Cultivation of Spice Crops

Related titles

Agriculture and Intellectual Property Rights: Economic, institutional and Implementation Issues in Biotechnology
V Santaniello, et al.

Controversial Drug Plants
Vasudevan Nair

Cultivation of Medicinal and Aromatic Crops (Rev. Ed.)
A A Farooqi and B S Sreeramu

Herbal Home Remedies
Vasudevan Nair

Indian Medicinal Plants: A Compendium of 500 Species
P K Warrier, V P K Nambiar and C Ramankutty

Intellectual Property Rights in Agricultural Biotechnology
F H Erbisch and K M Maredis (Eds.)

Nursery Manual for Forest Tree Species
D M Bhat, Vaidya S Swamy and N H Ravindranath

Plant Tissue Culture: Basic and Applied (Forthcoming)
Timir Baran Jha and Biswajit Ghosh

Vermicology: The Biology of Earthworms
S A Ismail

Concepts in Biotechnology (Rev. Edn.)
D Balasubramanian, et al.

This is Biology: The Science of the Living World
E Mayr

Travels in the Genetically Modified Zone
Mark L Winston

Universities Press Dictionary of Biology (3rd Edn.)
R Hine (Ed.)

Universities Press Dictionary of Biotechnology and Genetic Engineering (Rev. Edn.)
Mark L Steinberg and Sharon D Cosloy

Joint Forest Management: Spread, Performance and Impact
N H Ravindranath and P Sudha

Plant Tissue Culture and Biotechnology: Emerging Trends
P B Kavi Kishore

The Story of Our Food
K T Achaya

Cultivation of Spice Crops

A A Farooqi • B S Sreeramu
• Srinivasappa

Universities Press (India) Private Limited

Registered Office
3-5-819 Hyderguda, Hyderabad 500 029 (A.P.), India
Email: hyd2_upilco@sancharnet.in

Distributed by
Orient Longman Private Limited

Registered Office
3-6-752 Himayatnagar, Hyderabad 500 029 (A.P.), India

Other Offices
Bangalore / Bhopal / Bhubaneshwar / Chennai
Ernakulam / Guwahati / Hyderabad / Jaipur / Kolkata
Lucknow / Mumbai / New Delhi / Patna

First Published 2005
ISBN 81 7371 521 1

Typeset by
OSDATA, Hyderabad 500 029

Printed in India at
Orion Printers Private Limited, Hyderabad 500 004

Published by
Universities Press (India) Private Limited
3-5-819 Hyderguda, Hyderabad 500 029

Contents

Preface

India is the land of spices and has been the major producer of spices like black pepper, cardamom, ginger, turmeric, chilli, coriander, cumin, celery, fennel, fenugreek, garlic, etc., since time immemorial. India produces about 36.68 lakh tonnes of various spices from 28.31 lakh hectares of land. The value of spices produced is worth about Rs.7000 crores, of which the export component is approximately Rs. 1612.07 crores.

One or more spices are grown in almost all the states of the country. The major spice-producing states in India are Andhra Pradesh, Kerala, Gujarat, Rajasthan, Maharashtra, West Bengal, Karnataka, Tamil Nadu, Orissa and Madhya Pradesh. The North-Eastern region and the Andaman and Nicobar Isands have also been identified as potential areas for spice cultivation. While the cultivation of black pepper and small cardamom are mainly confined to South India, ginger and turmeric are grown in many states. Large cardamom is mainly confined to Sikkim, whereas cumin, coriander and fenugreek are grown in the northern states on a considerable scale, both for use within the country and for export.

The intensive research carried out actively on these groups of crops since 1971 has resulted in a marked improvement in the agro-techniques as well as in the release of many high yielding varieties. Thus, in order to increase productivity and reduce the cost of production and make them more remunerative to the farmers and others interested in these groups of crops, there is a need to extend this knowledge.

As the information on these crops, particularly about their cultivation, is very scattered and scanty, an effort has been made to compile all the possible pooled information available in various books, bulletins, journals and other sources. The authors have no pretensions to expertise in this vast field, nor do they claim an exhaustive treatment of the subject in this book. They do hope, however, that this book will be highly useful to students, teachers, extension workers and growers of spice crops.

As it is their maiden attempt in this direction, there may be many shortcomings. Any feedback from the enlightened farmers in this regard is welcome so that in future editions such mistakes can be avoided.

The authors record their sincere thanks to the authorities of University of Agricultural Sciences, Bangalore, for providing the support necessary to publish this book.

The secretarial help rendered by Mrs N Radha and Mrs Rekha in preparing the manuscript of this book is gratefully acknowledged.

December 2004
Bangalore

A.A.FAROOQI
B.S.SREERAMU
K.N.SRINIVASAPPA

Introduction

pices constitute an important group of agricultural commodities, which, since antiquity, have been considered indispensable in the culinary arts for flavouring foods. Some are used in pharmaceutical, perfumery, cosmetics and several other industries, and others possess colourant, preservative, antioxidant, antiseptic and antibiotic properties. Besides, they also play quite a significant role in the national economy of India and also in those of various other spice-producing, exporting and importing countries of the world.

India has been a leading spice-producing, consuming and exporting country of the world since the recorded history. During 2000–2001, India has earned foreign exchange worth Rs. 1,61,206.65 lakhs through the export of about 230000 tonnes of spices, as compared to Rs. 2,02,508.63 lakhs and Rs. 1,75,802.20 lakhs earned during the years 1999–2000 and 1998–1999 through the export of 2,36,142 and 2,31,389 tonnes of spices, respectively. (Spices Board, 2001). In addition, huge quantities of spices are consumed within the country for seasoning of foods and for several other purposes mentioned above. These figures indicate the economic importance of spices in our country and the world.

Definition of Spices and Condiments

According to the latest available International Organisation for Standardisation (ISO) report, Budapest, Hungary, there are about 109 spices grown in different parts of the world. India grows about 63 of these. But the commercial cultivation is limited to about a dozen of spices. Both ISO and ISI (now BIS) experts have concluded, after considerable deliberations, that there is no clear-cut division between 'spices' and 'condiments' and they are now clubbed together. The term 'Spices and Condiments' applies to such natural plant or vegetable products or mixtures thereof, used in whole or ground form, mainly for imparting flavour, aroma and pungency to food and also for seasoning of foods and beverages like soups, etc.

Spices are natural plant products used to improve the flavour, aroma, taste and colour of food products, They are also used in beverages, liquors, pharmaceutical, cosmetic and perfumery products. From time immemorial, India has been known as the 'Land of Spices'. No other country in the world has such a diverse variety of spice crops as India. Indian spices are renowned for their excellent aroma, flavour and pungency, not easily matched by any other country. Even in minute quantities, spices are a real delight to the senses, making food more palatable, tasty and easily digestible. While, their usage is common in one form or the other in every household, their demand in the food processing industry is increasing at a rapid pace across the world. Thus, the demand for spices in recent years has been continuously on the rise.

Classification of Spices and Condiments

The list of botanical and English names of 63 spices classified by the Bureau of Indian Standards (BIS) and the ISO as officially recognised spices and condiments, is given in Table 1. The table also covers the part of the plant that is used as a spice or condiment, its botanical and common English names as well as its name in Hindi.

There are as many as 86 important spices and condiments, officially recognised by both the ISO and the BIS. The BIS holds the ISO Secretariat for 'Spices and Condiments' on behalf of the Government of India. The ISO has now approved 86 spices that have been published in an ISO document.

There are over 190 ISO Technical Committees (TCs). Of these, TC 34 sub-committee 7 deals with the formulation of International Standards (ISO) for individual spices, spice mixtures (curry powder), their processed products like their essential oils and oleoresins as well as their methods of testing. These standards cover physical, chemical and microbiological parameters of both whole and ground spices.

World Trade in Spices—Present Status

The status of the world spice trade in 1998 was 026,076 tonnes which is valued at over US $ (000) 2,33,8541.

The comparative importance of individual spices in world trade in decreasing order is illustrated in Table 2

Importance of individual spice requirements

The importance of individual spice requirements differs from one importing country to another. However, in most world markets, pepper is almost invariably the principal or the most important world spice imported in terms of both its volume and value. The only exceptions are Finland, Saudi Arabia and Kuwait (importers of small cardamom), and the Democratic Republic of Yemen (importers of ginger and chillies). Next in order of importance is the capsicum group, consisting of paprika (the leading item), chillies and cayenne pepper.

Table 1 *Nomenclature of spices and condiments*

Sl. No.	Botanical name	Part used	Common name (English)	Common name (Hindi)
1.	*Acorus calamus* Linn.	Rhizome	Sweet flag	*Bach*
2.	*Allium cepa* Linn.	Bulb	Onion	*Piyaz*
3.	*Allium sativum* Linn.	Bulb	Garlic	*Lasan*
4.	*Alpinia galanga* Wild (Linn.)	Rhizome	Greater Galanga	*Bara kulanjan*
5.	*Amomum subulatum* Linn.	Fruit & seed	Greater Indian Cardamom, Large Cardamom, Sikkim Cardamom, Nepal Cardamom	*Bara elaichi*
6.	*Anethum graveolens* Linn.	Fruit and leaf	Dill seed	*European Dill*
7.	*Apium graveolens* Linn.	Fruit	Celery seed	*Ajmud*
8.	*Amoracia rusticana* Gaetn. Syn *Cochleria amoracia* Linn.	Root	Horseradish	-
9.	*Brassica juncea* Linn Cezn & Cosson.	Seed	Indian mustard	*Rai*
10.	*Brassica nigra* Linn. W.D.J.Koch	Seed	Black mustard	*Asa raiorBanarasirai*
11.	*Capparis spinosa* Linn.	Unopened flower bud	Caper	*Kabra*
12.	*Capsicum annum* Linn.	Fruit	Chillies or capsicum	*Lal-mirichi*
	Capsicum fruitescens Linn.	Fruit	Bird eye	*Lal-mirchi*
13a.	*Carum bulboccstanum*	Fruit	Black caraway	*Kalajira*
13b.	*Carum carvi* Linn.	Fruit	Caraway seed	*Shajira*
14.	*Cinnamomum aromaticum* C.G.Ness	Bark	Cassia or China cassia	*Jungli dalchini*

(continues...)

(... *contd.*)

Table 1

Sl. No.	Botanical name	Part used	Common name (English)	Common name (Hindi)
15.	*Cinnamomum mantii* (Blume) C.G.Ness	Bark	Batavia cassia or java cassia	*Jungli dalchini A.Q.*
16.	*Cinnamomum tamala* (Buch Ham) Nees & Eberm,	Bark and leaf	Indian cassia lignea	*Tejpet*
17.	*Cinnamomum zeylanicum* Blume	Bark	Cinnamon	*Dalchini or A.Q.*
18.	*Coriandrum sativum* Linn.	Leaf and seed	Coriander	*Dhanya, Dhania*
19.	*Crocus sativus* Linn.	Stigma	Saffron	*Kesar, Zafran*
20.	*Cuminum cyminum* Linn.	Fruit	Cumin seed	*Jira, Zira*
21.	*Curcuma longa* Linn.	Rhizome	Turmeric	*Haldi*
22.	*Elettaria cardamomum* Linn. (C.Sprengel) Bullock	Fruit and seed	Cardamom	*Choti elaichi*
23.	*Eugenia caryophyllus* (C.Sprengel) Bullock Harrison, Syn *Syzygium aromaticum* Merril. Perril	Unopened flower bud	Cloves	*Laung or loving*
24.	*Ferula alliacea* Boiss, *syn Ferula rubricaulus* Boiss	Oleogum resin from	Asafoetida (*Hingra*)	*Hing*
25.	*Ferula asafetida* Regel	Oleogum resin from rhizome and thickened root	*Hingra*	*Hingra*
26.	*Foeniculum vulgare* Mill.	Fruit	Fennel	*Bari-saunf or Saunf*
27.	*Garcinia indica* (Gaertner) Desrous-Seaux	Pericarp-lobes	Gambodge	*Aradal*
28.	*Garcinia indica* Choisy	Peel of fruit	Kokam	*Kolam*
29.	*Hyssopus officinalis* Linn.	Leaves	Hyssop	*Zujah-Yabair*

(*continues...*)

(... contd.)

Table 1

Sl. No.	Botanical name	Part used	Common name (English)	Common name (Hindi)
30.	*Juniperus communis* Linn.	Berry	Juniper	*Aaraar*
31.	*Laurus nobilis* Linn.	Leaf	Sweet bay or true laurel leaves	*Indian Bazar Hab-el ghar*
32.	*Levisticum officinale* W.D.J.Koch	Fruit and root	Lovage	
33.	*Marjorana hortensis* Moench	Leaf and flowering top	Sweet marjoram	*Marwa*
34.	*Mangifera indica* Linn.	Unripe fruit, Dried raw peeled slices or powder	Mango	*Amchur*
35.	*Mentha arvensis* Linn.	Leaf	Japanese mint	*Pudina*
36.	*Mentha longifolia* (Linn.) Hudson	Leaf	Mint	*Jangali pudina*
37.	*Mentha piperita* Linn. emend. Honds	Leaf	Peppermint	*Peppermint or Vilayati pundina*
37a.	*Mentha spicata* Linn. Syn *M.viridis* Linn.	Leaf	Spearmint	*Pahari pundina*
38.	*Murraya koenigii* (Linn.) Sprengel	Leaf	Curry leaf	*Kathnim or kurry patta*
39.	*Myristica fragrans* Houttuyn	Aril	Mace	*Jaipatri or jalavatri*
40.	*Myristica fragrans* Houttuyn	Kernel	Nutmeg	*Jaiphal*
41.	*Nigella sativa* Linn	Fruit	Black cumin	*Kalonjii*
42.	*Ocimum basilicum* Linn.	Leaf	Sweet basil	*Balui tulsi*
43.	*Origanum vulgare* Linn.	Leaf and flowering top	Oregano	*Sathra*
44.	*Papaver somniferum* Linn.	Seed	Poppy seed	*Ajim, khaskhas*
45.	*Petroselinum crispum* Nyman ex. A.W.Hill Syn *P. Sativum* Miller	Seed	Parsley	*Ajmood*

(continues...)

(... *contd.*)

Table 1

Sl. No.	Botanical name	Part used	Common name (English)	Common name (Hindi)
46.	*Pimenta dioca* (Linn) Merril	Fruit, leaf	Pimento or allspice	*Aril*
47.	*Pimpinella anisum* Linn.	Fruit	Aniseed	*Saunf*
48.	*Piper longum* Linn.	Seed and leaves	Long pepper	*Pipal or Pipli*
49.	*Piper nigrum* Linn.	Berry	Black/white pepper	*Gol mirch, Kali mirch*
50.	*Punica granatum* Linn.	Seed (dried with flesh)	Pomegranate	*Anardana*
51.	*Rosmarinus officinalis* Linn.	Leaf	Rosemary	*Rusmari*
52.	*Salvia officinalis* Linn.	Leaf	Sage	*Salviar seyfakuss*
53.	*Satureja hortensis* Linn. or *Satureia hortensis* Linn.	Stem, leaf and flowering top, seed	Summer savory	–
54.	*Sinapis alba* Linn. Syn *Brassica alba* or *B.hitra*	Seed	White mustard or yellow mustard	*Sufaid rai*
55.	*Tamarindus indica* Linn.	Ripe fruit or pod	Tamarind	*Imli or Amli*
56.	*Thymus serpyllum* Linn.	Leaf and flowering top	Wild thyme	*Jangli Banajwain*
57.	*Thymus vulgaris* Linn.	Leaf and flowering top	Common thyme	*Banajwain*
58.	*Trachyspermum ammi* Linn.	Fruit	Bishop's weed Ajowan	*Ajowan*
59.	*Trigonella foenum–graecum* Linn.	Seed	Fenugreek	*Metha danao or Methe*
60.	*Vanilla fragrans* (Salisbury) Ames, *Vanilla planifolia* (Andrews)	Pod	Vanilla	*Vanilla*
61.	*Vanilla pompona* Schneiden	Pod	Vanilla	*Vanilla*
62.	*Vanilla tahitensis* Moore	Pod	Tahiti vanilla	*Vanilla*
63.	*Zingiber officinale* Linn.	Rhizome	Dried ginger	*Adrak or Adhrak*

(*Source*: Pruthi, J.S., 2000)

Nutmeg, mace, cinnamon and cassia also feature prominently in the import of spices into the industrialized markets of Western Europe and North America. Pimento (Allspice) is the major item in the socialist countries of Eastern Europe and the USSR.

The other volume items of significance are ginger, turmeric and spice seeds e.g. coriander, juniper, anise, caraway and cumin. In terms of volume, world imports of spice seeds amount between 55,000 and 60,000 tonnes, annually, but in general, they are low-value items. On the other hand, vanilla, saffron and cardamom are high–unit value spices but their trading volume is low. Another high value item is clove, but this is a special case, as over 75–80% of world clove exports are imported into Indonesia for their '*Kretek*' cigarette industry.

Table 2 *Comparative importance of individual spices in world trade*

World Rank	Spices
I	Black and white pepper
II	Capsicums (Chillies and Paprika)[1]
III	Seed Spices
IV	Cinnamon and Cassia
V	Turmeric
VI	Ginger
VII	Nutmeg and Mace
VIII	Cardamom
IX	Curry powder
X	Clove[2]
XI	Pimento (Allspice)
XII	Vanilla
XIII	Saffron[3]

[1]Figures include trade in paprika, chillies and cayenne pepper, whole and ground;

[2]Excluding imports into Indonesia (about 12,000 to 15,000 tonnes annually), Hong Kong and Singapore;

[3]In tonnes only (and not in thousand tonnes).

Source: International Trade Centre, UNCTAD/GATT, Geneva, Switzerland, 1991

Trends in individual world markets

In general, there has been an increase in the value of world imports of various spices. However, the rate of increase has varied across countries and from spice to spice. Of all the markets, FR Germany, Japan, Netherlands, United Kingdom, Singapore and certain Middle Eastern countries, particularly Saudi Arabia, have registered impressive growth rates. Also, in the USA, the largest importer of spices, has shown is a moderate increase in imports of spices.

The 1998 import of spices by European Union (15 countries) amounted to about 250,987 tonnes valued at US $ (000)794,907. Other countries such as Poland, the Russian Federation and Switzerland together imported about 65,309 tonnes of spices valued at US $ 21,900.

Germany continues to be the largest single market in Europe, accounting for one-third of the total imports of spices into Western European countries. Next comes Netherlands and United Kingdom, Sweden and Finland, after the Middle East.

The United States is the World's largest single market for spices. Imports of spices (excluding spice herbs, mustard and sesame seeds) into this country were 1,72,421 tonnes in 1998 valued at US $ (000) 4,78,455.

The Canadian market for spices is relatively small, i.e., 52,155 tonnes valued at US $ 20,673 in 1998.

In the Asian and Pacific Region, the major consumer of spices is Japan, the third largest world market, followed, at some distance, by Singapore and Malaysia. Japan imported 124,632 tonnes of spices, valued at UD $ (000) 18569 in 1998. Although domestic consumption of spices in Hong Kong and Singapore is very small, the importance of Singapore in the export trade in spices is still of considerable significance. It is estimated that from 20–25% of the total pepper entering the international trade, is being handled by the Singapore. This can be attributed to the excellent communication, finance and shipping facilities available in Singapore. The other factor is the ability of the exporter to provide both the quality and timely delivery that the importer requires. Although Indonesia is one of the major producers of a number of spices, yet it is also a substantial importer of cloves for the '*Kretek*' cigarette industry.

Forms of spices and spice products traded

Although spices are traded in a variety of forms, it is estimated by ITC (1986), that by far the bulk of spices (about 90%) is traded in the whole form (unground), as has been the case since ancient times. The remainder 10 to 20% consists of (i) ground spices, (ii) curry powder (iii) spice essential oils and (iv) spice oleoresins. Important ground spices include paprika, pepper, turmeric, curry powder and a number of other spices mixtures. Retail consumer-packs of spices entering international trade are very small at present. In recent years, trade in processed spices (i.e. ground spices, mixed spices or both) packed in bulk containers is gradually gaining momentum (20% of total exports in 1997–1998).

Present Status of Indian Spice Industry

Area and production

Some of the important spices of India are pepper, cardamom, chilli, ginger, turmeric, coriander, cumin, fennel, fenugreek, celery, saffron, tamarind and garlic. Spices produced and exported in small quantities are aniseed, bishop's weed (*ajowan*), dill seed, poppy seed, *tejpat*, curry leaves, cinnamon, *kokam* and a few other culinary herbs. Certain spices, though cultivated in the country, are imported for want of sufficient production. Such spices are clove, nutmeg, mace, star-anise, allspice and some of the herbal spices like rosemary, thyme, marjoram, oregano, chive, parsley, sage, savory, tarragon and basil. Commercial cultivation has still not commenced in vanilla and paprika, but there is immense potential for their production and export. The area and production of important spices in India are given in Table 3.

Kerala retains the lead in black pepper production in the country, contributing 96% of area and 97% of production. The black pepper area in India increased from 0.8 lakh hectares in 1950–51 to 1.92 lakh hectares in 1990–00. Similarly, production increased from 20,500 tonnes to 58,290 tonnes from 1950–51 onwards, mainly due to the lucrative prices prevailing for black pepper in the domestic and international markets. The annual growth rates observed for area, production and productivity of black pepper for the last decade were 1.3, 3.3 and 1.5%, respectively. However, it is paradoxical to note that in spite of highly favourable climate, improved varieties and high-tech production technologies, the productivity of black pepper in the country is very low (303 kg/ha) compared to the other pepper producing countries (3 to 4 t/ha).

Black pepper production in the country during 2002 was estimated at 65,000 tonnes while the global production was assessed to be about 3,10,000 tonnes. However, the Jakarta-based International Pepper Community had projected Indian pepper production to be around 80,000 tonnes during 2002. Vietnam, with a production of 55,000 tonnes (2001), is a strong competitor for India in the international trade of black pepper. If Vietnam, presently the second-largest producer of black pepper, maintains its current rate of growth in production, there is no doubt that it might overtake India in a couple of years. The production and productivity of Indian pepper is highly influenced by the oscillating prices, which were quite pronounced recently. The cost of one kg black pepper was only Rs. 30 in 1990–91, but went up to Rs. 170 during 2000–2001 and nose-dived to Rs. 55–65 in January–February 2002. At its peak, in November 1999, black pepper was sold at Rs. 270/kg.

Table 3 *Area under spice cultivation in India during the IX Five Year Plan ('000 ha)*

Spices	1996–97 (base year)	1997–98	1998–99	1991–2000	2000–01	2001–02	Average annual growth rate (%)
Pepper	180.26	181.53	238.32	247.85	257.77	268.08	8.80
Ginger	70.29	75.57	77.61	80.71	83.94	87.30	4.44
Chillies	944.20	840.60	892.20	927.89	965.00	1003.60	1.43
Turmeric	135.20	139.70	155.80	162.03	168.51	175.25	5.37
Coriander	425.70	656.50	546.50	568.36	591.09	614.74	8.05
Garlic	94.30	108.80	114.40	118.98	123.74	128.68	6.50
Cumin	307.05	288.83	264.02	274.58	285.56	296.99	–0.50
Fennel	25.90	28.28	18.32	19.05	19.81	20.61	–2.81
Fenugreek	36.32	77.40	82.67	85.98	89.42	92.99	26.38
Celery	4.60	4.59	4.60	4.78	4.98	5.17	2.40
Clove	2.30	2.30	3.17	3.30	3.43	3.57	9.97
Nutmeg	6.30	6.47	6.47	6.73	7.00	7.28	2.94
Cinnamon	0.83	0.73	0.73	0.76	0.79	0.82	–0.01
Others	111.96	113.00	112.36	116.85	121.53	126.39	2.47
Total	2372.21	2524.30	2517.17	2617.86	2722.57	2831.47	3.63

(*Source:* DA&SD, Calicut)

Table 4 *Production of spices during the IX Five Year Plan ('000 tonnes)*

Spice	1996–97 (base year)	1997–98	1998–99	19991–2000	2000–01	2001–02	Average annual growth rate (%)
Pepper	55.59	57.33	65.99	69.00	79.00	85.32	9.06
Ginger	232.51	252.11	263.17	284.22	306,96	331.52	7.30
Chillies	1066.40	870.10	921.30	995.00	1074.60	1160.57	2.30
Turmeric	528.90	549.20	598.40	646.27	697.97	753.81	7.36
Coriander	255.50	337.70	290.00	313.20	338.26	365.32	8.41
Garlic	437.90	484.40	517.70	559.12	603.85	652.15	8.30
Cumin	117.91	116.27	109.72	118.50	127.98	138.22	3.40
Fennel	28.98	37.18	24.18	26.11	28.20	30.46	3.47
Fenugreek	51.06	65.33	80.00	86.40	93.31	100.78	14.88
Celery	6.00	6.5	4.50	4.86	5.25	5.67	0.31
Clove	1.78	1.66	2.80	3.02	3.27	3.53	17.19
Nutmeg	1.49	1.53	1.53	1.65	1.78	1.93	5.34
Cinnamon	0.37	0.36	0.36	0.39	0.42	0.45	4.26
Others	20.61	21.35	30.82	33.95	35.95	38.82	14.39
Total	2805.00	2801.02	2910.47	3141.04	3396.80	3668.55	5.57

(*Source:* DA & SD, Calicut)

Cardamom, the 'Queen of Spices', enjoys a premium preference in the international market and is relished for its distinct enriching properties. The data on area under cardamom cultivation in India reveal that there was not much fluctuation during the last four decades. The highest area of 1 lakh hectares was recorded during 1985–1986, which was reduced to 72,400 hectares in 1999–2000. Cardamom production declined to as low as 3,500 tonnes in 1998–1999 and rose to 10,450 tonnes in 2000–2001. The record price of cardamom was realised during 2001–2002, which was as high as Rs. 680/kg during January, 2002. The price of cardamom exhibits a firm trend, largely on account of the sharp decline in production in Guatemala.

India is the foremost producer and exporter of chillies. Among the various spices produced in the country, per capita consumption is the highest for chillies. Trends in area, production and productivity show a marked increase from 1950–1951 to 1999–2000. During 1999–2000, the area under the crop was 9.15 lakh hectares with a production of 10.18 lakh tonnes. The price of chilli was Rs. 32.01/kg during 2000–2001 which is slightly less than the average of the last three years (Rs. 35.44/kg).

Turmeric is a multipurpose crop valued for its colouring pigment, spicy flavour and medicinal properties. The area and production of turmeric in the country showed increasing trends during the last five decades. The highest area (1.69 lakh hectares), production (6.98 lakh tonnes) and productivity (4,130 kg) was recorded in 2000–2001. The annual growth rates in area and production for the last ten years work out to 3.7% and 9.1%, respectively. The price of turmeric had plummeted to around Rs. 26 kg during 2000–2001, while the average price for the last three years has been Rs. 33/kg.

Garlic is a major spice grown in Madhya Pradesh, Gujarat, Orissa and Rajasthan. The crop is grown over an area of 1.14 lakh hectares with a production of 4.95 lakh tonnes (1999–00). The price of garlic has nose-dived to Rs. 9.46/kg during the current year from the average of Rs. 17.58/kg for the last three years.

Ginger is one of the oldest known spices in the world and Indian ginger is highly valued in the global market because of its characteristic lemon-like flavour. Being an annual crop, the area and production are highly influenced by price fluctuations. The crop exhibited an annual growth rate of 4.6% in area, 7.4% in production and 2.7% in productivity. During 1970–71, the area under ginger was only 21,590 hectares which went up to 77,600 hectares in 1999–2000 and 83,940 hectares in 2000–2001. Ginger prices have remained almost constant at Rs. 35/kg during the last three years.

Vanilla is another important spice crop gaining momentum at the international level, as a result of the global movement to return to natural products. The present global demand for vanilla is around 32,000 tonnes as against the production of just 2,000 tonnes, creating a shortage of 30,000 tonnes. The climatic conditions prevailing in South India, particularly Kerala, are highly favourable for vanilla cultivation. Another important positive aspect is the lucrative price of vanilla in the domestic and global markets. At the present price of Rs. 800/kg, of raw beans, any farmer can enthusiastically venture into vanilla cultivation. The area under vanilla cultivation is, at present, quite meagre.

Nutmeg, cinnamon, clove, *tejpat*, tamarind and *Garcinia* are the important tree spices grown in India. The area under these crops are registering an upward trend mainly because of the attractive prices, both from domestic and international markets. *Garcinia* is a tree which is gaining importance due to the presence of an antiobesity factor and pro-hydroxy citric acid in the fruit rind. The area under nutmeg is 10.210 hectares with an annual production of 2,250 tonnes; while for clove, the area and production are 3,250 hectares and 2,940 tonnes, respectively. Cinnamon occupies an area of 730 hectares with a production of 370 tonnes. All other tree spices put together register an area of 14,190 hectares with an annual production of 5,560 tonnes.

The production of spices in the country has increased by 2.2 lakh tonnes during the first three years of the IX Five Year Plan.

Among the spices grown in the country, seed spices account for 37% and 17% of the total area and production, respectively. Seed spices contribute to approximately 24% of export and 10% of the national income from spices. Out of the 17 seed spices produced in the country, coriander, cumin, fennel, fenugreek, dill seed, celery and poppy seeds are important in selected pockets. The average annual growth rate was only 3 per cent, which was far below the estimate. The main reason was the decrease in the production of chillies. Since chillies is the highest in quantity among spices grown, the change in production will be reflected significantly in the total spices production of the country. The production of all the major spices except pepper and chillies has increased up to the targeted level during the initial years of the plan period.

Constraints faced in production

1. Lack of adequate infrastructure for large-scale production and distribution of quality planting materials of the released varieties.
2. Low productivity due to cultivation of varieties of poor genetic potential with regard to yield and quality parameters.
3. Non-manuring, under-manuring or imbalanced manuring is generally followed.

4. Non-adoption of recommended cultural practices, as well as soil and water conservation measures.
5. Non-adoption of integrated pest and disease management practices.
6. Inadequate extension network for effective transfer of technology.
7. Frequent fluctuations in prices of the commodities and the absence of a support price.

Export

The changed economic order in the context of globalisation and liberalisation of world trade in agriculture has opened up new vistas of growth. The spices sector is one of the key areas in which India has an inherent strength to dominate the global markets. Recorded history elucidates the extreme fascination of the rest of the world for the fabled wealth of India—the variety of spices. As a result of the recent WTO regime, quality competitiveness has emerged as the prime mover of international food marketing. The present dispensation, of course, widens the access to global markets, but at the same time raises many challenges for survival of the spice industry.

In India, spice exports have been consistently moving up during the last decade. The Spices Board under the Ministry of Commerce has the mandate for the export of spices from the country. Spices account for about 16% in quantity and 19% in value of the total horticultural exports from India. India's share of the world spices trade is estimated as 40–50% by volume and 25% by value. During 1999–2000, Indian spices exports rose creating an all-time record in terms of value. Thus, for the last couple of years, India has been enjoying a commanding position in the global market. However, in 2000–2001 the scenario changed due to the fall in the export of black pepper. According to the estimates of the Spices Board, Indian spice export during 2000–2001 was estimated at 230,000 tonnes valued at Rs. 1612.07 crores as against 236,142 tonnes valued Rs. 2025.09 crores in the previous year, showing a decline of 20% in value.

During 1999–2000, 49% of the world trade in spices was Indian spices. The country enjoyed monopoly in the export of spice oils and oleoresins, with a contribution of 16% of the total export earnings from spices during 1999–2000; this was followed by chillies (12%). The United States of America is the major importer of Indian spices. In 1999–2000, 38% of our spices export earnings came from the USA. Other countries importing Indian spices are the United Kingdom, Japan, Germany, Canada, Netherlands, Russia, Sri Lanka, UAE, Malaysia, etc.

Table 5 *Major importers of Indian spices (percentage to India's spices export earnings)*

Countries	1997–1998	1998–1999	1999–2000
United States of America	30	30	38
United Kingdom	6	6	6
Japan	5	5	5
Germany	4	4	4
Canada	3	3	4
Netherlands	2	2	3
Russia	4	2	3
Sri Lanka	3	5	4
UAE	5	4	3
Malaysia	2	2	2
Others	36	37	28

(*Source:* Spices Board, Cochin)

Indian pepper exports, which account for a major part of the global share, seems to be facing a stiff threat from the new entrants, especially Vietnam. During 2000, the world pepper output was estimated at Rs. 2,63,000 tonnes. This better supply position and the competition in the international markets have put pressure on the price. As a result of this, black pepper export contributed only 20% of the total export earnings during the fiscal year 2000–2001, a reduction of 55% was registered in the black pepper valued at Rs. 326.33 crores as compared to 42,806 tonnes valued at Rs. 884.88 crores in 1999–2000. The way to meet the challenge is to develop a market-oriented production strategy, that is, to increase productivity, encourage value additions and adopt organic cultivation.

With better production and the resultant decline in the price, Indian cardamom (small) export was increased by 70% in quantity and 77% in value. Export of large cardamom also increased by 36% in quantity and 63% in value.

Better production in the competing countries like China and Pakistan and the low volume of import by Bangladesh has resulted in the decline in export of chillies during 2000–2001. However, due to better production and the resultant price fluctuation, India was able to achieve the export target set by the Spices Board. The export of ginger and turmeric also registered a decline in both quantity and value.

Seed spices account for 20% of the total foreign exchange earning, contributing to a volume of 13% of the total spice export. Among seed spices, the export of cumin marked a record increase with 125% in quantity and 243% in value during 2000–2001 as against the export during 1999–2000. This achievement was mainly due to the poor production in competing countries like Iran, Syria, Turkey, etc. The export of value-added products like spice oils and oleoresins has been steadily increasing during the past years and during 2000–2001 India exported 3625 tonnes valued at Rs. 364.05 crores.

Demand driven spice development

The spice industry is entering into a new era of diversified use of spices, particularly medicinal and neutraceuticals. The antioxidant properties of curcumin from turmeric, which prevent the entry of the cancer virus, is well documented. The phenomenal growth of curcumin exported from India clearly proves the acceptance of such functional properties in addition to its wider use as food colourant. *Garcinia* is another spice which has gained a lot of importance due to its functional properties by the presence of hydroxy citric acid which is found to be an 'appetite suppressant' helping to reduce weight. The significant growth in the export of this product in the last five years shows its the effectiveness in reducing obesity.

Piperine from black pepper, is a potent inhibitor of drug metabolism. It acts as a bio-availability enhancer. Bioperin is a pharmaceutical formulation of piperine in the US. Piperine enriches the availability of drugs such as vascicine, spartein, sulpha diazine, rifampicin and phenytoin. Zingiberine from ginger is also considered as a potent bio-active material.

Chillies, that are usually regarded as an irritant spice, have counter-irritant properties also at very low concentration and they are being used in skin ointments. They can reduce the retention time of faecal matter in the rectum and can help in preventing rectum/colon cancer.

Garlic, onion and fenugreek are well known for their cholesterol reducing and antidiabetic properties. Also, dill and fennel are used as digestents. The exploration of medicinal and neutraceutical properties of our traditional spices will go a long way in the development of the Indian spice industry.

Crop diversification

Our spice farmers are accustomed to selling the produce in the primary form. With respect to export, about 85–90% of exports also go out as primary produce. No doubt, there exists a good scope for value-added products in the European and American markets. Presently we are exporting the bulk of our value-added products to the Middle East where the price is quite low. A few organic products fetch a premium price, higher by 100% or more.

Table 6 *Export of spices from India (Quantity: tonnes, value: Rs. in lakhs)*

Item	1998–1999		1999–2000		2000–2001	
	Quantity	Value	Quantity	Value	Quantity	Value
Pepper	34864	63811.28	42806	88488.14	19250	32632.75
Cardamom (S)	475	2521.21	646	3201.84	1100	5654.70
Cardamom (L)	1424	1190.88	1211	1696.89	1645	2768.75
Chillies	61253	21661.11	64776	25065.89	61000	19523.50
Ginger	8778	4064.79	8773	3060.15	6580	2295.40
Turmeric	36522	12454.98	35556	12168.69	34500	9106.00
Coriander	20685	4588.99	13973	3496.77	11700	2742.50
Cumin	10723	6010.92	6145	3428.17	13800	11743.50
Celery	3981	969.08	3497	1059.55	5250	1700.50
Fennel	5279	1538.11	2953	1441.53	4000	1778.00
Fenugreek	10082	1914.91	10901	2095.41	9050	1787.50
Other seeds (1)	2001	749.11	2349	1026.46	2425	896.25
Garlic	4068	741.02	8542	1326.87	11000	1040.30
Other Spices (2)	19077	7659.48	22012	8850.38	35000	14490.00
Curry powder	5210	3596.78	5814	3460.60	6200	3997.00
Mint oil	4207	12252.18	2820	10310.71	3875	12645.00
Spice oleoresins and other oils	2750	30077.37	3368	32330.58	3625	36405.00
Total	**231389**	**175802.20**	**236142**	**202508.63**	**230000**	**161206.65**1)

(1) Includes aniseed, ajwan seed, dill seed, poppy seed, mustard etc. (2) Includes tamarind, asafoetida, cinnamon, cassia, *kokam*, saffron, *etc* for export data. (*Source:* Spice Board, Cochin)

In general, if the price rises 20 to 30% above the conventional rate, the number of consumers will drastically reduce. Hence, the price should be affordable to the users. The cost of certification for organic spices is prohibitive and beyond the scope of the average Indian farmer now.

In January 1993, the European Union Council Regulation pertaining to production and commercialisation of organic products came into existence. For a non-member country of the EU, there are two ways to export organic products to Europe. One is to get included in the 'Third Countries List', whereby the EU commission will do the evaluation or to get an import authorisation from the importing country to the effect that production and inspection were carried out as per the EU regulations.

Denmark eagerly plans to convert 40% of its agriculture to organic by 2010. About 10% of land is already under organic cultivation in Australia and Sweden. The annual growth rate of organic market in the US is around 20%.

Looking at the world figures, it is obvious that organic market is leaving the niche and entering the mainstream.

The Spices Board of India has already launched a number of programmes to create awareness for organic spices production in the country. Training, indigenous certification programmes, sourcing organic spices to international buyers, etc., are some of the activities of the Board in this field.

The Board has also launched a scheme to assist groups of farmers/ societies in getting organic certification from international agencies, by providing 50% of the cost of inspection and certification charges.

Quality Assurance Mechanism

Even though India has been in the forefront for export of spices for several centuries, it was not until the mid-80s that the word 'quality' assumed importance. The imposition of quality standards by importing countries resulted in efforts for quality assurance. The steps were taken by the Spices Board to develop and inculcate good post-harvest management right from the stage of harvesting—particularly in seed spices. The right stage of harvest decides the quality, *viz.*, size, colour, fibre content, volatile oil, chewing quality (fennel), splitting (coriander), etc. Clean drying, proper winnowing to remove dirt, chaff, plant parts, birds' droppings, rat excreta, hair, etc., followed by storage in hygienic and well-ventilated godowns ensures a good product. At the international level there is a stringent quality standard which has to be met by exporting countries. The specifications of American Spices

Trade Association (ASTA) and the International Organisation for Standardisation (ISD) guidelines, with mutually agreed terms are followed. Japan and many other countries use the ASTA specifications; India, Malaysia, Spain, Indonesia, Sri Lanka and Hungary have their own quality standards for exporting spices. ASTA specifications are for unprocessed spices imported into the USA, and they regulate the cleanliness of these spices. ASTA specifications place limits on extraneous matter (insects, insect excrement, stones, stems, sticks, moulds, etc.) and set standards in sampling and testing procedures. Imported spices not meeting these specifications are reconditioned at the port of entry, while local spices need to be reconditioned (using fumigants) before they are processed for use in a product.

The US Federal Specifications establish quality standards for whole and ground spices. Specifications include total ash, acid-insoluble ash, volatile oil, moisture, colour and granulation, and these are standardised for the major spices. These specifications have maxima or minima values that differ considerably between importing countries. Standards for defective spice (stalks, stems or other contaminants) are also established. The ISO has specifications for most spices, such as extraneous matter, moisture content, total ash, acid-insoluble ash and other chemical characteristics. Based on the ISO, Britain and Canada have established their own standards. Spices that are irradiated, reheated with ethylene oxide or contaminated with aflatoxin are prohibited under the Japanese Sanitation Law.

In India, quality standards are established for unprocessed and processed spices that particularly regulate moisture content, volatile oil, total ash, acid-insoluble ash, starch and extraneous matter. Grades are also established with specifications. Many spices, such as turmeric, black pepper, ginger and chilli peppers, have individual grade specifications that differentiate them from similar spices. Even curry powders are graded based on the amount of spices and salt they contain. Similar specifications are established for spices in many other countries.

Grinding and sterilising procedures decrease the volatile oils in spices. Excessive heat volatilises and dissipates the essential oils in ground spices, and high humidity tends to cake them. Exposure to light, humidity variations, air and certain metals can discolour many spices such as paprika, turmeric or the green leafy spices. Dry, ground chilli peppers turn from a natural green or red colour to an olive or dirty reddish-brown colour when exposed to light. Flavour and aroma losses, as well as insect and rodent infestation, occur when spices or spice extracts are stored in tightly closed containers in cool, dark, dry conditions below 4.4°C and 60% humidity. Some spices need storage at low temperatures of about 2–5°C to prevent mould infestation (capsicum peppers), colour deterioration (paprika) and to avoid rancidity (in high fixed oil seeds, such as sesame seeds). Colder temperatures also help preserve the volatile oil flavour and aroma, freshness and sanitary quality; refrigeration slows microbial growth. Whole spices tend to have better shelf life than ground spices. Ground spices for minimally processed foods such as salad dressings, condiments or 'sprinkle on' seasonings should be well cleaned and sterilised.

High levels of moisture in ground or whole spices indicate mould and microbial growth. During storage, insects breed on spices in varying degrees, depending upon storage conditions, where they are harvested, transportation contamination and the extent of cleaning. Filth levels include foreign materials such as insect fragments (moths, mites, beetles), small stones, metal fragments and glass pieces. Insects and mould growth can change the colour and, to some extent, the flavour of the spice.

Microbiological requirements for 'clean' spices include counts for total bacteria, yeast, mould, coliforms and food pathogens such as *Eschersia coli* and *Salmonella*. High microbial counts are caused by contamination during growing and post-harvest handling. Spore-forming bacteria, such as the *Bacillus* species or *Aerobacter aerogenes* found in the soil can be transferred to the spice during the drying process, particularly to 'under the ground' spices such as turmeric, ginger, galangal or garlic. The type and amount of moulds and bacteria on a spice depends upon the type of spice and the conditions under which it is harvested and dried. *Staphylococcus* and *Streptococcus* species of bacteria predominate, but pathogenic bacteria tend not to exist on spices. Spices that show strong antimicrobial properties tend to have low counts of microbes. Moulds, such as *Aspergillus,* that produce the toxin Aflatoxin and *Penicillium*, are found on certain species including red pepper, fenugreek and ginger, so there are specification limits for these toxins. Moulds tend to multiply during the drying process and storage. The control of insects and microbes is important in receiving quality spices. Spices need to be free of microbes so as to reduce the initial bacteria or mould content in processed foods. Spice extractives and sterilised spices tend to meet these objectives. Organic foods are becoming more popular, and for food to be labelled as organic, irradiation is not permitted. Spices that are not sterilised can be potent sources for microbial contamination and growth.

Marketing

The marketing of spices is very important in their trade. This is in the hands of several intermediaries in the country. There is normally no government intervention in this trade, with regard to the marketing procedure or in the maintenance of a reasonable price for the farmers. Farmers dispose of their produce either locally to a village shop, co-operative society or even to a terminal market. The choice depends on a number of factors such as loyalty or financial obligations to a particular trader, the distance to the terminal market, the prevailing market price, etc. Marketing the produce to pre-harvest contractors is also in vogue in certain areas. In this case, the farmers receive an advance payment immediately after the acceptance of the offer. Collection centres operate within a short distance of the growing areas and also in almost every town. The merchants purchase from these centres when the price is quite attractive in the terminal markets. Commission agents receive the goods and sell them to the exporters through brokers and dealers. The exporters grade the finished product according to grade specifications and finally consign the packed material to the shipping agents. The important channels in the spices marketing are:

1. Producer–Village merchant/Wholesaler–Commission agent/ Exporter-Foreign importer
2. Producer–Village merchant/Wholesaler-Commission agent/ Wholesaler of consuming market–Retailer–Consumer.

Of these two, the more important one for black pepper is the former channel, as the major share of the produce is being exported and the internal price is solely dependent on foreign demand.

At present there is no regulated market for spices in Kerala. However, the co-operative marketing societies in Kerala handle these commodities, but they are also not functioning very well. The main reason for their failure is their poor financial resources that makes them incapable of providing short term/long term loans to the farmers in time. In other states like Karnataka, where regulated marketing is in vogue, the farmers do not get their legitimate share.

Normally, the farm price a grower gets is much lower than the retail sale price in the assembling or distributing markets. This wide range of disparities in the price received by the producers is mainly due to:

1. Expenditure incurred in marketing.
2. Margin of profit enjoyed by the pre-harvest contractors/ intermediaries.

3. Deduction on account of high moisture percentage.
4. Incidence of various taxes and other charges.
5. Transportation costs, etc.

Challenges

- A major issue that needs to be addressed is the low productivity of spices in the country. In black pepper, our national average is only 316 kg/ha whereas, in Thailand it is 3,352 kg/ha. Guatemala produces cardamom at the rate of 250 kg/ha as compared to 135 kg/ha produced in India. Ginger is about 3391 kg/ha in India as against 8,116 kg/ha in Indonesia. At present we do not have high yielding varieties, appropriate production technologies and a highly conductive climate for spices production.
- Quality is the key to good marketing of spices. Our motto should be 'clean spices' rather than 'cleaned spices'. In order to compete and retain our position in the world spices market, our ability to meet the quality expectations in the areas of pesticide residues, mycotoxins and microbial load should be strengthened.
- Broad-basing the spices export basket by product diversification to meet global quality standards.
- Price competition in the global market is another major challenge for the Indian spice industry. A realistic price reduction to offer healthy competition to our major competitors like Vietnam (for black pepper), Guatemala (for cardamom), etc., should be seriously considered.
- Challenges with respect to the WTO regime and IPR issues.
- Many Indian spices are much valued in the world market because of their intrinsic qualities. However, the import of low grade spices to India and re-export under the Indian label should be regarded very seriously.
- Promotion of organic spices production, so as to cash in on the growing demand for organic produce in the world market. The huge expense involved in the process of certification remains a major challenge to be addressed.

Research Support

Quite a few spices research schemes were initiated in the 1940s. Research on spices was initially limited to formulate cultural practices for major spices like black pepper, cardamom, ginger and turmeric at a few centres under the Department of Agriculture in the Southern states comprising Kerala, Karnataka, Tamil Nadu and Andhra Pradesh.

Since 1949, there have been organised attempts to conduct spices research in the country and to formulate specific programmes for the development of spices. The Government of Madras initiated the first pepper research scheme at Panniyur, Kerala. It was followed by cardamom research schemes at Mudigere (Karnataka) and Pampadumpara (Kerala) in 1951. The Spices Enquiry Committee set up by the Government of India in 1953 made recommendations to commence specific research activities and the opening of new research centres on spices. Accordingly, research centres were set up at Chethalli (Karnataka), Dergaon (Assam) and Sirsi (Karnataka) for black pepper; Kandaghat (Punjab), Targaon (Maharashtra) and Thodupuzha, Ambalavayal (Kerala) for ginger and turmeric; and Burliar (Tamil Nadu) for minor spices. The pre-plan period included research schemes for cardamom in Chennai (Tamil Nadu), ginger in Kerala and Orissa and turmeric in Orissa. These programmes came to a natural end with the termination of their respective schemes. The first five year plan stressed research on pepper, cardamom, clove, nutmeg and minor spices. During the second and third plans, research on spices spread to a number of states including Orissa, Andhra Pradesh and West Bengal. Schemes for minor spices and vanilla were also initiated during this period.

To increase the production and productivity of spices, adequate research support is needed. The research and development work in spices is carried out by Indian Council of Agricultural Research (ICAR), State Agricultural Universities (SAUs), Spices Board, Directorate of Arecanut and Spices Development, Council of Scientific and Industrial Research (CSIR) and a few other agencies. The Indian Council of Agricultural Research is the apex body that conducts, coordinates, manages and funds various crop research programmes. The research and development of spices received new impetus and vitality when ICAR commenced concerted research through the All India Coordinated Research Project on Spices. The establishment of the All India Coordinated Spices and Cashew Improvement Project (AICSCIP) by the ICAR in 1971 (IV Plan) was a major step in spices research in the country

During the fifth plan period, ICAR recognised the need to initiate concerted research on major spices like black pepper, cardamom, ginger, turmeric and tree spices. A regional station under CPCRI (ICAR) was started on 10th November, 1975 at Calicut, Kerala, to carry out research exclusively on spices. The Cardamom Research Centre at Appangala, Coorg, Karnataka, originally established by the Mysore Government to initiate research on cardamom, was taken over by CPCRI in 1974 from the Indian Institute of Horticulture Research,

Bangalore. This was followed by the setting up of the Indian Cardamom Research Institute in 1976 by the Spices Board to work exclusively on cardamom, with four regional stations.

The CPCRI (RS), Calicut, was upgraded and designated as National Research Centre for Spices (NRCS) in April 1, 1986, by merging the CRS, Appangala, Karnataka. The AICRPS, functioning from 1971, was bifurcated into two independent projects, one on spices and other on cashewnut, in September 1986. Based on the recommendations of the QRT and the parliamentary committee (Rajya Sabha), the NRCS was upgraded to a full-fledged institute and called the Indian Institute of Spices Research from July 1, 1995.

The AICRPS is the largest spices research network in the country, providing the much needed R & D base. Research on 12 mandate crops is carried out under the AICRPS in 20 coordinating centres, located in 15 states with the respective Agricultural Universities, representing the major agro-climatic regions.

Highlights of spice research

- The IISR & AICRPS centres hold about 10,500 germplasm accessions in black pepper, cardamom, ginger, turmeric, tree spices, seed spices and vanilla.
- The concerted efforts made during the last three decades resulted in the release of many high yielding varieties.
- Packages of technologies have been developed, such as production (rapid multiplication) of planting materials, seed germination technology, optimum time of planting, spacing, high density planting as well as high production technology(ies).
- Manurial recommendations for different spices have been standardised for obtaining optimum yield and quality.
- Integrated plant nutrient management systems are being evolved, weed management and water requirement have been standardised for a few spices.
- Advances have been made to locate sources of resistance/ tolerance to biotic and abiotic stress.
- Technology for management of pests and diseases has been developed.
- Integrated disease management has been developed against major diseases.
- High quality lines in black pepper, cardamom, ginger and turmeric have been identified.
- Advances have been made in biocontrol of diseases in major spices.
- The production of planting materials have been undertaken.

Research attainments

A considerable increase in productivity has been reported in research stations especially in pepper, ginger, turmeric and coriander.

The critical need is effective transfer of technology. Attempts are being made to identify main production constraints and work out appropriate and economic strategies.

Table 7 *Potential for productivity increase at the national level (kg/ha)*

Crop	National	Progressive farmer	Research station	Foreign
Pepper	315	2000	2445	2925 (Malaysia)
Cardamom	154	1625	450	250 (Guatemala)
Ginger	3477	5500	8250	–
Turmeric	3912	6200	10700	–
Coriander	591	–	1900	515 (Morocco)
Cumin	578	–	2000	–

(*Source:* Directorate of Economics and Statistics, New Delhi and FAO STAT)

An exclusive nucleus seed multiplication programme to supply quality planting materials to farmers is under implementation. Research results are available on the rapid multiplication through tissue culture of black pepper, cardamom, clove, nutmeg, cinnamon, ginger and turmeric.

Biotechnology research has great potential in spices

Potential applications of plant biotechnology in spices

- Micro-propagation and rapid clonal multiplication of high yielding, elite genotypes to generate adequate good quality and disease-free planting materials. E.g. cardamom, vanilla, black pepper.
- Exploiting somaclonal variation and utilisation of techniques like somatic cell hybridization, anther culture, embryo rescue, etc., for crop improvement. E.g. bold rhizome types in ginger.
- *In vitro* selection for resistance to biotic and abiotic stresses. E.g. soft-rot resistant lines in ginger.
- *In vitro* selection for resistance to biotic and abiotic stresses. E.g. soft-rot resistant lines in ginger.
- *In vitro* conservation and safe exchange of germplasm.
- Production of flavour and volatile constituents in culture.
- Synseed technology available for many spices.

Table 8 *Main production constraints in spices*

Spices	Constraints	Research attainments
Black pepper	*Phytophthora* foot rot	Management of the disease through package of practices is standardised. Tolerant lines to the disease are developed (P24), biocontrol is being recommended
	Stunt disease	No effective control measures except uprooting of diseased plants. Causative organism is virus/mycoplasma, insect, micronutrient deficiency
	Slow decline	Effective management method(s) are available except on source of resistance to *Radopholus similes* nematode.
	Low productivity	High Production Technology (HPT) is demonstrated
Small cardamom	'*Katte* disease', 'Vein clearing (*Kokke kandu*) virus diseases,	Virus etiology is established. '*Katte*' escapes are identified. '*Katte*'-tolerant lines are multiplied
	Low productivity	Removal of diseased plants is advocated. No effective management measures are available. The spread has been checked by social action.
Ginger Turmeric	Rhizome-rot	High Production Technology (HPT) is demonstrated. *Pythium* sp. involved, soil solarisation.
	Bacterial wilt	Causitive organisms are identified. Solarisation of soil prior to planting is effective. Biocontrol is being tried out.
	Nematodes infestation	*Ralstonia solancearum* is the causal organism. No resistant variety is available. *Meloidogyne* sp. and *Pratylenchus* sp. identified.
	Low curcumin	High curcumin lines have been identified.
	Low productivity	Integrated plant nutrient management (IPNM) has been developed.

(*Source:* Spice Board, Cochin)

Table 9 *Improved variety and hybrid released so far*

Spices		Hybrids /varieties released
Black pepper	12	Panniyur-1 (H); Panniyur-2 (V); Panniyur-3 (H); Panniyur–4 (V); Panniyur–5 (V); Panniyur-6 (V); Panniyur-7 (V); Subhakara (V); Palode–2 (V); Sreekara (V); Panchami (V); Pournami (V)
Small Cardamom	8	Mudigere–1 (V); Mudigere–2 (V); PV–1 (V); CCS–1 (V); ICRI–1 (V); ICRI–2 (V); ICRI–3 (SKP–14) (V); ICRI–4 (V)
Cinnamon	2	IISR Navasree (V); IISR Nithyasree (V)
Ginger	5	Suprabha (V); Suruchi (V); Suravi (V); Varada (V); Himgiri (V)
Turmeric	16	Co-1(M); Krishna (V); Sugandham (V); BSR-1 (V); BSR-2 (V); Suvarna (V); Roma (V); Suroma (V); Rajendra Sonia (V); Suguna (V); Sudarshana (V); Ranga (V); Rasmi (V); IISR Prabha (V); IISR Prathibha (V); Megha Turmeric-1 (V)
Coriander	17	Gujarat Coriander-1 (V); Co-1 (V); Co-2 (V); Co-3 (V); Gujarat Coriander 2 (V); Rajendra Swati (V); RCr-1 (V); RCr-41 (V); Sadhana (V)' Swathi (V); CS-287 (V); Sindhu (V); UD 20 (V); DH-5 (V); CIMPO S-33 (V); Hisar Anand (V); Pant Dhania (V);
Cumin	5	S-404 (V); MC-43 (V); Gujarat Cumin-1 (V); Gujarat Cumin-2 (V); RZ-19 (V)
Fennel	6	S-7-9 (V); PF-35 (V); Gujarat Fennel-1 (V); Co-1 (V), Gujrat Fennel-2 (V); Rajendra Saurabh (V)
Fenugreek	10	Co-1 (V); Co-2 (V) Rajendra Kanti (V); RMt-1 (V); Lam Selection-1 (V), RMt 303 (V), RMt 143, Hisar Sonty (V) Gutarat Methi-1 (V); Rajendra Khushba (V)

H-Hybrid, V-Variety, M-Mutant
(*Source:* Spice Board, Cochin)

Table 10 *Methods of plant propagation in spices*

Spices	Method of propagation	Advantages
Black pepper	Single-noded rooted cutting using bamboo method	Multiplication rate of 1:40 per year. Good anchorage due to presence of two root systems
Cardamom	Trench method of sucker production	Multiplication rate of 1:20 per year
Clove	Inarching on clove seedlings	Earliness, dwarfness and high productivity
Nutmeg	Epicotyl grafting, top working	Female plants are propagated. Conservation of male plants to female plants. Rapid multiplication of elite plants per unit area
Cinnamon	Cuttage, air layering (rapid multiplication techniques)	Earliness, more plants per unit area
Cassia	Air layering, cuttage	True to type plants
Allspice	Cuttage, layering	Management of rhizome-rot
Ginger and Turmeric	Rhizomes	Crop rotation, solarisation and disease-free rhizomes

(*Source:* Spice Board, Cochin)

Some of the salient achievements of biotechnology in spices include micro-propagation and clonal multiplication in cardamom, black pepper, ginger, turmeric, cinnamon and vanilla. *In vitro* selection for resistance to biotic and abiotic stresses especially rhizome-rot in ginger is in progress. Protocols are also available for direct regeneration through tissue culture in long pepper, chaba pepper, kaempferia, curry leaf, tamarind, coriander, fennel, cumin, dill, lavender, aniseed, thyme and large cardamom. Synseed technologies are also available for cardamom, ginger, turmeric and pepper.

Quality maintenance in Indian spices

India has a global monopoly in spices oils and oleoresins. During 1998–1999 India exported 2625 tonnes of oils and oleoresins valued at Rs. 293.10 crores against 2351 tonnes valued at Rs. 222.21 crores during 1997–1998. There are AGMARK specifications for various Indian species and their value-added products. The Bureau of Indian Standards has published a series of specifications to maintain and improve quality of Indian spices. High curcumin lines in turmeric, high piperine lines in black pepper, high essential oil lines in clove, cinnamon and nutmeg, etc., are in the pipeline at the Indian Institute of Spices Research, Calicut.

Table 11 *Salient achievements in biotechnology in spices*

Cardamom	Clonal multiplication, field evaluation of tissue cultured plants, inflorescence culture, regeneration of plantlets from callus, somaclones evaluated
Black pepper	Clonal multiplication, callus culture and regeneration of plantlets, micro-propagation of related species of *Piper*, *Piper longum, P.chaba; P.betle, P.colubrinum* and *P.barberi*, Synseeds
Ginger	Clonal multiplication, field evaluation of tissue culture plants, *In vitro* rhizome formation, regeneration of plantlets from callus, *In vitro* selection for resistance to diseases, *in vitro* polyploidy, inflorescence culture and *in vitro* development of fruit, synseeds
Paprika	Plants regenerated from shoot and leaf explants
Curry leaf	Protocols isolated. Micro-propagation protocols standardised
Turmeric	Micro-propagation, plant regeneration from callus, micro-propagation of related species and genera, synseeds
Tree spice	Micro-propagation, *in vitro* proliferation of mace in nutmeg
Vanilla	Seed and embryo culture
Herbal spice	Micro-propagation and callus cultures

(*Source:* Spices Board, Cochin)

Table 12 *Bio-control of pests and diseases in spices*

Spice	Pests/diseases	Bio-control agent tried
Black pepper	*Phytophthora capsici*, scale insects, top shoot borer, nematodes	*Trichoderma* sp. in combination with VAM. Natural enemies viz., *Encarsia lounsburyi*, *Apanteles cypris*, an important parasite, *Verticillium chlamydosporium*, *Trichoderma* sp., VAM
	Pollu beetle	An entomophagous mite and nematode
Cardamom	Rhizome-rot and damping-off	*Trichoderma* sp. and *Paecilomyces lilacinus*
	Root knot nemadote	*Paecilomyces lilacines*
Ginger and turmeric	Rhizome-rot	*T. harzianum, T. hamatum, T.viride*
Coriander	Wilt	*T. viride, T. harzianum, Pseudomonas fluoresences*
Fenugreek	Root knot	*T. viride*

(*Source:* Spices Board, Cochin)

Environmentally friendly spices culture

The use of plant protection chemicals has led to residual toxicity in spice products. Moreover, there is a high premium for organic spices. Biocontrol of pests and diseases is another area where useful information is being generated.

A fungus, *Trichoderma* sp, in combination with VAM suppresses *Phytophthora capsici*, the causative organism of foot-rot in black pepper. Biopesticides to manage pests of spices are also being developed. There has been a major break-through in the generation of new technologies.

Spice Oils and Oleoresins

As raw spices do not provide uniformity in flavour, strength and colour, spice oils and oleoresins are used in flavour formulations. The spice oleoresins and oils score over the respective raw materials in their flavour concentration, solubility, stability, uniformity and hygienic quality and are hence preferred in modern food industries. Spice oils are volatile substances which are mostly terpenic in nature and are obtained by the steam or hydrodistillation methods. Oleoresins are obtained by organic solvent extraction of the powdered spices and they contain essential oils fraction, bite principles, colour and resinous matter.

Spice oils

In most spices, the odour or aroma factor is attributed to the volatiles (essentials oils) present in them. This 'essential oil' can be obtained by steam distillation of the ground spice. The oil is made up of hydrocarbons like terpenes, sesquiterpenes and their oxygenated derivatives.

Steam distillation is a gentle process, in which the oil components from the ground spices are entrained by steam at a temperature less than 100°C. With a rapid flow of steam, large quantities of oil can be collected.

Production

The spice is ground before distillation, using a suitable mill. The crushed material is distilled as quickly as possible to avoid any loss of oil. The distillation of oil should be carried out using stainless steel instruments to avoid metallic and other contaminations which are undesirable in materials used for food flavouring. The ground spice is kept in a cage with a perforated bottom. From an external source, steam is let in through the bed of the material. As the steam passes, it carries with it the volatile components of the spice by entrainment. The mixture of steam and oil passess through a water-cooled condenser, gets condensed and the mixture of water and oil enters the oil-water separator situated at the other end of the condenser. Since most of the essential oils are lighter or heavier than water, they are separated out and the oil can be collected easily.

Precautions have to be taken regarding the rate of flow of steam, condensate temperature, rate of distillation, etc., to ensure the best results. The distillation time should be sufficient to ensure complete recovery of all the important constituents of the oil.

The oil collected from the separator is filtered and dried over anhydrous sodium sulphate. The oil is packed in brown glass or tinned metal containers. The containers are filled to the brim to exclude air, since oxygen can cause deterioration. The oils are then stored in a cool place.

Supercritical fluid extraction technology (SCFE)

The SCFE system is the modern technology of making oils and oleoresins through carbon-dioxide processing. This system provides a modern cost-effective technology for value addition in the processing of various agricultural commodities like spices. The CO_2 extraction of essential oils utilises liquid carbon dioxide at low temperature (0 to + 10°C) and high pressure (8 to 80 atm) as a selective solvent. Under these conditions, CO_2 liquifies to become a non-polar solvent which is odourless, tasteless, colourless, easily removed without residues and non-flammable. Selectivity is demonstrated by the extraction of the under-graded essential oil plus the lighter molecular weight fractions of the resin, imparting the character of an 'absolute' to the extract. The liquid and supercritical gaseous extraction of spices and other aromatic plants offers many interesting and novel products to the food product developer. This technology is preferred worldwide for commercial-scale extraction because it offers a:

Superior product

- Delicacy and freshness is close to natural
- High potency of active components
- Excellent blending characteristics
- Longer shelf life
- Free of biological contaminates

Super technology

- Simultaneous fractionation of extract
- Pollution-free process

- Provides solution to international concerns
- No residual solvent
- No residual pesticide

Yields

The volatile oil content of different spices varies widely. It may be as low as 0.1% and as high as 18%.

Quality evaluation

Ideally, the quality of spice oils should be assessed organoleptically by experienced, well-trained panels. But this method is slow and impractical when quick evaluations are needed by industry.

Physico-chemical properties like density, solubility, refractive index and optical rotation are used to characterise essential oils, but they do not provide complete information on the quality. Thin-layer chromatography, gas liquid chromatography, infrared spectroscopy, mass spectrometry, etc., are modern methods used for the objective assessment of quality of essential oils.

Spice oleoresins

The oleoresins represent the total flavouring components of spices in a very concentrated form. They are very close to the flavour of the whole ground spice. Spice oils only give the aroma of the spice, whereas spice oleoresins represent the total flavour of the spice. Oleoresins are prepared by extracting the ground spice with solvents like alcohol, acetone, hexane, methylene chloride, ethylene dichloride and ethyl acetate. Care should be taken to use pure solvents (food grade) that do not leave behind high boiling residues, which affect the flavour adversely. The major advantages with oleoresins are: flavour stability, less storage space requirement and uniformity in flavour strength. The main steps involved in the oleoresin process are:

1. Grinding the spice
2. Extraction
3. Distillation of the miscella (extract)
4. Blending the finished product

The ground spice is placed in extractors, which are preferably made of stainless steel. The bed of material rests on a perforated false bottom. The solvent is let in at the top. It percolates down the bed of material, carrying with it the solubles like volatile oil, fatty oils, colours, resins and pungent principles. The contact time between the solvent and the spice particles should be sufficient to ensure diffusion and extraction of the flavour constituents. About 4 to 5 volumes of the solvent are required to get most of the solubles out of the ground spice. This volume can be reduced by using countercurrent extraction, where dilute extracts can be used for extracting fresh material.

The concentrated extract (miscella) is distilled to remove the solvent. A vacuum is used towards the end, to prevent damage by heat and to ensure the complete removal of the solvent. Since food laws require the oleoresin to contain not more than 30 to 60 ppm of solvent residue, the final traces of solvent have to the removed carefully.

For obtaining a standard quality oleoresin, it is important that the choice of the solvent, particle size of the spice and the ratio of solvent to spice are properly determined. Attempts have been made to prepare oleoresin by using the super-critical extraction technique. The process is yet to be commercialised.

The finished product (oleoresin) is a dark viscous liquid. The flavour or colour strength of the oleoresin can be adjusted to the required standard by dilution with permitted dilutents. It can be fortified with separately distilled essential oils to achieve a balance between pungency and aroma. The yield of oleoresin and the main constituent present in them, from some spices, are given in Table 13

Table 13 *Yield of oleoresin and the principal constituent*

Spice	Yield of oleoresin (%)	Principal constituent in oleoresin
Pepper	10–12	Piperine (35–60%)
Ginger	4–7	Gingerene (25–30%)
Chillies	12–16	Capsaicin (2–4%)
Turmeric	6–9	Curcumin (20–30%)
Coriander	18–20	D–linalool (1.5–2.0%)

(*Source:* Spices Board, Cochin)

As mentioned earlier, the oleoresins are too viscous and concentrated to be used as such in flavour blends or food products. It is customary to disperse them on solid or liquid media to dilute them and make their use more easy.

Standards for spice oleoresins have been given by the Essential Oil Association of America. Other organisations like the Indian Standards Organisation and the International Standards Organisation have also prescribed such standards.

Applications of spice oils and oleoresins

Spice oils and oleoresins can be used to advantage wherever spices are used, except in those applications where the appearance of the filler aspect of the spice is of importance.

In addition to the benefits of standardisation, consistency, and hygiene afforded by spice oils and oleoresins, there is a huge potential in their use for new product development. New flavours and fragrances are constantly being sought to entice the consumer. This applies equally to food products, medications, as well as other non-food products.

Food

Processed meats

The use of spices, particularly pepper, in the manufacture of meat products, is traditional for imparting flavour and maintaining the quality of the products. The typical seasoning mix for fresh sausages, for example, consists of pepper, capsicum, ginger, nutmeg, plus herbs. For dry sausages and pickled meats, cardamom and coriander are also used.

The move to use oleoresins has been accelerated by the increasing size of manufacturing plants, where the use of spice extracts benefits production quality, as well as easy handling and cost savings.

The above-mentioned spices are used in the dispersed form of their oleoresin, while cardamom and coriander are used in the form of their oils.

Fish and vegetables

The seasoning mix for both fish and vegetables, and particularly pickled or brined products such as herrings, contain a wide range of spices and herbs. The use of oleoresins, particularly dispersed oleoresins on a soluble base, will provide a means of easier preparation, reduced handling and costs.

Soups, sauces, chutneys and dressings

The increasing demand for convenience products available in the form of a dry mix for ready reconstitution, has caused a rapid move from conventional seasoning towards dispersed or encapsulated oleoresins and oils.

Oleoresins of celery, pepper, capsicum, etc., are used in conjunction with the oils of onion and garlic. Coriander and ginger extracts are used in barbeque sauces. The use of the lesser known but highly useful spices, such as fenugreek, broaden the new product opportunities.

Cheeses and dairy products

The use of spices in cheeses is established in Germany, including 'Quark'. Spice extracts are unlikely to be used in these products as spices provide the flavouring and visual impact.

However, spice oils and oleoresins have significance in processed cheeses and savoury spreads.

Baked foods

The use of cardamom baked goods in Scandinavia and Germany is traditional. The baking industry generally uses ginger, cinnamon and nutmeg. The move from the spice to their oleoresin has been effectively taking place for many years for ease of handling and simplicity in manufacture. The use of spice extracts in cake fillings, biscuits, and snack products is also steadily increasing.

Confectionery

The use of spices and spice extracts in the confectionery area is rather rare, but the demonstration of the use of such material as cardamom oil and other extracts in toffees, chocolates, and others, has shown that they provide a very novel and pleasing confectionery ingredient which is new to the market.

Snacks

Flavouring is an essential component of the appeal of snack products, and unusual because the flavour is often applied on the surface, either by spray coating or dusting. For this purpose, the seasoning mix has to be capable of being applied in spray form or powder. Oleoresins of pepper, chilli and celery, are widely used. Turmeric and chilli extracts are used to provide colour.

Beverages

Spice oils are used for the preparation of soft beverages, for example, ginger oil in the preparation of ginger beer, etc. Some of the less well-known spice extracts can be used to produce very pleasing soft drink products. But, these are not yet widely known outside of the countries in which they originate

Cosmetics

The use of spice oils in the preparation of creams, soaps, shampoos, lacquers, lipsticks, etc., is well known. However, some of the materials available from India are not yet widely used, nor are they recognised as providing a new dimension to cosmetic products.

The growing preference for herbal, spicy and spicy coniferous products like shampoos and hair tonics are noted, yet such extracts as those of cardamom and fenugreek are little heard of. The use of lesser known spice extracts can provide new product appeal.

Perfumes

The perfumery industry uses a wide range of essential oils and oleoresins from sources far and wide, yet some of the lesser known oils and oleoresins are hardly used. An examination of the wide range of those available from India could well provide a new basis for products of appeal.

Hygiene products

Products like toothpastes, mouthwashes, etc., depend on essential oils to provide their pleasing flavour, making them not only acceptable, but pleasant to use.

In cleansing materials, detergents, etc., spice oils provide the aromatic appeal in otherwise uninteresting sometimes offensive notes associated with some of the base products.

Aerosols

The use of aerosols worldwide is increasing at a significant rate in products such as air freshners, polishes, lacquers, and many cleansing agents, and waxes, etc. All of these are perfumed with essential oils for pleasant and fresh aroma.

The range of spice oils from India can contribute to new product development.

Pharmaceuticals

Both oils and oleoresins are widely used in pharmaceutical products, to provide either a pleasant taste or aroma to render them pleasing and easy to use, which would otherwise be difficult to accept. These include medications, skin creams, cold remedies, etc.

Value-added Products

Most staple foods are bland, although they provide important sources of nutrition and are essential for growth and maintenance. Food is made palatable and enjoyable by using spices and herbs which are valued for their flavour and colour. The use of whole spices in food processing is limited to the extent of providing 'eye appeal' to the final product and not imparting full characteristic flavours. The flavour moiety is well protected from evaporators and in whole spices by the cells. Whole spices release flavour slowly; ground spices release the flavour more quickly.

Need for value-added products

1. Ground spices have a limited shelf-life and the flavour loss/modification is considerable during grinding/storage. They are of poor hygienic quality and hence find restricted use in processed foods.
2. Dispersed spices are prepared by blending spice oils and oleoresins with carriers like dextrose, salt, rusk, etc. They tend to lose flavour during storage due to the large internal surface area on which the aromatic components are spread.
3. The characteristic flavour (fresh)/aroma is absent in dried/dehydrated spices like ginger, green chilli, garlic, curry leaf, etc.
4. Spice oils do possess good flavour but are often incomplete and unbalanced. Some flavour components are readily oxidised. They are quite concentrated and hence difficult to disperse in product mixes.
5. Spice oleoresins represent the total flavour profile of the spice being concentrated (10–50 times the spice). They are viscous and difficult to incorporate in to the end product. Solvent residues (high boiling fractions) contribute to off-odours and off-flavours. Doubts have also been raised regarding the safety-in-use of residual organic solvents in spice extractives.
6. Processing steps like ultra-high temperature and microwave demand stable flavours.
7. The growing demand for convenience foods needs suitable stable flavouring materials.
8. Better flavour quality can be obtained by suitable blending of naturals and synthetics.

Irradiated spices

Whole or ground spices are unhygienic since they are contaminated with a heavy load of saprophytic and thermophilic bacterial spores, moulds and yeasts (10^2–10^7/g). Gamma irradiation serves as an effective means of reducing bacterial load in natural spices. A dose of 3 to 10 KGY can reduce the viable cell count to an acceptable level of $<10^4$/g and the flavour quality remains unaffected.

Liquid water-soluble emulsion

The oil or oleoresin can be extracted in glycerol, isopropanol, propylene glycol, etc. Although the extraction is poor, the shelf-life of the emulsion is generally better. Solubilised spices are prepared by mixing them with a polysorbate ester (Tween 80) or by emulsifying them with plant gums.

Encapsulated flavours

One way by which the loss of volatiles can be obviated is to encapsulate the flavouring components by using an impermeable shell, which locks in the aromatics and thereby considerably increases their shelf-life. Encapsulation is essentially a method of providing a 'wall' material around the 'core' flavour, thus protecting the flavour from deterioration/oxidation. In this technique liquid droplets/particles are trapped in a continuous film of a polymer-celled wall material. The selection of the wall material is governed by its final end use. There are several methods of encapsulation like spray drying, molten extrusion, molecular inclusion, gelatin coating, spray cooling, etc. Encapsulated flavours are particularly useful in dry mixes and convenience foods.

Spray drying

This is the most widely used method for the encapsulation of flavours. The method involves two unit operations, viz., emulsification of the spice flavour with an aqueous solution of plant gum/modified starch and dehydration in a suitable drying chamber. The 'core' flavour is homogenised with gum/starch solution and the emulsion is pumped in controlled amounts to the spray dryer's centrifugal atomiser. The dried particles are carried away in the stream of air and separated in a cyclone collecting chamber. There are several processing parameters which decide the product quality. These are viscosity of the emulsion, flavour load, degree of homogenisation, in-feed temperature, in-feed flow rate, operating variables of the spray dryer, etc.

Molten extrusion

Otherwise known as Sunkist fixation method, this was originally employed for the encapsulation of citrus flavours. Extrusion as applied to flavour encapsulation means that a flavour emulsion is forced through a die. A mixture of glycerol and corn syrup solids is heated to a molten state and cooled to a semiplastic mass when the essential oil is added and emulsified. The hot emulsion is extruded into a cold solvent when the filaments or standards harden. These filaments are impact-broken, separated from the solvent, dried under vacuum conditions and suitably packed after the incorporation of an anti-caking agent.

Molecular encapsulation

The encapsulation occurs on a molecular basis with beta-cyclodextrin. The reaction between beta-cyclodextrin and a suitable guest molecule takes place in a solution. Unlike spray drying, molecular inclusion techniques have the disadvantage that some of the chemical constituents cannot be encapsulated. The use of beta-cyclodextrin is approved of by some countries like Japan, Hungary and the Netherlands.

Spray cooling

A variant of the spray-drying process is known as 'spray cooling'. The product is prepared by blending an essential oil/oleoresin with a fatty carrier (hydrogenated fat). It is formulated for use in products such as mayonnaise, bakery, etc. In the hydrogenated fat, the product may be in the form of a solid block, which must be melted before use or it may be spray coated to obtain a free-flowing powder. To obtain a spray-cooled spice, the blend of spice oil/oleoresin and the fatty carrier is melted to 45–120°C and atomised in a stream of chilled inert gas (nitrogen) when minute fatty globules are formed and instantly 'set', thereby protecting the entrapped flavour. An antioxidant may be added to delay the onset of rancidity on storage. The use is limited to fatty products such as dry cake mixes, potato chips, etc., since it is water insoluble. Other methods of encapsulation do not find use in food applications.

Organic Cultivation

In recent years, organic agriculture has been gaining considerable importance. Many farmers today show interest in organic cultivation. Several of them have begun switching to this traditional method of cultivation as a means to produce safe foodstuffs and preserve the environment. The concept of sustainable farming has finally caught on in India.

World demand for organic foods

The world demand for organically produced foods is growing rapidly in developed countries in Europe, in the USA, Japan and Australia. The current estimated share of organic foods in these countries is approximately 1 to 1.5 per cent which is expected to increase to 10 per cent in the next couple of pentinniums. The market size of organic spices in the US is 30,000 tonnes, Europe 21,000 and Japan 6,000.

The worldwide food trends are changing with a marked health orientation. Since organic foods are free from chemical contaminants, the demand for these products should steadily increase in the new millennium. According to the ITC, UNCTAD/GATT, more than a 130 countries produce certified organic foods. Many of them are from Asia and Africa.

Internationally, there is a definite shift towards traditional/ethnic medicines. Since spices form part of many of these medicines, the demand for organically produced spices should grow.

Organic cultivation is nothing new to India. The country has always been practising the traditional ways of using indigenous technologies and inputs, mostly in line with modern organic farming principles. The per capita consumption of fertilisers and pesticides in India is far below that of the other developed countries. This means that it is very easy for Indian farmers to embrace organic spice farming in its true sense.

The Spices Board, India, has taken a major initiative in promoting the production and export of organic spices on a large scale.

The Indian initiative

The export of organic spices from India has started in right earnest. The country at present exports around 50 tonnes of different varieties of organic spices. These exports will get a significant boost in the coming years as more farmers are switching to organic methods.

The Spices Board, India, has prepared a document on the production of organic species. It features the organic concepts, principles, basic standards, production guidelines, documentation, inspection and certification. The document has been published after approval by the National Standards Committee constituted by the members of IFOAM in India.

Research programmes on the organic cultivation of important spices have commenced. The work is being carried out at the Spices Board's Indian Cardamom Research Institute at Idukki District in Kerala. Besides organising demonstrations to educate and motivate prospective organic spice growers, the Board is simultaneously involved in conducting training programmes for existing spice growers, on organic principles and practices.

Developing organic spices

The Spices Board encourages non-governmental organisations and farmers' groups to promote and adopt organic farming techniques. Infrastructure development for production of organic input is underway in prominent growing areas. The development of vermi-compost units and the supply of post-harvest aids for upgrading the quality of the produce at the farm level are also in progress.

Certification of organic spices

Certification for organic spice farms is an essential pre-requisite for marketing their produce, especially in the international markets. The Government of India is in the process of setting up indigenous certification programmes, since there is no domestic certifying agency in India. However, the offices of certain international certification agencies do operate in India.

For timely inspection at optimum cost, the Board has access to the services of trained inspectors from both governmental and non-governmental organisations. These inspectors were trained in 1998 by the members of IFOAM in India. They will commence work once the domestic certification body is set up and approved by the competent authorities.

Help in sourcing organic spices

The Spices Board provides assistance to overseas buyers in sourcing organic spices from India. The country has the capability to produce and supply a variety or organic spices. If firm orders are placed well in advance, farmers can go ahead with confidence, cultivate squeaky-clean spices organically, and deliver them on time.

What follows is a list of agencies/farmers involved in the production and export of organic spices/spice products from India. The products covered are black pepper, white pepper, cardamom, clove, nutmeg, mace, ginger, turmeric, vanilla and seed spices. Many of these farms are certified, while others are either in the process of getting certification, or have initiated the necessary action.

Ajowan

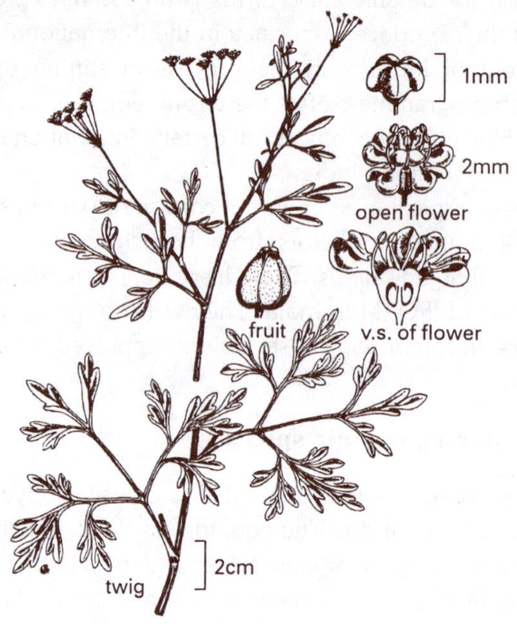

Plant Profile

Family	:	*Umbelliferae/Apiaceae*
English name	:	Bishop's weed
Indian name	:	*Ajowan* (Hindustani / Hindi), *Jowan or Joan* (Bengali),
	:	*Yavan* (Gujarati), *Jawind* (Kashmiri), *Oma* (Kannada), *Omum* (Malayalam), *Onva* (Marathi), *Juani* (Oriya), *Ajowain* (Punjabi, Urdu), *Ajamoda, Aavanika (*Sanskrit), *Omum* (Tamil), *Vamu* (Telugu).
Species	:	*Trachystermum ammi* (L.); Sprague (Syn. *Carum copticum* Heim; *T. copticum* Link)
Distribution	:	Egypt, Iran, Iraq, Afghanistan, Pakistan and India
Uses	:	Flavouring, beverages, pan mixture, medicine

The scientific name of *ajowan* is *Trachypermum ammi* (L) Sprague (Syn. *Carum copticum* Heim, *T.copticum* Link) and it belongs to the family *Umbelliferae/Apiaceae*. It is an annual herbaceous plant grown for its herb and greyish-brown fruits (seeds) which constitute the spice. The trade name *ajowan* or *ajwain* is based on the Indian name which is derived for *adarjawan*. It is known as 'Bishop's weed' in English and *ajowian* in Bengali and Hindi.

Ajowan is grown for its seeds (fruits) the composition of which is as follows; moisture-8.9%; protein-15.4%; fat (ether extract); 18.1%, crude fibre; 11.9%, carbohydrates-38.6%, mineral matter (total ash), 7.1%, calcium-1.42%, phosphorus-0.30%, iron-14.6 mg/100 g, vitamins (per 100 g) thiamine-0.21 mg, riboflavin-0.28 mg; nicotinic acid 2.1 mg; carotene; 71 mg. Its calorific value is 379.

The steam distillation of crushed *ajowan* seeds yields 2.5 to 4% essential oil (*ajowan katel*). The *ajowan* seed oil is pale yellowish-brown in colour, contains twenty-seven compounds and has a characteristic thyme odour with a sharp burning taste. The characteristic odour of the oil is due to the phenols mainly, thymol (30-60%) and some carvacrol. On standing, the major portion of thymol gets crystallised which is sold in Indian markets as '*ajowan kaphul*' or '*sat ajowan*'. The remainder of the oil is called thymene, which has utility in the perfume industry (mainly in soaps) due to the presence of the gamma-terpinene (30-35%) in it. The other major components of the oil are p-cymene (50-55%) γ and β-pinenes (4-5%), dipentene (4-6%) and a small amount of camphene and myrcene. The aqueous solution of thymol is an excellent mouth wash and thymol is a constituent of many toothpastes and is useful for curing a variety of fungal infections of the skin. The p-cymene in oil has germicidal properties. The aqueous portion, left behind after the separation of the essential oil, is known as *omum* water, which is used in gripewater preparation for children to treat flatulence.

The seeds of *ajowan*, like other spices, are used in small quantities for flavouring numerous food, as antioxidants, preservatives in confectionaries, beverages and *pan* mixture, in medicines, for the manufacture of essential oil for ultimate use in perfumery, essences, etc. Tasty pickles, mouth-watering *chats* and delicious vegetable dishes are dull without *ajowan*—the spice for strength and vigour. *Ajowan* is also a must in most households after a heavy meal to aid digestion.

Ajowan has long been used in indigenous medicine as a carminative, from the time of ancient Indian physicians like Charaka and Sushruta. Even Greek physicians like Dioscrides and Gelan used it in various carminative medicines. The seeds are also used for treatment of diarrhoea, dysentery, atonic dyspepsia, cholera, colic, skin diseases, hysteria, flatulence and indigestion. A tablespoon of these seeds with a little rock salt, asafoetida and myrobalan is a household remedy for indigestion and relieving flatulence. *Ajowan* is an effective remedy for common cold used traditionally over the ages. It has the remarkable power to open up clogged and congested nasal passages. A tablespoon of seeds crushed and tied up in a cloth bundle can be used for inhalation. Moreover, the seeds are effective for curing bronchitis and asthma. They are also used in lotions and ointments in cosmetic industries.

Ajowan oil, both pure and dethymolised, is used as an antiseptic, aromatic stimulant and carminative. It possesses antimicrobial properties. The oil is useful as an expectorant in emphysema, bronchial pneumonia and other respiratory ailments. It also seems to possess an antidiuretic effect. It is beneficial in the treatment of migraine and rheumatic pains. The IC_{50} of essential oil from *T. ammi* against P 388 mouse leukaemia cells was found to be 12.2 mg/ml.

The material left over after distillation is rich in protein (15-17%) and fat (20-25%) and is used as cattle feed. The leaves are used as a vermicide. Even the roots of the *ajowan* plant are reported to be diuretic and carminative.

Origin and distribution

It is a native of Egypt and is cultivated in the Mediterranean region and Southwest Asian countries like Iran, Iraq, Afghanistan, Pakistan, India and even in Egypt in Africa. In India, it is grown largely in the states of Uttar Pradesh, Bihar, Madhya Pradesh, Gujarat, Punjab, Rajasthan, West Bengal, Tamil Nadu, Maharastra and Andhra Pradesh.

Description of the plant

Ajowan is an erect, glabrous or minutely pubescent, annual herbaceous plant growing to a height of 70-90 cm. It has soft fine hairs. It has many branched striated leafy stems, feather-like leaves, 2-3 pinnately divided. The segments are linear and the ultimate segments are 1.0-2.5 cm long; the flowers are in terminal or seemingly lateral, pedunculate, compound umbels and white in colour. The minute grayish-white fruits are ovoid in nature, muricate, aromatic cremocarps with distinct ridges and tubercular surface and are 1-seeded.

Climate

The crop is tolerant to drought and has wide climatic adaptability. The area should be free from severe frost during winter months. It is mainly grown as a winter-crop in subtropical and temperate climate. It is generally grown in October-November and harvested in February-March in North India, and sown in May-June in South India.

Soil

Ajowan grows on any soil type from heavy clay to silt loams, but performs best in humus-rich loam or clayey loam soils. It is grown as a rain-fed crop in heavy soils, whereas it requires irrigation in light textured soils. It is very sensitive to water-logging and needs good drainage in case of medium to heavy soils. Black cotton soil is well suited for its cultivation.

Varieties

The following three improved varieties have been released in this crop.

Gujarat Ajowan-1

It is tolerant to drought, flowers in 110 days after sowing. Each plant produces about 291 umbels with 13 umbellets in an umbel and bears 21 seeds per umbel. It has a 1000 seed weight of 90 g. The seeds are bold with an essential oil content of 3.61% and matures in 176 days. It has yield potential of 1475 kg/ha.

Lam Sel.1

This is an erect type, grows to a height of 99 cm, matures in 120 days and produces good quality seeds. Flowering starts 83 days after sowing, produces 85 umbels per plant and has a yield potential of 1140 kg/ha.

Lam Sel.2

This is a spreading, bushy type. Flowering starts in 90 days after sowing and matures in 132 days, it produces 96 umbels per plant and yields 1170 kg/ha.

The other varieties of *ajowan* include S25, S95, S47, S84 and IC3743, which are reported to yield essential oil upto 3.9% in the plains and upto 4.9% in cooler areas. These selections exhibit considerable variations in agronomic characters.

Cultivation

Ajowan is generally grown as an annual crop from the fruits (seeds), and is sown directly in the field, following the line sowing method. The field is ploughed repeatedly during September-October. Farmyard Manure (FYM) @ 10-15 t/ha is incorporated into the soil before the final ploughing. The seeds are sown broadcast or drilled in rows 45 cm apart; the seed rate is 3-4 kg/ha for the broadcast method of sowing and 1.5 kg/ha for drill sowing. The seeds germinate in 7-14 days. The broadcast as well as the line sown crops should be thinned to a spacing of 35 cm within 30 days of sowing. Thinning should be carried out when the plants attain a height of 20-25 cm.

Manures and fertilisers

A fertiliser dose of 20-80 kg N, 30-50 kg P_2O_5 and K_2O and 50 kg S should be applied per hectare for obtaining the best yields. The full dose of P_2O_5 and K_2O should be incorporated into the soil prior to sowing. Nitrogen is applied in three equal split doses, the first being at the time of sowing, and the subsequent two doses being at 30 days' intervals.

Irrigation and interculture

Irrigation is provided immediately after sowing if there are no rains and later at 6-7 days' intervals up to one month and further once in 15 days is sufficient. Weeding is generally done twice whenever the weeds are noticed.

Crop rotation

Ajowan is grown often in rotation with sorghum and pearl millet.

Pests and diseases

Pests

Aphids and ants are the major insect pests of this crop. To control aphids, Rogor (0.1%) is sprayed. The ant menace can be minimised by mixing about 10 kg of 6% Heptachlor per hectare to the soil.

Diseases

Collar-rot and root-rot caused by *Sclerotium rolfsii* is observed in this crop. The disease can be controlled by spraying 0.2% Mancozeb. Powdery mildew is a major disease of this crop, which can be controlled by spraying Contaf (0.1%) at weekly intervals. In warehouses, the cold treatment of *ajowan* seeds at 2°C for 16 days is reported to destroy the fungus.

Harvesting and yield

The plants start flowering in 2 months' time. The crop is harvested 120-140 days after sowing when the flower heads turn brown or the mature umbels can be cut with a sickle manually. The harvested crop is then tied in small bundles and staked for drying, keeping the bundles upside down and then threshed on a clean floor and winnowed to separate the clean seeds.

An average yield of 1100 to 1400 kg of seeds can be obtained from one hectare, which in turn will give 30-35 kg of *ajowan* oil per hectare.

Grading

Indian standards are specified for three grades of *ajowan* viz., special, good and fair.

Packaging and storage

The *ajowan* seeds which are free from leaf, chaff, stones, etc., are packed in clean gunny bags of 40-45 kg capacity and stored in hygienic warehouses or godowns. The seeds are normally stored for 6-8 months, but can be kept upto a year or so.

Adulterants

Commercial samples of *ajowan* are often found adulterated with earthy matter; the seeds are also adulterated and known in commerce as '*Makara*'.

Value-added products

The major value-added products of *ajowan* are the oil of *ajowan*, thymol, dethymolised oil or thyme and fatty oil.

Distillation of oil

The dried seeds are crushed and distilled to obtain the essential oil. Hydro or steam distillation or use of subcritical liquid carbon-dioxide is used. The seeds lose the essential oil when stored for a long time. On an average, the dry seeds yield 2.5-4% essential oil.

Allspice

Plant Profile

Family	:	*Myrtaceae*
English name	:	Allspice, Pimenta, Pimento, Jamaica Pepper
Indian name	:	*Gandamenasu, Sarwasambaru* (Kannada), *Kappalmulaku* (Malayalam), *Kattukkaruve* (Tamil)
Species	:	*Pimenta dioica* (Linn.) Merr. Syn. *P. officinalis* Lindl, *Euginia pimenta* DC.
Distribution	:	West Indies, Central America, Southern Mexico, Honduras, Guatemala, Cuba, Costarica, India.
Uses	:	Condiment and Flavouring agent

Allspice (*Pimenta dioica* (Linn.) Merr.Syn. *P.officinalis* Lindl, *Eugenia pimenta* DC, belonging to the family *Myrtaceae,* is a minor tree spice. The spice is also called 'pimenta'. The term pimenta had its origin in the initial incorrect belief of the early Spanish exporters who thought these berries were similar to pepper. Then came the English expression 'Jamaica pepper' and the German 'Nalkenpfeffer'. It is now officially recognised as 'pimenta'. The allspice or 'pimento' of commerce is the dried immature fruit. The name 'Allspice' came from the fact that the spice is said to possess the characteristic flavour and aroma of cloves, nutmeg and cinnamon all combined in this one spice. Allspice is available both in whole or ground form. The analysis of ground berries of Allspice shows the following composition: moisture-8.8%, protein-6.0%, fat (ether extract)-6.6%, fibre-21.6%, carbohydrates-52.8%, total ash-4.2%, calcium-0.8%, phosphorus-0.1%, sodium-0.08%, potassium-1.1%, iron-7.5 mg/100 g, vitamin C (ascorbic acid)-39.2 mg/100 g, vitamin B_1 (thiamine)-0.1 mg/100 g, vitamin B_{12} (riboflavin) 0.06 mg, niacin-2.9 mg and vitamin A-1445 IU. The food energy or calorific value is 380 calories per 100 g. The characteristic odour of allspice is due to the presence of an essential oil (3.3–4.5%), concentrated mainly in the pericarp. In addition, it contains inercitannic acid (over 8%) responsible for the astringency, a soft resin with a burning taste, fixed oil (5.8%), proteins (5.8%); crude starch (20%) and traces of alkaloid.

The berries are used as a condiment and flavouring agent in ketchups, soups, sauces, pickles, canned meats, sausages, gravies, relishes, fish dishes, pies, puddings, preservatives, etc. They are used as a flavouring ingredient for wines and as a perfume in soap making. It is an important ingredient of whole mixed pickling spice and spice mixtures, *viz.,* curry powders, mincemeat spice, poultry dressing, frankfurter and hamburger, etc.

Allspice is used as an aromatic stimulant in digestive troubles. The powdered fruit is used in flatulence, dyspepsia and diarrhoea. Earlier, it was used in medicine as an adjuvant to tonics and puratives. It is considered to be carminative. It is an anodyne against rheumatism and neuralgia. Pimenta berry oil is used for flavouring condiments and food products and in perfumery, soap and pharmaceutical preparations.

Origin and distribution

Allspice is indigenous to West Indies and tropical Central America and grows semi-wild in Jamaica (West Indies), which is the major producer. It is also reported to occur in Southern Mexico, Honduras, Guatemala, Cuba and Costa Rica. In India, allspice is cultivated in

West Bengal, Bihar and Orissa. The cultivation is slowly increasing in Kerala, Karnataka and Tamil Nadu. It is found to grow well and fruits heavily in Bangalore. Its cultivation has been recommended in the hilly districts of Karnataka along the river valleys.

Description of the plant

Allspice is a small, medium-sized evergreen tree which grows upto a height of 8 to 10 m. The flowers are small and whitish in colour with a peculiar aroma. They are present in groups of cymes. They are structurally hermaphrodite, but functionally dioecious and are scientifically called '*Polygamo dioecious*'. Those trees which bear no fruit are male trees wherein the flower will have above 100 stamens, while the flowers bearing female trees have around 50 stamens. The receptacle has four cream-coloured calyx lobes, spreading at the anthesis and persistent in the fruit. There are four petals, which are whitish and quickly decidous. The style is white with a yellow stigma. In females the style is slightly shorter and the stigma longer than in the barren trees. The ovary is inferior and 2-celled, usually with 1 ovule in each cell. The plants flower during March-June and the fruit, which is a berry (4-7 mm in diameter), matures 3-4 months later. For spice purposes, it is picked when it is fully developed, but still green. The fruit has two kidney-shaped seeds. The male trees flower early. These plants normally do not bear fruit in the plains.

Soil and climate

The plant grows in a wide range of soils including poor soils with good drainage. It is not suited to red soils with inadequate moisture. Pimenta grows from sea level to an altitude of 1065 m above MSL; however, it also grows well below 300 m. An annual rainfall of 100 to 200 cm or more with a mean monthly temperature upto 27°C are ideal. The performance of allspice in the plains is poor and fruiting is not observed. At higher, cooler and wetter areas, the incidence of rust disease is severe.

Varieties

There are no named varieties but there are few local varieties available in the Nagarcoil area, Kallar and Burliar. IISR, Calicut has a germplasm collection of 137 indigenous types.

Cultivation

Propagation

Seed propagation is the common method of propagation of allspice. To obtain the seeds, the ripe fruits are collected from high yielding and regular bearing trees. The seeds are extracted after soaking the fruits overnight in water, rubbing in a sieve and washing them with clean water. The seeds are dried in the shade. The seeds should be sown immediately or else germination is reduced. The seeds are sown in nursery beds, boxes, pots, basins or polyethylene bags (30 × 15 cm). Beds of 1.2 m width are prepared by using light soil and incorporating organic matter or a mixture of sand and coir dust or coir dust alone. After sowing, the nursery beds are mulched to hasten germination. Dried leaves, straw, paper and damp sacks are used as mulch. Watering is done using a fine spray. Germination takes place between 9 and 10 days or sometimes 15 days after sowing.

Vegetative propagation by bottle grafting, budding, approach grafting, air layering and top working are also possible in this crop. Propagation by micro-propagation is also reported. Shoot tip cultures of allspice could be established and activated to grow. Callus cultures from leaf as well as shoot explants could also be established.

Land preparation

The land selected for planting should be cleared of wild vegetation. In slopy areas contour planting is suggested. Pits of 60 cm^3 are dug at the recommended spacing and they are filled with well rotten FYM or compost and top soil.

Planting

Six to ten-month-old seedlings, 25-30 cm in height are ideal for field planting. The seedlings are planted at a spacing of 6m × 6m. If the soil is poor in fertility, closer spacing is suggested. To avoid losses, instead of one seedling, three are planted in each pit. When grafted plants are used (278/ha), it is suggested to plant one per pit. To ensure sufficient 'males' as pollinators, it is necessary to plant female and male in the ratio of 8:1. Care should be taken at the young stages by providing shade and regular irrigation to reduce casualties in the main field. Allspice can be grown as an intercrop in arecanut, coconut gardens and coffee estates.

Manures and fertilisers

Not much work has been done in this area. However, the tentative, fertiliser schedule per plant is the application of 10 kg compost along with 20:18:50 g of N, P_2O_5 and K_2O during the first year, and 20 kg compost with 40:36:100 g of N, P_2O_5 and K_2O during the second year. The dose is gradually increased to 50 kg compost and 300:250:750 g of N, P_2O_5 and K_2O per tree upto 15 years after planting. In some areas it is recommended to apply 80 kg FYM and 300:250:700 of NPK per plant from the 15th year onwards. The fertiliser is applied in two splits, one in May-June and another in September-October.

Irrigation

The plants are watered during the dry months in the early stages of growth, upto 2-3 years after planting.

Weeding

As the crop is widely spaced, weed problems are common, especially in the early stages. The base of plants should always be kept clean by regular weeding and mulching. Weeds are controlled chemically by using a herbicide like Gramaxone (Paraquat).

Regulation of flowering and fruit set

Generally, allspice is a shy bearer of flowers/fruits. In order to enhance the seed set, the tree basins should be drenched with Cultar (Paclobatrazol) at low concentrations. The plants are reported to show a positive response to flowering and fruit setting.

Pests and diseases

Pests

Infestation by the tea mosquito bug (*Helopeltis antonii*) has been observed on tender shoots and leaves. The pest infestation results in drying of the tender shoots and formation of necrotic spots on the leaves. Spraying with Quinalphos (0.05%) will control the pest. Scale insects (*Odonaspis pimentae* and *Lalveia primitiva* var. *pimentae*) are the other pests which are reported to cause damage to allspice. The above control measure holds good for the scale insects also. The other pests include termites and white ants, white fly (*Aleurodicus pimentae*), red-banded thrips (*Selenothrips rubrocinctus*), bag worm caterpillars (*Oeceticus abboth*), feeder beetle (*Prepodes* and *Pachanaeus* sp.), etc.

Diseases

Leaf-spot (Cylindrocladium quinqueseptatum)

Grey to brownish-grey discolourations appear on the margins of the leaves which extend inward causing leaf rot. The affected leaves drop off causing varying degree of defoliation. The lower surface of the infected leaves show a downy growth due to the production of abundant conidia. The disease is generally noticed during July-September.

Leaf rust (Puccinia psidii)

It is the most serious disease in Jamaica which results in the defoliation of the young leaves; successive rust attack results in the death of the trees. To successfully control the disease, the young leaves and inflorescences are sprayed with Dithane M-45 or Menzate D.

In addition, *Pestalotiopsis* also causes leaf rot. The brown lesions of *Pestalotiopsis* exhibit a dark brown colour advancing into the margins. The fungi produce conidia in acervuli on the upper surface of the affected leaves which could be seen as black dots. In severe cases, both these fungi cause die-back. Spraying the foliage with Bordeaux mixture (1%) or 0.1% Carbendazim during May-June checks this disease. Die-back or canker or fibre blight (*Ceratocystis fimbriata*) is the other disease reported on this crop. This can also be controlled by 1% Bordeaux mixture spray before the onset of the monsoon.

Harvesting and yield

The male trees flower early compared to the female ones and the usual flowering time in India is during March to June. It takes 5-6 years for flowering under good management. The tree takes 20-25 years to come to the full bearing stage and continues upto a good age. The fruits are ready for harvest after 3-4 months of flowering. The berries grow in clusters, and are best used as the spice when they are green, fully matured but not ripe. The unripe berry is more spicy and somewhat peppery in taste. They are manually gathered by climbing up the tree on a ladder. After harvesting, the ripe berries are separated from the green ones.

The berries are spread out in the sun and turned over with a wooden rake, so that they dry uniformly. Drying takes three to twelve days. A good dry wind accelerates the drying process. The end product should be bright brown in colour. The completion of drying is confirmed by sharp, dry and crisp rattling sounds when a handful is shaken close to the ear. Then the berries are cleaned by winnowing and stored after removing any dust. A well-grown tree yields 20-25 kg of dry berries per year. Yields as high as 50 to 60 kg of dry berries have also been reported. The berries are light and about 14,000 berries weigh 1 kg.

Pimento should be stored in gunny bags lined with polyethylene in covered premises, well protected from the sun, rain and excessive heat.

Allspice generally does not set seed under all agro-climatic conditions. Combined application of Indole Acetic Acid (IAA) and Benzyl adenine (BAP) (50:5) results not only in fruit setting but also in producing large-sized fruits.

Grades/types

Pimento is marketed as whole or ground pimento. There are four major grades/types of pimento that are traded. They are, Mexican, Guatemala, Honduras and Jamican pimenta. Jamaican pimenta has the best appearance, aroma, flavour and volatile oil content (4.0-4.5%). The dried berries range in size from 6.5 to 9.5 mm in diameter and there are 13 to 14 berries per gram.

Value-added products

The following products are manufactured from pimenta.

Pimenta berry oil

The essential oil (3.3 to 4.3%) obtained by steam distillation of the crushed dried berries is known as pimenta berry oil. The oil has resin, protein, pentosans, starch, traces of alkaloids, pigments and minerals. It is yellow to yellowish-red in colour. It darkens with age and possesses the characteristic colour and warm spicy and sweet flavour of allspice. It contains eugenol (65-80%), methyl eugenol (8%), beta-caryophyllene (4.2%), humulene (2.7%) and cineole (2.3%).

Pimenta leaf oil

Steam distillation of dried pimenta leaves yields 1.8-3.8% of an oil, which is the pimenta leaf oil. The leaf oil contains eugenol as its main component (58-62%), it also contains myrcene (14-17%) and 1, 8-cineole (6.6-8.4%) but has an inferior odour and flavour to that of the berry oil. The leaves also contain tannin which may be used for tanning purposes.

Pimenta oleoresin

The berries are processed on a small scale in some of the importing countries for the manufacture of oleoresin.

Pimenta bark and wood

Pimenta bark also contains tannin and a small quantity of essential oil. The wood from the tree trunk is dark to light salmon pink in colour with a very firm, hard, close texture and a smooth surface. The wood is used for making sticks, umbrella handles and cart shafts.

Anise

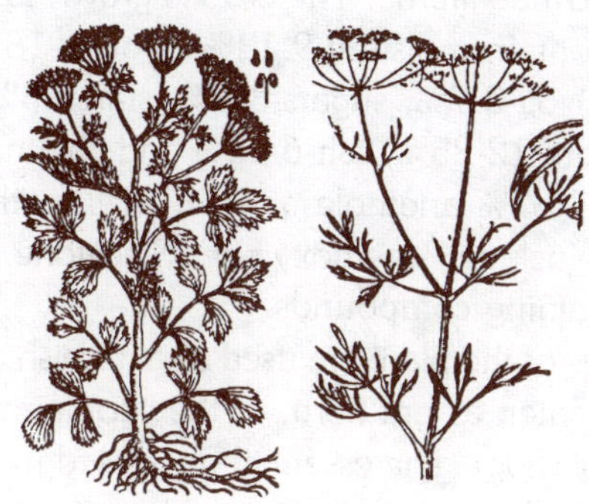

Plant Profile

Family	:	*Umbelliferae/Apiaceae*
English name	:	Anise
Indian name	:	Imported or *Valuili Saunf* or *Aawonf, Badian* (Hindi), *Muhuri, Mitha jira* (Bengali), *Anisi, Sowa* (Gujarati), *Sompu* (Kannada), *Shombu* (Malayalam), *Somp, Badishep* (Marathi), *Sop* (Oriya), *Valaiti Sounf* (Punjabi), *Shetapusapa* (Sanskrit), *Shombu* (Tamil), *Kuppi Soptu* (Telugu)
Species	:	*Pimpinella anisum* L. (Syn. *Anisum vulgare* Gaertner; *A. officinalis* Moench)
Distribution	:	Central, South and South-East European countries, Egypt, Syria, Iran, Mexico, South America, Turkey, Russia, India
Uses	:	Flavouring salads, Perfumery, Soaps, Toilet articles, High-grade cattle feed,

Anise, botanically known as *Pimpinella anisum* L. (Syn. *Anisum vulgare* Gaertner, *A.officinalis* Moench.), belongs to the family *Apiaceae/Umbelliferae*. The crop is grown for its seeds, which are reported to contain moisture-9-13%, protein-18%, fatty oil-23%, volatile (essential) oil-2-7%, sugars-3-5%, starch-5%, N-free extract-22-28%, crude fibre-12-25%, ash-6-10% and choline. The volatile oil contains about 84-87% anethole and 12-15% methyl chavicol. The other compounds include P-methoxy phenyl acetone, terpenes, choline and sulphur containing compounds.

The fresh leaves of the plant are used as a garnish and for flavouring salads. They are eaten as pot herb, as they contain essential oils and vitamin-C (8.7mg/100g). The essential oil is used in perfumery, soaps and other toilet articles like perfuming sachets, in dental preparations and in mouth washes. The seeds are used for flavouring food, confectionery, bakery products, beverages, in processed food anisette and in liquor industries. The oil is used externally as an insecticide against small insects and also has fungicidal properties. Medicinally, the oil is used as an aromatic carminative, to relieve flatulence and as an ingredient of cough lozenges in combination with liquorice. It is a mild expectorant and is used as an antiseptic and also against cholera. It also finds use in preparation of gripewater. The fruit is considered to be a mild expectorant, stimulating, carminative, diuretic and diaphoretic and is used in flatulent colic, in the preparation of asthma powders and in veterinary medicine. The alcoholic extract of aniseed possesses fungicidal properties. The residue left after extraction of oil may be used as a high grade cattle feed. It contains 17-19% of protein and 16-22% fat, in addition to carbohydrates.

Origin and distribution

The plant is native to the East Mediterranean region. It is extensively cultivated in central, south and southeast European countries and also in Egypt, Syria, Iran, Mexico, South America, Turkey and the USSR. In India, it is grown to a small extent in Rajasthan, Punjab, Orissa, Malwa tract, Indore, Lucknow and Delhi. Aniseed is being imported from France, Spain, U.K., etc.

Description of the plant

Anise (2n = 18) is an annual pubescent aromatic herb, which attains a height of 60 to 80 cm under cultivation in India. The basal leaves are long and bisected with long petioles, whereas the leaves on the stem are pinnately decompound, they emit the characteristic anise aroma when crushed. The flowers appear in umbels after 120 days of sowing. They are small, white in colour, borne centripetally in the umbels. The flowers are pollinated by little bees (*Apis florea*). The fruit is small, ovoid and laterally compressed, grey to grayish-brown in colour, 3.0 to 5.0 mm in length with a short pedicil attached. Five longitudinal ridges are visible on each pericarp. It has a characteristic agreeable odour and a pleasant aromatic taste.

Climate

A fairly warm weather during sowing with a temperature of 20-22°C is desirable. Low temperature and higher atmospheric humidity are conducive for vegetative growth, while the crop requires warm sunny weather during the seed formation and development stage. Frost in the early stages is deleterious for growing seedlings.

Soil

The crop is grown on a wide range of soils, ranging from sandy loam to clay loam, having a pH of 6-8.5 and moderately fertile. However, higher seed yields have been reported over heavier, moderately rich black cotton soils. A poor drainage of the land as well as stagnating water damages the crop stand and reduces its yield.

Varieties

A selection EC-22091 from French Anise has been developed at IARI, New Delhi, and has shown good growth as a *rabi* crop. Its flowering and maturity are uniform and it is a good yielder. It contains over 2% essential oil. It is suitable for cultivation in the plains.

Cultivation

Propagation

Anise is grown from its seed as a cold weather crop in India. Since the seeds are small, they are mixed with sand for even distribution in rows. The seedlings can be raised in beds and transplanted. However, the *in situ* crop is found to perform better than the transplanted crop.

Land preparation

The seeds are small and need meticulous land preparation for obtaining a good germination and high seedling stand. The land should be ploughed and harrowed 2-3 times to bring it to a fine tilth. Further, it should be quickly levelled to conserve soil moisture.

Sowing

The seeds are sown at a spacing of 45 × 20 cm at a 1 cm depth in late October or early November. Deeper sown seeds reduce the percentage of germination. The germination of seeds is rather tricky and is often uneven. To obtain uniform germination, the seeds should be soaked in water overnight, and moist seed sowing is recommended followed by light irrigation. The seed rate is between 7 and 10 kg/ha depending upon the soil moisture and the quality of the seeds. The germination commences in 7-8 days and is over within 10 days.

Manures and fertilisers

The crop is moderate in its requirement of manures and fertilisers. FYM at 10 t/ha is applied at the time of land preparation. An application of 40-80 kg N, 30-60 kg each of P_2O_5 and K_2O/ha is followed. Out of the total quantity, 50% N and full dose of P_2O_5 and K_2O are applied when the seedlings are 20-30 days old. The remaining 50% N is applied at the full vegetative stage after another month.

Irrigation

For better growth and higher seed yield, anise needs good irrigation. Daily irrigation is done till the seeds germinate. Later, irrigation is given at intervals of 3-4 days depending on the soil moisture conditions. It does not appear to withstand too much of moisture. At the flowering and grain-filling stages, the crop needs copious application of water.

Interculture

Weeding, preferably manual weeding, has to be carried out at monthly intervals and the field should be kept weed free. The crop needs at least 3-4 weedings during the crop cycle.

Crop rotation

Crop rotation of anise with sacred basil and senna has been tried and these combinations produce better returns.

Pests and diseases

Pests

Semilooper (Thysanoplusia orichalcia)

This is a defoliating insect and can be controlled by spraying 0.2% Monocrotophos or Endosulphon.

Besides, the cigarette beetle, *Hasioderma serricorne* (F) has been reported as a storage pest of aniseed.

Diseases

Altenaria leaf blight

This is a fungal disease characterized by the blightening of leaves and can be controlled by spraying 0.25% Mancozeb twice at an interval of 10 days.

Harvesting and yield

The umbels come to maturity from March to mid-April after 170-175 days of sowing. The terminal umbels mature first and the lateral ones 15 to 20 days later. When 80-90% fruits begin to turn grayish-green in colour, the tops of the plants are cut along with the branches and are tied in bundles. The bundles are stacked in a central place upto 2 m in height, keeping the fruiting heads downwards, towards the centre. Later, the bundles are threshed and the seeds are cleaned, bagged and stored.

A good crop yields 7 to 8 quintals of seed per hectare.

Processing

Aniseed should be crushed and steam distilled immediately to obtain higher oil yield and also to save on fuel and time in distillation. The anise oil is colourless or a pale yellow liquid having the characteristic odour and taste of the fruit. The fresh seed crop generally yields a volatile oil of about 15 kg/ha at 2.1-2.4% recovery of oil.

Adulterants

Aniseed is often adulterated with exhausted fruits, fine earth, other small seeds and fruits, star anise, etc. Ground aniseed is sometimes found adulterated with ground fennel.

Anise oil is frequently adulterated with the lower priced star anise oil. The other adulterants of the oil are turpentine oil, cedar wood oil, capaiba and gurjan balsam oils and synthetic anethole made from pine oil.

Black Cumin

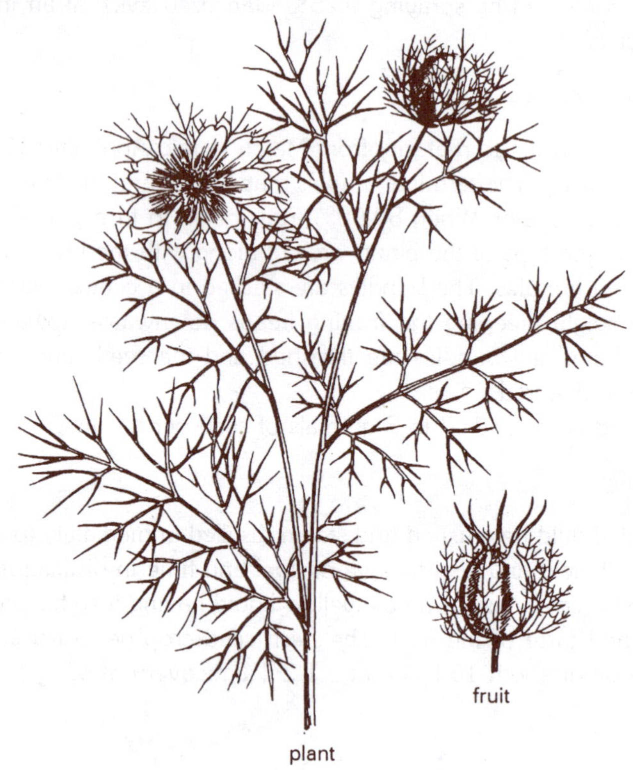

Plant Profile

Family	:	*Umbelliferae*
English name	:	Black cumin
Indian name	:	*Kalonji, Kala jira* (Hindi Punjabi and Urdu), *Kaljira (*Bengali), *Karunshiragam* (Malayalam), *Kalanjire* (Marathi), *Krishna-jirake* (Sanskrit), *Karun-jiragam* (Tamil), *Nulakilakara* (Telugu)
Species	:	*Nigella sativa* L.
Distribution	:	Egypt, Syria, Greece, Arabian countries, India
Uses	:	Culinary, Food flavouring, Beverages, Medicine

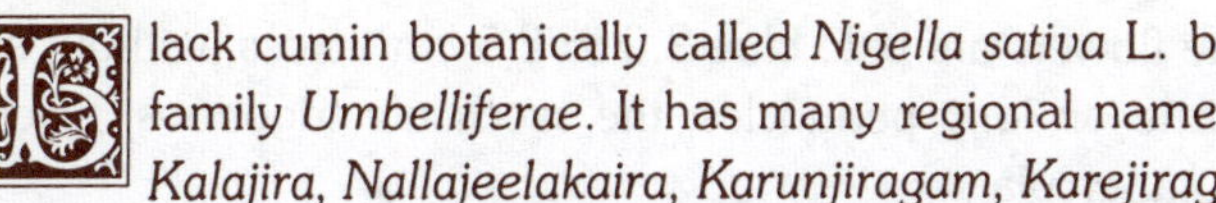

Black cumin botanically called *Nigella sativa* L. belongs to the family *Umbelliferae*. It has many regional names like *Kalonj, Kalajira, Nallajeelakaira, Karunjiragam, Karejirage,* etc., and it is also called as small fennel. The dried seed like fruits are parts of economic importance in black cumin. Black cumin is a seed spice-cum-medicinal crop. An analysis of the Black cumin seeds gave the following values; total ash-3.8%-5.3%, ash insol-in acid-0-0.5%, volatile oil-0.5-1.6%, ether extract (crude oil)-35.6-41.6%, alcoholic acidity as oleic acid-3.4-6.3%.

The seeds contain mainly volatile (0.5-1.6%) and fatty oils (31%), a bitter principle nigellin, tannins, resins, proteins, reducing sugars, saponins, and arabic acid and other alcohol-soluble organic acids. Free amino acids like cystine, lysine, aspartic acid, glutamic acid, alanine, tryptophan, valine and leucine are present in the dormant seeds.

Black cumin seed is an expensive spice in India, extensively used for culinary purposes and for flavouring foods and beverages. The seeds are considered carminative, stimulating, diuretic, emmenagogue, galactogogue and are used in the treatment of mild cases of puerperal fever. The paste of the seeds are externally applied for eruptions of skin. Alcoholic extracts of the seeds show antibacterial properties against *Micrococcus pyogenes* var. *aureus* and *Escherichia coli*. The fatty oil obtained by the expression of the seeds is reported to be used for edible purposes. The seeds are often scattered between folds of linen or woollen clothes to preserve them against insect attacks. They can be used as a stabilising agent for edible fats.

Origin and distribution

Nigella is a small genus of annual herbs found in Southern Europe and Western Asia, but chiefly in the Mediterranean region. Black cumin is a native of the Levant (Eastern Mediterranean region). It is being grown in Mediterranean countries like Egypt, Syria, Greece and also the Arabic countries and India as a minor spice. Reliable data relating to the area and production of *Nigella sativa* in India is not available. As the seeds are being collected mostly from plants growing wild in forest areas (for use as flavouring material and for medicinal purposes) till recent times, the crop was not cultivated on any considerable area. But now states like Himachal Pradesh, Uttar Pradesh, Bihar, Assam, Punjab, Madhya Pradesh and West Bengal are cultivating this crop over small areas.

Description of the plant

Black cumin is a small herb growing to a height of 45 cm. The leaves are pinnatisect and 2.5-5.0 cm long, cut into linear-lanceolate segments. The flowers are pale blue 2.0 to 2.5 cm across, without an involucre, borne on long peduncles; the seeds are trigonous, black, rugulose and tubercular.

Climate

Black cumin is a *rabi* season crop. It requires a mild, cool climate with plenty of sunshine throughout its life cycle. The crop can be profitably cultivated in the winter season. In West Bengal and Uttar Pradesh, the crop sown in November and December, gave the maximum seed yield. Preliminary trials conducted in Bangalore have shown similar results. However, rainfall of high intensity for prolonged periods, has a detrimental effect on growth and maturity of the crop.

Soil

The crop is hardy and grows on a variety of soils ranging from alluvial sandy loams to red sandy loams. Light friable soils with good aeration, proper drainage and rich organic matter are preferable.

Varieties

A local selection from a collection of *Nigella* showing superior growth and yield attributes has been released as Azad Kalaunji (Mangral) from the Vegetable Research Station, Kalyanpur, Uttar Pradesh, in 1998 by the State variety release committee for general cultivation. Baliabil, Kalapani, Chhagalia, Bengerbhita, Sadullabari and Sukhchare are some of the local cultivars grown in Assam.

The other improved varieties are listed below.

NS - 44

It is a high yielding variety of black cumin in Jabalpur. It yields 4.5 to 6.5 quintals per hectare and matures in 140-150 days.

NS - 32

It is a high-yielding variety producing 4.5 to 5.5 quintals per hectare and matures in 140-150 days.

Kala Jeera

This one also a high-yielding potential variety. It yields 4-5 quintals per hectare and matures in 135-145 days.

Cultivation

Propagation

The crop is commercially propagated through seeds. The seeds are small and should be mixed with sand/leaf compost for even distribution in the rows.

Land preparation

The land is prepared by repeated ploughing and planking and brought to a fine tilth and then divided into plots of a convenient size. FYM @ 15-20 tonnes is added to the soil at the time of the last ploughing or at least one month before sowing.

Sowing

About 5-8 kg of seeds are required per hectare. The seeds are sown in the first fortnight of November or December. Late sowing results in a poor yield. The seeds are treated with a fungicide like Bavistin (2-3 g/kg of seeds) before sowing and are sown thinly and at shallow depths, (1-1.5 cm) in 20-25 cm wide rows. The seeds are very light, a 1000 seeds weigh around 2 g. The seed germination is very high and they germinate in 14-15 days. The seedlings are thinned out within the row three weeks after germination, to maintain a spacing of 20-25 × 5-10 cm. Sometimes, the seeds are also broadcasted. If broadcasting is done, seeds should be mixed with soil with the help of rakes so that all the seeds are well covered with the soil.

Manures and fertilisers

From studies conducted at the Division of Horticulture, GKVK, Bangalore, it has been found that the application of N, P and K at the rate of 50, 40 and 30 kg/ha, respectively, is optimum for obtaining the maximum seed yield. Half the dose of N along with the full dose of P_2O_5 and K_2O is applied to the soil at the time of sowing and the remaining quantity of N is applied after 30-40 days.

Irrigation

The land should be irrigated properly before sowing to obtain uniform germination of the seeds. Later, irrigation can be given at intervals of 7-10 days. Protective irrigation at 30 days after sowing and at flowering is a must.

Weeding and hoeing

One good weeding and hoeing with a hand hoe (*khurpi*) is required at 20 days after sowing when the plants are 5 cm tall. Another weeding is done 50-60 days after sowing or as and when required.

Pests and diseases

Pests

There are no reports of any pest of serious nature on this crop. However, stray incidents of the leaf webber have been observed and the insect can be controlled by spraying Monocrotophos (0.1%).

Diseases

Damping-off and root wilt are the two fungal diseases observed in this crop. Damping-off of the seedlings occurs when the soil moisture level is very high due to excessive irrigation or water-logged conditions. The damping-off disease can be managed by providing proper drainage and also by drenching the plants with a solution of Bavistin (0.1%).

The root wilt disease is characterised by the blackening and death of the roots. The aerial parts of the plant remain green to pale green but the leaves and branches droop down and subsequently the plant dies. This disease can be controlled by treating the seed with Bavistin before sowing, removing and destroying the affected plants, drenching the soil with 0.1% Bavistin which should be repeated after 15-21 days and by giving light irrigation at increased irrigation intervals.

Harvesting and yield

The crop is ready for harvest in 110-135 days. At the mature stage, the plants as well as the capsules (pods) turn yellow in colour. The capsules when split between the fingers discharge deep black coloured seeds with pleasant characteristic aroma. The crop can be harvested at this stage. While harvesting, the whole plants are cut and dried in the shade. The seeds are separated by threshing followed by winnowing of the dried herb. The seeds are dried under sunlight until the moisture level reaches 8-10%. Later, they are stored in airtight containers or polybags in a cool and dry place. The seed yield per hectare ranges from 700 to 1000 kg.

Distillation of oil

The seeds are steam distillated to obtain a yellowish-brown volatile oil with an unpleasant odour. The oil contians 45-60% carvone, d-limonene and cymene. A carbanyl compound, nigellone, which was shown to protect guinea-pigs against histamine-induced bronchospasms, has been isolated from the oil. Its therauptic use in cough and bronchial asthma has also been recorded.

Black Pepper

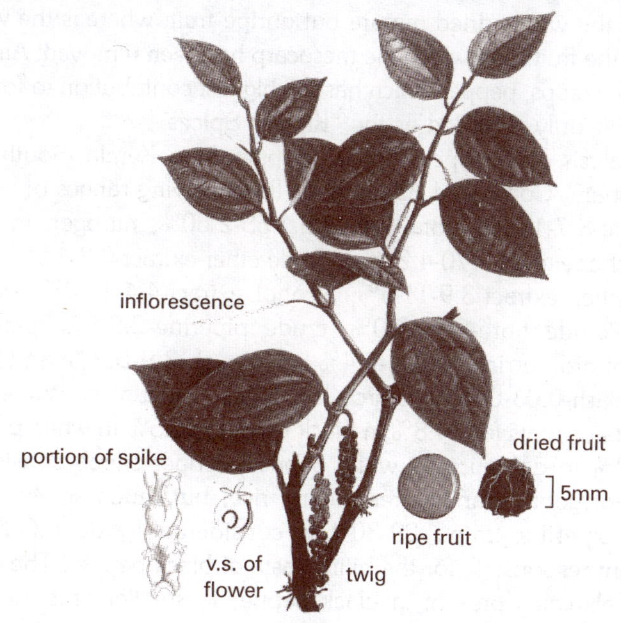

Plant Profile

Family	:	*Piperaceae*
English name	:	Black Pepper
Indian name	:	*Kali Mirch* (Hindi), *Kare Menasu* (Kannada), *Kurumaluku, Nallamulaku* (Malayalam), *Maricha Ushana, Hopusha* (Sanskrit), *Milagu* (Tamil), *Miriyalu* (Telugu).
Species	:	*Piper nigrum* L. *P. betle* L. *P. methysticum* Forst (Kava) *P. cububa* P. *longum* P. *retrofractum*
Distribution	:	Western Ghats of India, North Myanmar, Indonesia, Malaysia, Brazil, Madagascar, Sri Lanka, Vietnam, Thailand, China.
Uses	:	Spice and condiment.

lack and white pepper are obtained from a climbing vine called *Piper nigrum* L., which belongs to the family *Piperaceae*. It is one of the oldest and the world's most important spices. Black pepper is the whole dried mature but unripe fruit, whereas the white pepper is the fruit from which the mesocarp has been removed. Among all the spice crops, pepper which has the highest contribution to foreign exchange is duly regarded as the "King of Spices".

An analysis of 23 types of black pepper from Kerala, South and North Kanara, Coorg and Assam gave the following ranges of values – moisture-8.7-14.1%, total nitrogen-1.55-2.60%; nitrogen in non-volatile ether extract-2.70-4.22%, volatile ether extract-0.3-4.2%, non-volatile ether extract-3.9-11.5%, alcohol extract-4.4-12.0%, starch-28-49%, crude fibre-8.7-18.0%, crude piperine-2.8-9%, piperine (spectrophotometrically)-1.7-7.4%, total ash-3.9-5.7% and acid insoluble ash-0.03-0.55%. Starch is the predominant constituent of pepper. It accounts for 34.8% in black pepper, 56.5% in white pepper and 63.2% in decorticated white pepper. Pepper starch consists of minute polygonal granules resembling rice, but much smaller. The alkaloid piperine (m.p., 29-30°) is considered to be the major constituent responsible for the biting taste of black pepper. The other pungent alkaloids present in black pepper in smaller amounts, are chavicine, piperidine and piperetine. The characteristic aromatic odour of pepper is due to the presence of a volatile oil in the cells of the pericarp. On steam distillation, crushed black pepper yields 1.0-2.6% (upto 4.8%) of the volatile oil, the yield depending greatly upon the age of the dried fruits subjected to distillation.

Pepper is used for a variety of purposes. It is more frequently and regularly used than any other spice, as a condiment. It stimulates the digestive organs and thus produces an increased flow of saliva and gastric juices. It is used for seasoning many dishes and it imparts its characteristic blend of flavour and pungency to the cuisine.

The ancient Aryans considered pepper as a powerful remedy for various disorders of the anatomical system and prescribed it as an effective cure for dyspepsia, malaria, delirium tremors, haemorrhoids, etc. The Egyptians used it for embalming. The Asians are said to have used it as an aphrodisiac. However, its value as an essential preservative for meats and other perishable foods has been known for centuries. It is, therefore, largely used by meat packers and in canning, pickling, baking, confectionery and the preparation of beverages. It is also used as flavouring in culinary seasonings of universal use and an essential ingredient of numerous commercial foodstuffs.

Origin and Distribution

Pepper has its origin in the Western Ghats of India. It occurs wild in the hills of Assam and North Burma. It is a perennial climbing plant, mostly cultivated as a mixed crop in coffee, citrus and arecanut plantations in Kerala and Karnataka. Besides India, pepper is now grown in Indonesia, Malaysia, Brazil, Madagascar, Sri Lanka, Vietnam, Thailand, China and the Federated State of Micronesia.

Area and Production

The total world area under pepper in 1996 was 4,03,930 hectares and the production was 1,80,289 tonnes. Pepper is found throughout the Malabar coast, extending inwards to the cultivated areas on both the sides of the Western Ghats. In India, pepper is cultivated in over 1,95,050 hectares with an annual average production of 5340 tonnes, which accounts for one-third of the world's production. Kerala has the largest share of pepper production accounting for 96%, followed by Karnataka (3.9%) and Tamil Nadu (0.1%). Pepper is also grown in Pondicherry, Maharashtra, Goa, Andhra Pradesh, Orissa, West Bengal, some Northeastern states and the Andaman and Nicobar Islands. Nearly 80% of the pepper produced in the country is exported, contributing to about 37% of the total export earnings from spices. The main buyers of Indian pepper are the USSR, USA, Canada, Italy and Czechoslovakia. India exports nearly 26,346 tonnes of its production, earning an amount of, Rs. 19,64,709,000 (1996) in foreign exchange annually.

India enjoyed the top position in world pepper production till the 19th century, but lost its place to some South-east Asian countries. Now, on an average, under Indian conditions, we get only 1 kg green berries per vine, which in turn gives 250-350 g of dry berries, whereas the South-east Asian countries produce about 5 kg green berries per vine (1.25 kg dry berries).

Description of the Plant

Pepper is a large genus, with over 1,000 species, in the family *Piperaceae*. They are mostly herbaceous or woody climbers or shrubs distributed in the tropics of both hemispheres. The other economic species of the genus *Piper* are *P. betle* L. (Betel vine) and *P. methysticum* Forst (Kava). In addition to *P. nigrum* which now provides the true pepper of commerce, some of the species that have also been used as spices and as pepper substitutes are *P. cubeba* (Cubeba or tailed pepper), *P. longum* (Indian long pepper). *P. retrofractum* (Java long pepper), etc.

Piper nigrum is a climbing evergreen plant, growing to a height of 10 m or more. The vines branch horizontally from the nodes and do not attain much length. The pepper branches are dimorphic comprising: (i) the orthotropic vegetative climbing branches with 5-12 cm long internodes which give the framework of the plant, their stems are swollen at the nodes when young but become woody on aging; and (ii) a plageotropic fruiting branch develops from the auxillary bud present at each swollen node beside a leaf. The short adventitious roots produced at the nodes enable the plant to cling to the standard (climbing support). The leaves are alternate and simple, dark green and shiny above and pale green underneath. They are smooth and entire, broadly lanceolate but there is wide variation in leaf shape. The petioles are 2-5 cm long, and are grooved above.

The spike is a catkin type of inflorescence, which emerges opposite the upper leaves, on the plagiotropic branches. It is 3-15 cm long with 50-150 minute white to pale yellow flowers. The flowers are either unisexual with monoecious or dioecious forms, or bisexual, as seen in many cultivars. Under intense shade conditions, the hermaphrodite type produces more female flowers and less bisexual flowers. The flowers are cross-pollinated and thus much variation occurs among the progeny. Moreover, hermaphrodite flowers are protogynous; i.e., they attain receptivity well in advance of the pollen maturity, which results in uneven fertilisation of flowers and non uniform, incomplete filling of the spikes.

Flowering begins at the base of the spike and continues to the tip over a week. It is found that bagged inflorescence produces fruits, indicating that the hermaphrodite cultivars (Eg: Balankotta, Kalluvalli) are self-fertile and self-pollination can occur without the aid of rain or wind. The pollen is in glutinous masses of several grains, which is broken up by light showers and the grains are trapped in the papillae of the stigma. But wind pollination has not been found to be very efficient. It seems that pollination is confined mainly to individual spikes. The fruit is a one-seeded green berry, sessile, small, usually globose and sometimes elongated or oval. The seed is surrounded by a thin, soft pericap. During ripening, it becomes yellow and turns light red afterwards. The period taken from flowering to maturity is about 6 months.

Climate

Pepper is essentially a crop of the humid tropics and requires adequate rainfall and humidity. However, it thrives well even in the hot and humid climate of the foothills of the Western Ghats. It grows well at 500-1500 m elevation from the mean sea level. It requires a temperature of 10-40°C, below which the pollination and fruit set will be adversely affected. An optimum rainfall of 1250-2000 mm per annum distributed throughout the year is desirable. Though it is a shade-loving vine, too much of shade will affect flowering and fruiting. To avoid sun-scorching, it should be planted towards the eastern slopes avoiding the southwest sunlight.

Soil

Pepper thrives best on virgin soils rich in humus and naturally well drained; red lateritic soils or alluvial soils rich in humus are also highly suitable. Heavy clays and sandy soils should be avoided. In Kerala, pepper is also grown on the raised bunds of water-logged paddy fields. A pH range of 4.5 to 6.5 is ideal.

Varieties

The germplasm collection of black pepper at IISR is 2,776 (2,492 indigenous, 5 exotic and 279 wild). In India, out of about 100 varieties which are under cultivation, only ten have assumed commercial importance. The distinguishing characters of some of the important varieties are as follows.

Cheriyakaniyakkadan

The leaves are small and elliptic; the spikes are of medium length, closely set with medium-sized dark-green fruits. It is a popular type and bears regularly, the yield is heavy (42% dried pepper) and of high quality, it is wilt resistant and popular in north and central Travancore (Kerala).

Cheriakodi

The leaves are narrow, and dark green; the spikes are short, with dark or pale-green fruits, which are the smallest among all the types. The plant is a dwarf and sturdy type, bearing in alternate years, and of a high quality. It yields 38% dried pepper and is popular in north Travancore and north Malabar (Kerala).

Daddagya

The leaves are broad and the spikes are long, and curved. The fruits are large among Mysore types. It is a uniform yielder, esteemed for making white pepper; it yields 38% dried pepper. It is popular in North Kanara (Karnataka).

Kalluvally

It is a promising cultivar of north Kerala, being a hardy and regular yielder. The leaves are medium, ovate, elliptic and the spikes are medium and are the twisting type due to very thick setting. The berries are small to medium, heavy and have high driage (about 40%). It is a regular bearer and reportedly tolerant to water stress and diseases. Studies have shown that it is moderately tolerant to *Phytophthora* wilt. Though it is a very promising cultivar, its distribution is rather restricted to settlement areas of the submontane pepper tracts of Wynad and Cannanore districts.

The Kalluvally found in many areas in Cannanore and north Wynad does not conform to the above description. More than one cultivar is known by this name and some of these are rather poor yielders. They do not seem to be as hardy as the name indicates. The Kalluvally types are also found to differ in quality factors. For example, three Kalluvally types gave the following values; oleoresin (%) 8.8, 10.9 and 8.4; piperine (%) 4.2, 4.6 and 5.4, essential oil (%) 3.4, 0.4 and 3.0 and starch (%) 31.5, 29.0 and 20.7.

Kaniakkadan

The name Kaniakkadan appears to have been derived from Kaniakkar, a tribal sect inhabitating the hilly Western Ghat areas of the present day Idukki district. There are four different Kaniakkadan types. These are Cheriyakaniyakkadan, Valiakaniakkadan, Karutha Kaniakkadan and Valutha Kaniakkadan, of which the first two are more important.

Cheriyakaniyakkadan is a popular cultivar of the eastern parts of Kottayam and Quilon districts. It is also commonly found in many areas of north Kerala. It is a bisexual type, with small elliptical leaves, medium-long spikes and medium-sized berries. It is an average and regular bearer and has 30% driage. It gives 9% oleoresin, 3.9% piperine and 3.7% essential oil.

Valiakaniakkadan has slightly larger leaves, longer spikes, and bold and heavy berries, and its berry setting is good. It is a moderately high yielding type.

The other two Kaniyakkadans are also moderate yielders.

Arakkulamunda

This is a moderately good and regular bearer from the central area of Kerala. It derives its name from a village Arakkulam in the Thodupuzha taluk of Idukki district. It is commonly found in the settlement areas of north Kerala. It yields regularly and comes to maturity earlier than most other cultivars. The spikes are medium-long and the berries are bold and heavy. This cultivar gives 9.8% oleoresin, 4.4% piperine and 4.7% essential oil.

Balancotta

This cultivar is confined to north Kerala and grows very vigorously. It has the largest leaves among all the Kerala cultivars. The vines are large, growing to the top of the large supporting trees like mango and jackfruit. The spikes are medium long to long, the setting moderately good, and the berries are bold and pale-green. The variety is early to medium in duration, the berries give about 30% driage. It gives 9.3% oleoresin, 5.1% essential oil and 4.2% piperine. Its oil content is high and so has a very high flavour quality.

Balancotta is reportedly tolerant to moisture stress and is also somewhat tolerant to *Phytophthora* wilt. In the Malanad areas of North Kanara, this cultivar is grown under the name Malabar Pepper, especially in Sagar, Koppa, Sirsi and adjoining areas.

Karimcotta

The leaves are large, dark green and the spikes are short and curved with closely-set, large, dark-green berries. It is a hardy, regular bearer and a good yielder. It yields 42% dried pepper. It is popular in the north Travancore and north Malabar areas of Kerala.

Karimunda

The name Karimunda might have come from the bluish-black tender shoots and dark-green leaves and berries. It is the most popular cultivar grown throughout Kerala. This had been the traditional cultivar of central Kerala comprising Palai, Ponkunnam and the adjoining areas. The cultivar is bisexual and is characterised by small more or less oval leaves and short to medium-long, closely-set spikes. The spikes are 4 to 10 cm long and even more in certain cases, with a mean length of around 6.5 cm. In central Kerala, most of the vines have short, well-filled spikes, but the spikes found in the Idukki district and also in parts of Wynad are much longer. It is a prolific and regular bearer, having medium-sized berries of good driage (35%) and yields good quality pepper. This cultivar has about 11% oleoresin, 4.4% piperine and 4.0% essential oil content.

Its flowering time is in May-June, coinciding with the monsoon. The irrigated plants flower almost continuously. This cultivar is of medium maturity, and is suitable for intercropping and also for high density monocropping.

Kottanadan

It is the most popular pepper cultivar in South Kerala. This is a vigorous growing bisexual variety with large, broad, ovate leaves, long spikes, high fruit set and medium-sized berries. The cultivar is a high and regular yielder. Studies have shown that Kottanadan has the highest oleoresin content (17.8%). It has a high piperine content too (6.6%), though the volatile oil is comparatively less (2.5% to 4.2%). It gives about 37% driage and produces high quality, heavy pepper.

In the Wynad area, Kottanadan is grown as Aimpiriyan, the name being derived from the fact that berries are arranged in 5 distinct rows on the spike. In south Kerala, it flowers by April-May and is ready for harvest by January. In Wynad, it matures late and comes to harvest only by April.

Kumbhakodi, a cultivar grown in certain areas of Quilon, seems to be a variant of Kottanadan/Aimpiriyan.

Kuthiravally

A south Kerala type, now found in many pepper-growing areas throughout Kerala, Kuthiravally is a moderately high yielder, but an alternate bearer. The ovate leaves are medium-large, the spikes are very long and slender (like a horse's tail, hence the name Kuthiravally) and the flowers are bisexual. The berries are medium-large, having high driage (over 39%). It yields about 15% oleoresin, 6.0% piperine and 4.5% essential oil. Kuthiravally thus produces high quality pepper.

A cultivar called Thommankodi, grown in certain areas in eastern parts of Calicut (Koodaathai, Anakkampoyil areas), seems to be a variant of Kuthiravally. In Thommankodi, the spikes are slightly shorter and the berry filling somewhat better than those of Kuthiravally. A quality analysis of this cultivar has not been carried out.

Aranavalan, a cultivar found in the Malayatoor-Kaladi areas, also seems to be a variant of Kuthiravally.

Malligesara

This is the common cultivar of North Kanara district of Karnataka. In the Malnad areas of North Kanara, this is usually intercropped with arecanut. Two types of Malligesara are known. Karimalligesara and Bilimalligesara, which can be differentiated based on the anthcyanin colouration of the emerging shoots. In Karimalligesara it is purple-white and in Bilimalligesara it is pale-green. It is a moderate yielder, having medium-large leaves and spikes.

Narayakkodi

This is another central Kerala cultivar from the Kottayam-Champakkara-Mallappally tract, but is now found in many settlement areas throughout Kerala. The leaves are small to medium, ovate and have a characteristic twisting that results from a thick berry setting. The persistent stigmatic base looks like a 'pin' and hence the name 'Narayakkodi' (Narayam is a long iron-nail like tool used in the olden days for writing on palm leaves). It is a regular average yielder, having high driage (37.5%). It gives 11% oleoresin, 5.4% piperine and 4.0% essential oil.

The farmers in central Kerala are of the opinion that 'Narayakkodi' has a longer life span than 'Karimunda' and is also much less affected by diseases. Studies conducted in the Central Plantation Crops Research Institute, Regional Station, Calicut, have shown that Narayakkodi is more tolerant to foot-rot caused by *Phytophthora* than most other cultivars.

The following are some of the varieties and hybrids released from the Pepper Research Station, Panniyur, Taliparamba, Cananore district, Kerala and National Research Centre for Spices (NRCS), Calicut.

Panniyur-1

This is the first released black pepper and is evolved through hybridisation. This was one of the progeny of Uthiran Kotta x Cheriyakaniyakadan and was released in 1971. This variety out-yielded all the existing varieties at that time. The high productivity of this variety was due to the recombination of favourable and complementary yield components from the diverse parents. The chief attributes of this variety are the vigorous nature of the vine, large cordate leaves, non-pigmented growing tip, long spikes and bold berries. It performs well only under adequate sunlight. It gives an average yield of 2.2 kg green pepper per vine and 1,242 kg/ha dry pepper. Its potential yield is 8800 kg per hectare. The percentage of bisexual flowers is 99.92 per cent, female (%) 0.07 and male (%) 0.01. The setting percentage is 96 and the dry recovery is 35.3%. The contents of piperine, oleoresin and essential oil are 5.3, 11.8 and 3.5%, respectively.

Panniyur-2

This was a clone from the open-pollinated progeny of the cultivar Balankotta and released in 1989 for high yields. This cultivar is popular in the northern districts of Kerala and the South Kanara district of Karnataka. It tolerates shade better than the other cultivars. Panniyur-2 is vigorous in growth and has a non-pigmented growing tip. It gives a yield of 4.5 kg green pepper per vine with a dry recovery of 35.7%. The average dry pepper yield is 2,570 kg and the potential yield is 3,313 kg/ha. The piperine and oleoresin content of the berries is 6.6 and 10.9%, respectively.

Panniyur-3

This hybrid is a progeny of the cross Uthirankotta x Cheriyakaniyakkadan made in 1973. It was released in 1989 for its high yields. Its characteristics are similar to Panniyur-1, but it has faint pigmentation on the growing tip. The percentage of bisexual and female flowers are 99.9 and 0.1%, respectively. The setting percentage is 89.2. The average yield of green berry per vine is 4.4 kg and the dry recovery is

27.8%. Similarly, the average yield of dry berry is 1,953 kg and the potential yield is 3,269 kg/ha. The berries have fairly high content of piperine (5.2%) and oleoresin (12.7%).

Panniyur-4

There are a number of clones cultivated in Kerala under the name Kuthiravally. One such type maintained in germplasm at the Pepper Research Station, Panniyur, gave high yields. It was included in the replicated trials during 1975 to 1985 at this Centre. Besides high yields, this clone had been able to give comparatively good yields in unfavourable seasons. This was released as Panniyur-4 in 1989. It comes to harvest later in the season than most other cultivars. This variety has long and curved bracts in some of the basal flowers (at fruiting age) in some spikes, which is a varietal character. The yield per vine is 2.3 kg green pepper and the dry recovery is 34.7%. The variety can yield 1,277 kg dry pepper and the potential yield is 2,443 kg/ha. The berries contain 9.2% oleoresin.

The varieties Panniyur 5, 6 and 7 have also been released and their characteristics are given below.

Variety	Pedigree	Avg. yield (kg/ha)	Poten-tial yield (kg/ha)	1*	2**	Essential oil
Panniyur 5	Open polli-nated seed-ing of Peru-mkodi	1098	2587	5.3	12.3	3.8
Panniyur 6	Clonal selection from Karimunda	2127	3354	4.9	8.3	1.3
Panniyur 7	Selection from open pollinated Kaluvally	1410	2720	5.6	10.6	1.5

1* Piperine (%) 2** Oleoresin (%)

Subhakara and Sreekara

These are high yielding varieties selected from Karimunda clones and released in 1990 from NRCS, Calicut. The vines are slender and have characteristic dark green leaves and berries. The number of spikes produced in a season far exceeds that in other cultivars. Unlike most other cultivars, these varieties produce off-season flowers. The important characteristics of these two cultivars are as follows.

	Flower types			1	2	3	Yield (kg/ha)		
	Bisexual	Female	Male				Avg.	Potential	Piperine (%)
Subhakara	99%	0.5%	0.5%	68.0	4.2	35.5	2352	4487	3.4
Sreekara	98%	1%	1%	63.4	4.8	35.0	2677	4200	5.1

1 Setting percentage

2 Green berry yield/vine (kg)

3 Dry recovery (%)

Panchami

This variety was released in 1991 from NRCS, Calicut. This is a high yielding clone already in cultivation and locally known as Aimperian. The percentage of bisexual, female and male flowers is 95.5, 4 and 0.5%, respectively. The mean green berry yield per vine is 5.2 kg and the dry recovery is 34%. The average yield is 2,928 kg of dry pepper and the potential yield is 6,528 kg/ha. The piperine content of the berries is 4.7%.

Pournami

This variety was released in 1991 from NRCS, Calicut. It possesses tolerance to nematode and drought and has a potential for high yields. It has been tested under the name 'Ottaplackal-1' (Coll.812). The spikes bear 84% bisexeral, 15% female and 1% male flowers. The average green berry yield is 4.7 kg per vine and dry recovery is 39%. The average dry and potential pepper yield per hectare is 2,333 and 5,356 kg, respectively. The berries contain 44% piperin, 13.8% oleoresin and 3.4% essential oil.

Plode 2

This is a clonal selection of Kottanad released from IISR, Calicut. The average dry berry yield is 2,475 kg/ha. The berries contain 15.5% oleoresin, 3.3% piperine and 4.8% essential oil. This variety is suited to southern districts of Kerala.

In addition to the above released varieties, several promising cultivars are in the pipeline for release from IISR, Calicut.

Name	Pedigree	1	2	3	4	Remarks
Coll No. 1041	Clonal selection from Cv. Thevan-mundi	4.76	7.3	3.1	2.0	High tolerance to *Phytophthora* foot-rot
OPK m	Open pollinated progeny from Karimunda	3.4	8.3	3.0	3.0	Vigorous vine with long spikes and big bulk density
HP-105	Narayakodi × Neelamundi	4.6	7.2	3.8	2.0	Hybrid–suitable for high altitudes
HP-813	Cholamudi × Panniyur-1	3.52	10.4	5.4	3.37	Hybrid–good yielder, suitable for high altitudes
HP-1411	Aimpiryan × Panniyur –1	3.65	11.0	3.8	4.0	Hybrid–vigorous vine, good spiking. Compact spike and bold berries

1 Average yield kg/vine (Fresh berry)

2 Oleoresin

3 Piperine

4 Essential oil

Cultivation

Propagation

Pepper can be propagated by seeds as well as by vegetative means.

Propagation by seeds

The fully ripe berries are collected, soaked in water overnight and later rubbed with a paste of cowdung. The treated seeds are sown in the nursery. The seeds are viable upto 20 days after the harvest. They germinate within a month in the nursery beds and will be ready for transplanting within 45 days. In India, transplanting is done in July-August.

The seedling progenies show a lot of variation, since pepper is a cross-pollinated crop. They take 7-8 years to come to the first bearing.

Vegetative propagation

By grafting

Single-node cuttings could be successfully grafted, either by means of side grafting or inarching. Varieties like Uthirankotta can be successfully top worked with bisexual types and made productive.

By cuttings

Propagation by cuttings is commercially preferred over seed propagation. In Karnataka, cuttings are taken from long shoots in the old bush as these possess small roots and establish easily and quickly. While selecting cuttings, utmost care is necessary to prepare the cutting from the pedigree vines producing hermophrodite flowers. One to one-and-a-half metre long cuttings gathered from the runner at the base of the plant are also used for planting. In Maharashtra these runners are ground layered by covering them lightly with compost or leaf mould.

When planting material is in short supply, single-node cuttings could be successfully used. In Trinidad, a simple method has been devised wherein the above cuttings, treated with IBA at 2 mg per ml of 50% alcohol in the quick dip method, resulted in 75% success and rooted within 21 days.

For the rapid multiplication of pepper, pits of 50 cm^3 spaced 50-100 cm apart are dug. They are then filled with FYM, top soil and 250 g of super phosphate. Further, sixteen rooted cuttings per pit are planted 12.5 cm apart in both the directions. Shade and irrigation is provided. Urea and superphosphate at 20 g per pit is given at monthly intervals. The shoots arising are allowed to coil on a peg without touching the ground. These coiled shoots are cut in the 2nd fortnight of February for obtaining rooted cutting. By this method, we can obtain 100 cuttings per pit in the first year and 150-200 cuttings in subsequent years.

The rapid multiplication method (bamboo method) is adopted in different research institutions in India. For better rooting and quick establishment, growth regulators can also be used. IBA (1000 ppm) enhances the rooting percentage and shoot growth.

When pepper is propagated by air layering the fruiting branches, the layers develop into non-climbing dwarf plants with a bushy habit. They start fruiting at an early stage and are prolific bearers.

Micro-propagation of black pepper *in vitro,* using shoot explants both from mature and juvenile plants, were standardised at the Indian Institute of Spices Research (IISR), Calicut. The multiplication rate is around six shoots per culture in 90 days. The protocols for plant regeneration were also standardised and the plants were successfully micro-propagated from the callus cultures derived both from leaf and stem explants.

Land preparation

The pits are prepared well in advance of planting at the base of the standards with a minimum size of 50 cm^3. The pits are filled with a mixture of top soil, 5 kg of well rotten FYM/compost, 2 kg neem cake and 150 g rock phosphate.

Planting and aftercare

The pepper vine requires standards for support and for partial fulfilment of shade, if it is taken up as a pure plantation. Straight, quick growing standards have to be set in the main field before planting the pepper vines. The common live standards used to train pepper vines are *Erythrina indica, Garuga pinnata, Mangifera indica, Anacardium occidentale, Grevillea robusta, Terminalia sp.*, etc.

When pepper is grown as a mixed crop in coffee and coconut plantations or areca gardens, there is no need to plant standards or shade trees. The coconut or areca trees themselves provide shade and serve as standards. While planting, groups of 3 to 4 rooted cuttings are planted per standard. Usually the planting of vines is done during June-July in pits of 50 cm^3, dug on the northern side of the standard 15 cm away from it. The cuttings are planted with two nodes below the ground level. The pepper vine grows rapidly and after one year reaches a height of 2 m. As the vines grow they have to be tied to the standards at an interval of 30 cm. Also, the vines are brought down to ground level after a year and a half. They are cut and buried in the ground surrounding the standard. This helps in putting forth more shoots and good root systems.

Manures and Fertilisers

The dose of manures and fertilisers to be applied per unit area varies according to the soil fertility. Manuring for pepper vines is to be done in basins in a semicircular band on the northern side of the standard around the plants, 10-15 cm deep and 50-75 cm radius, depending upon the growth of the plants. Cattle manure/compost/ green leaves should be applied at the onset of the Southwest monsoon. It is desirable to apply lime at the rate of 500 g per vine in April-May in alternate years.

The following is the nutrient dose for pepper (3 years and above):

	Per vine per year
FYM or Compost or green leaves	10 kg
Castor cake	500 g
N	100 g
P_2O_5	40 g
K_2O	140 g

A one-third dose should be applied for one year old plants and half dose for two year old plants. The fertilisers may be applied in two split doses, the first in May-June and the second in August-September.

Intercultivation

During the first year, a thorough digging is given once during August-September and again in October-November around the standard and the vines to remove the weeds. Mulching is very essential where pepper is grown with minimum shade, to conserve moisture, using banana trash, dried grass and other substances. Cover crops like *Calapogonium mucanoides* can be sown during April-May and during October-November and they are cut and ploughed into the soil when they flower. Sawdust, areca husk and straw can also be used for mulching. Other operations like tying the vines to the standard, pruning the unwanted growth and the regulation of shade should be carried out as and when necessary.

Pests and diseases

Pests

In India, 34 species of insects are recorded to be pests on black pepper and the important ones among these are listed below.

Pollu Beetle (Longitarsus nigripennis)

This is commonly known as the 'flea beetle'. It is small, shining and brownish-black is colour. The adult beetle lays eggs in small shallow depressions made on the rind of tender pepper berries. The eggs hatch into pale-yellowish grubs which, in turn, bore into and feed on the contents of the tender berries making them hollow. The external indication of infestation is the presence of dark, drying berries possessing a characteristic circular hole in the midst of green healthy berries. A single grub destroys 3 or 4 berries. Fully-fed grubs drop to the soil, construct oval-shaped earthen cocoons and pupate inside them. The total life-cycle of the pest is 40-50 days. It completes 4 generations between July and January. Egg-laying stops by December when the pepper berries mature. The extent of damage caused by the 'pollu' beetle goes up to 40% in certain endemic areas.

Cultural operations like raking the soil and regulation of shade by standards are reported to be beneficial in reducing 'pollu' beetle infestation. The application of insecticides into the soil can also be effective in controlling pupating grubs. Spraying vines with Endosulfan (0.05%), Quinalphos (0.025%), Dimethoate or Monocrotophos (0.05%) twice a year, during July and September, effectively controls the pest. The leaf extract of *Strychnos nux-vomica* can be sprayed on the berries as an antifeedent.

Top-shoot borer (Cydia hemidoxa)

The caterpillars of this tiny crimson and yellow coloured moth damage the terminal shoots by boring into them. The intensity of incidence of this pest is more during August to December, when the succulent shoots are available. As the caterpillars bore into the terminal shoot, the growing point is damaged and this results in drying of terminal portions of the vine. The pest takes about a month to complete its life-cycle. Spraying the vines with Endosulfan, Dimethoate, Phosphamidon or Monocrotophos at 0.05% concentration at the point of new shoot emergence has been found to be effective in controlling top-shoot borer infestation. *Hexamermis* sp. and *Apanteles cypris* are the bio-control agents, for this pest.

Leaf gall thrips (Liothrips karnyii)

Infestation by leaf gall thrips is more severe at high altitudes and also in nurseries on plains. The feeding activity of the thrips on the leaves causes the leaf margins to curl downwards and inwards resulting in formation of marginal leaf galls. Later the infested leaves become crinkled and malformed. Spraying with Monocrotophos, Dimethoate, Endosulfan (0.05%) or Phosphomidon (0.03%) during the emergence of new flushes reduces the infestation. *Montandoniola moraguesi* and *Androthrips flavipes* are the biocontrol agents for this pest.

Scale insects and mealy bugs

Eighteen genera/species of scale insects and mealy bugs, *Pseudococcus vigatus*, infest black pepper. The scale insects cause severe damage to black pepper at higher altitudes. The Mussel scale (*Lepidosaphus piperis*) and Coconut scale (*Aspidiotus destructor*) are important. They are sedentary, remaining permanently fixed to the plant parts and appear as encrustations, mainly on the stems, leaves and berries. They feed on the plant sap, cause yellowing and wilting of infested parts. In severe cases of infestation the affected portions of the vines dry up. The infestation is more severe during summer. Spraying with Monocrotophos or Dimethoate (0.1%) or Quinalphos (0.05%) during January-February at 15 day intervals checks the infestation. A total of

17 predators and 4 parasitoids of the scale insects have been recorded, of which *Chlocarpus* sp., *Encarsia lounsburyi* and *Pseudoscymnus* could be employed for the biocontrol of scale insects.

Nematodes

Plant parasitic nematodes viz., *Radopholus similis* (burrowing nematode) and *Meloidogyne incognita* (root-knot-nematode), cause considerable damage to pepper plantations. The initial symptom is the turning of a few leaves from normal green to yellow. With the gradual advancement of the disease, many leaves turn yellow followed by defoliation and die-back, resulting in loss of vigour and leading to the death of the vine.

The nursery mixture should be sterilised with Methyl bromide to denematise the soil. In the main field the pits are treated with nematicide (Phorate 10 G or Carbofuran 3 G) at the time of planting. The planting of nematode-free, healthy rooted cuttings is essential. Phorate 10 G or Carbofuran 3 G @ 3 g a.i. per vine is applied once in May-June and again in September-October to check the population of nematodes. The variety CLT-P-812 is resistant to the root-knot nematode.

There are a few species of mealy bugs which infest the leaves, stems and sometimes the spikes and berries. Spraying Dimethoate, Parathion-methyl and Quinalphos (0.1%) is an effective control measure.

Pterolophia griseovarla and larva of *Synegiu* sp. which feed on the spikes and foliage, are the new pests recorded on black pepper.

Diseases

In India, although 17 diseases affect black pepper, the following are serious ones.

Quick wilt or foot-rot disease (Phytophthora palmivora Var. piperis)

This is the most serious and dreaded disease affecting the pepper vine. Three types of symptoms, i.e., leaf-rot, collar-rot and root-rot, are generally observed in a quick wilt infected plant. The first visible symptom on the leaf is water-soaked lesions in the infected area. Within 2 days the water-soaked areas change colour and develop into necrotic spots surrounded by a yellow halo. The lesions expand rapidly and cover large areas of the leaf. Subsequently, the infected region shows a wet rotting and causes defoliation.

Collar-rot is the most fatal type of infestion in pepper. It appears on vines at the ground level and above, up to about one metre height. The initial symptom is the appearance of water-soaked lesions similar to those on the leaves. On the stem and branches, lesions the spread and the affected area gets rotten within 8-15 days. In the advanced stages, the affected tissues emit a foul smell. Infection of the collar region progresses towards the underground portions of the stem and also to the root system and causes their rotting as well.

In the case of root-rot, the degree of damage depends upon the type and the number of roots infected and the extent of rotting. The infection generally starts on the fine roots and progresses towards the main roots, reaching finally underground portions of the stem. By this time, the entire plant collapses.

The disease can be effectively controlled by the following practices.

- Before the onset of the monsoon, good sanitation measures should be taken in pepper gardens by removing all dried vines and leaves.
- Good drainage should be provided.
- After a few early showers from the Southwest monsoon, the soil should be drenched twice around the base of the vine at a radius of 30 to 45 cm with 0.1% solution of methoxy ethyl mercuric chloride (Agallol G 3%) or Copper oxychloride (0.2%).
- The vines should be sprayed and drenched with 1% Bordeaux mixture during the onset of the Southwest monsoon and once again during the peak period of monsoon.
- The main stem should be pasted up to 1 to ½ m above soil level with Bordeaux paste (10% Boreaux mixture) during the onset of the monsoon will protect the stems from infection.
- Metalaxyl 500 ppm, a systematic fungicide, potent upto 50 days, can effectively be used to check *Phytophthora* infection.
- Application (per vine) of neem cake (1 kg) + *Trichoderma harzianum* (50 g), Ridomil MZ (metalaxyl) + garlic and mustard seed extract (2 kg garlic + ½ kg mustard, crushed and volume made upto 200 lit) + mulching the wet soil with transparent polythene sheets during the hot summer is the most effective treatment. The biological control measure should be followed by spraying of Akomin 6.2% (Potassium phosphate) in August or 0.3% Potassium phosphate @ 3 l/vine and drenching @ 5 l/vine twice, before the onset of the monsoon and during the second week of August.
- Prior inoculation with VAM provides better protection for the root system against this disease.

The sources of resistance/tolerance for this disease are the P-24 and P-603 varieties and *Piper colubrinum* and *P. arboreum*:

Slow wilt

This is a complex problem involving fungi, nematode, soil and nutritional factors. The fungal pathogens associated with this disease are *Fusarium, Colletotrichum, Rhizoctonia bataticda, Diploida,* etc. The slow-wilt disease of pepper as the name suggests is identified by gradual appearance of wilting symptoms in the pepper vines. The symptoms usually appear after the Northeast monsoon. The leaves and tender shoots first lose their turgidity and become flaccid, as if the plant suffers from moisture deficit in the soil. Later as the disease progresses, the leaves become pale and then yellow due to infestation by the nematodes *Meloidogyne incognita* and *Rhodopholus similies* and finally they shed. The symptoms become more pronounced during summer. With the onset of rains, new flushes are produced and the wilting symptoms apparently disappear. But after the Northeast monsoon, or the following season, the symptoms reappear more severely. The process is repeated for one or two years. The root system of the affected plants withers and decays. After the first appearance of the disease, it takes two to three years for the death of the plant, during which period, the health, vigour and yield of the plants gradually reduce.

The following control measures have been suggested to protect possible invasion of the organisms.

- A soil drenching treatment should be given with 0.1% mercurial fungicide or 1% Bordeaux mixture @ 5-10 l/vine 2-3 times during the monsoon season to the root zone.
- Apply Phorate 10 G or Carbofuran 3 G @ 30 g/vine plus Bordeaux mixture spraying and drenching plus soil application of neem cake @ 2 kg/vine during May-June and October-November.

Pollu disease (Anthracnose)

This is a berry disease caused by *Collectotrichum gloeosporeoides*. The disease can be controlled by two rounds of spraying with 1% Bordeaux mixture or Captafol (1.0%) once before flowering and at berry formation stage or spraying 0.2% Carbendazim before the onset of monsoon and subsequently at monthly intervals till September.

Stunting disease

This disease is caused by the cucumber mosaic virus (CMV). The symptoms are: (i) stunting, crinkling, cupping and narrowing of leaves, (ii) scattered chlorosis flecks on the leaves, (iii) chloratic flecks along veins, leading to interveinal chlorosis, (iv) vein banding, severe interveinal chlorosis with dark green veins, (v) leathering of leaves with mosaic pattern and vein banding. No effective measure has been recorded against this disease.

Phyllody, caused by phytoplasmas and leaf-spot diseases, is also of increasing importance in black pepper. Yellow disease is reported to be a serious disease of black pepper in Indonesia.

Harvesting and yield

The spikes are ready for harvest when they start yellowing and in a spike 1 or 2 berries become red. At this stage, the spikes are removed from the vines. The spikes are kept as such for a day or so, thereafter the berries are removed by rubbing or scrubbing, and then dried in the sun. The spikes are sometimes dried directly in the sun for a few days on mats or on clean concrete floors. These are turned over, and later are removed when completely dry. The outer skin of the berries becomes dark brown to black and shrivels.

In Kerala, where 90-95% of pepper is grown, the plants flower in May-June. The crop takes 6-8 months from flowering to harvest. The harvest season extends from November to January in the plains and January to March in the hills.

The yield varies widely in different pepper producing areas depending upon: (i) elevation (ii) temperature (iii) distribution of rainfall (iv) soil fertility (v) cultural practices (vi) type or variety and (vii) age of the vine. In India, the yield varies from 110 kg to 355 kg/ha.

Commercial grades of pepper

- Malabar Black Garbled - light green in colour with 11.7% oleoresin content.
- Malabar Black Ungarbled - light green in colour with 10.1% oleoresin content.

Standard specifications for pepper

The Indian Government, through the Ministry of Agriculture and Food, has prescribed the obligatory grading and standardization of agriculture products under the label "AGMARK". This specifies limits for extraneous matter, light berries and moisture content. They provide a compulsory quality control and pre-shipment inspection apart from meeting the requirements of importing countries. For export, the American Spice Trade Association standards are generally followed, where moisture content, ash content, essential oils and starch content are considered.

	Moisture (%)	Ash (%)	Crude fibre(%)	Volatile oil (%)	Starch (%)
Black pepper	12	7	10.15	2 (minimum)	30 (minimum)
White pepper	13	3.5	5	1.5	52 (minimum)

Value-added products

The value-added products of pepper are dehydrated green pepper, freeze-dried green pepper, white pepper, green pepper in brine, pepper oil, pepper oleoresin, ground pepper, organic pepper, sterile pepper and canned tender green pepper.

White pepper

It is prepared by removing the outer pericarp (skin) of the berries by either water steeping, steaming or boiling and decortication. In the water-steeping method, fresh, ripe spikes or berries are packed in gunny bags or steeped as such in water tanks or running water for 7-10 days. The pin-heads and light berries, which float are separated. Once the skin is softened, they are rubbed by hand or tampered to remove their outer skin. The deskinned berries are sun-dried and sold as white pepper. The recovery of white pepper varies from 22-27% of the green pepper. White pepper can also be prepared from black dried pepper by steeping them in water for 15-20 days. However, a loss of about 20-25% of berries is incurred due to breakage in this method.

Pepper oil

The pepper oil is extracted by the steam distillation of powdered pepper. The quality and composition of the pepper oil varies considerably, generally due to variation in variety, grade, storage conditions and processing. To arrive at uniform quality in pepper oil, a proper blending of inputs is necessary. Structurally, the oil is richly carpeted with over 250 compounds comprising (%) monoterpene hydrocarbons (70-80), sesquiterpenes (20-30) and oxygenated compounds (>4). The monoterpene hydrocarbons oscillate as follows (%): α-pinene (1-19), β-pinene (5-35), sabinene (0-44.5), car-3-ene (trace-15), α-phellandrene (1-27) and myrcene (trace-10). The oil of pepper is occasionally adultrated with the lower-priced common adulterants like phellandrene, dipentene and caryophyllene, which are also natural components of the oil itself.

Pepper oleoresin

It is prepared by the solvent extraction of ground pepper. The production process uses a number of equipments like precleaners, pulverizers, extractors and solvent recovery units. Besides, for quality control measures, instruments like gas chromatography and ultraviolet spectrophotometer are used. The oleoresin of pepper can be prepared based on the customer's quality requirement.

Piperine

The alkaloid piperine (3-6%) is the major constituent responsible for the biting taste of black pepper. The other pungent alkaloids are chovicine, peiperidine and piperine.

Green pepper in brine

Bottled green pepper has great demand in non-traditional areas. The green colour is maintained under the high salinity of the steeping liquid. The minimum salt level should be 12%. The addition of a small amount of citric acid prevents the discolouration due to phenols.

Dehydrated green pepper

Keeping the freshly harvested, despiked pepper in boiling water for over 10 minutes deactivates the bleaching enzyme. Treatment with sulphur dioxide reduces the chances of darkening. As sun-drying destroys chlorophyll and the green colour, to make dehydrated green pepper, drying should be done in hot air or in a microwave oven.

Frozen green pepper

This is prepared using blast freezers. Such peppers, on thawing, are almost equivalent to the fresh material.

Other products like spice essences and emulsion, spice decoctions, encapsulated spices, fat-based spices, etc., are also prepared from black pepper.

Celery

Plant Profile

Family	:	*Apiaceae / Umbelliferae*
English name	:	Celery
Indian name	:	*Shalari, Ajmud* (Hindi), *Bandhuri, Chanu* (Bengali), *Bodiajmoda* (Gujarati), *Ajmoda* (Marathi), *Kernauli* (Punjabi), *Ajamoda* (Sanskrit), *Ajmoda* (Tamil)
Species	:	*Apium graveolens* Linn. *Apium graveolens* var. *rapaceum* (Mill), *Apium graveolens* var. *secalinum* and *smallage*
Distribution	:	India, Algeria, Asia, Baluchistan, Belgium, the Caucasus, Egypt, Ethiopia, France, Germany, Holland, Italy, Sweden, UK, USA
Uses	:	Culinary, Drugs, Flavouring, Food, Perfumery

The dried fruit of *Apium graveolens* Linn. (2n=22) (Family: *Apiaceae; Umbelliferae*) is known as celery. It is popularly known as *Karnauli* or *Ajmod*.

There are four known horticultural types of celery. *A. graveolens* var.*dulce* (Mill) Pers., whose leaves and stalks are consumed as salad, appetizer and as a flavouring in soups; and var. *rapaceum* (Mill) Gaudich commonly called celeriac, with dark green leaves is cultivated for root tubers which are cooked and eaten. The other two are var. *secalinum* and *smallage*. The main variety cultivated in India is *dulce*.

Commercially celery is available as celery seeds, celery flakes, vegetable, celery seed oil and celery seed oleoresin.

The celery seed has the following quality characteristics; moisture-5-11%, volatile oil-1.5-3, (average 2.4%), non-volatile ether extract-5.8-14.2% (average 9-4%), cold water extract-5.9-12.6% (average 8.4%), total ash-6.9-11.0% (average 8.8%), ash insoluble in acid-0.5-4.0% (average 2.5%).

Fresh celery leaves and stalk are used as a salad plant. They contain moisture-81.3, 93.5 per cent, protein-6.0, 0.8 per cent; fat-0.6, 0.1 per cent, fibre-1.4, 1.2 per cent, carbohydrates-8.6, 3.5 per cent, mineral matter-2.1, 0.9 per cent, calcium-0.23, 0.3 per cent, phosphorus-0.14, 0.4 per cent, Iron-0.06, 0.05 per cent, vit A-5800, 7500, I.U., vit B_1 – traces, vit C-62,6 mg/100 g calorific value; 64,18, calories per 100 g.

The seeds on steam distillation yield a pale yellow essential oil (1.75%) which contains limonene (80%) as a major constituent. The other constituents of the oil are: p-dimethyl styrene, n-pentyl benzene, caryophyllene, selinene, n-butyl phthalide and sedanolide along with sabinene, β-elemne, trans-1,2-epoxy limonene, linalool, isovaleric acid, cis-dihydrocarvone, trans-dihydrocarvone, terpinene-4-ol, cis-p-menth-2,8-dien-1-ol, trans-p-menth-2,8-dien-1-ol, terpineol, carvone, trans-carveol, cis-carveol, tans-anethole, trans-p-menth-1(7), 8-dien-2-ol, perillaldehyde and thymol. The seeds are also rich in vitamin B.

The dried ripe seeds are used as spice to flavour food and liquids. The seeds are stimulating and carminative, used as a nerve tonic in domestic medicine. It is also a remedy for rheumatism. The seed oil is used as a food flavourant, and in perfumery and pharmaceutical industries. The fatty oil obtained from the fruit is used as an anti-spasmodic and nerve stimulant.

Origin and Distribution

The native habitat of celery is the lowlands of Italy from where it spread to Sweden, Egypt, Algeria and Ethiopia, and in Asia, to India, the Caucasus and Baluchistan. Celery is also claimed to be mentioned for the first time as a cultivated food plant in France as early as 1623. The countries growing it are the USA, France, Germany, United Kingdom, Italy, Belgium and Holland. In India it is cultivated in the Northwestern Himalayas, the Punjab, Haryana and Western Uttar Pradesh in area of about 5000 hectares. Punjab contributes to about 90% of the total Indian production. Celery seed is in great demand both in India and abroad. Indian celery seed is exported mostly to the UK and USA and to a lesser extent to continental countries like France, Italy, the Netherlands, Germany as well as Australia and New Zealand. Currently, about 8,000 hectares are under celery cultivation. India exported celery seed worth Rs. 46 crores during 1991-92.

Description of the plant

Celery is a herbaceous annual or biennial erect herb growing to a height of 60-90 cm. It has a shallow taproot system. The stem is branched, succulent and ridged. The leaves are reduced, pinnate, large, deeply divided and coarsely-toothed at the apex. The leaflets are ovate to suborbicular, 3-lobed, 2.0-4.5 cm long. The inflorescence is a compound umbel. The flowers are small and white and the calyx teeth are obsolete. There are five ovate, acute petals, with their tips inflexed. The carpels are semi-terete, sub-pentagonal, the primary ridges being distinct and filiform. The fruit is a schizocarp with two mericarps, suborbicular to ellipsoid, 1-2 mm in diameter, aromatic and slightly bitter. Celery is naturally cross-pollinated but not self-incompatible.

Climate

In colder climate and on the hills, celery behaves as a biennial plant but in the plains, it behaves as an annual plant. Celery is a moisture-loving plant requiring a cold climate. It can be successfully cultivated as a winter crop in areas which are free from severe frost and have low atmosphereric humidity during February-March when the crop flowers. Northern and Central India including the hills, that have cold and dry climate are most suited for celery cultivation. It is sensitive to very low day temperature (5°-10°C) during vegetative growth causing early bolting. A combination of 12° to 15° and 22° to 25°C day-night temperatures gives 80% seed germination within 2 weeks.

Soil

Celery can be successfully cultivated on all soils except saline, alkaline and water-logged soils. However, loamy soils which are rich in organic matter and retentive of soil moisture are the best. Celery is not suited to clayey soils. It does not withstand water stagnation. Celery is sensitive to the extremes of soil reaction, a soil pH range of 6.0 to 7.5 is suitable for its cultivation.

Varieties

The important high yielding varieties recommended for cultivation under this crop include EC-99249-1 and RRL-85-1. The selection RRL-85-1 yields 2-3% yellow volatile oil. It has 16-20% fatty oil. 'Standard Bearer' and 'Wright Grove Gaint' are the varieties of celery recommended for cultivation by IARI, New Delhi, for salad purposes.

Cultivation

Propagation

Celery can be grown from seeds, by raising the seedlings in nursery beds. The seeds are quite small. An ounce contains about 70,000 seeds. The germination rate of seeds is about 50%. A seed rate of 1.5 kg per hectare is sufficient. The seeds are sown in boxes and covered with thin layers of soil and FYM mixture. The seedlings are ready for transplanting when they are about 10 cm tall (60-70 days after sowing).

Tissue culture

The major breakthrough in tissue culture research is the encapsulation of somatic embryos from hypocotyl segments, thus leading to the production of artificial celery seeds.

Land preparation

The land is brought to a fine tilth and laid into plots of convenient sizes for irrigation. It is preferable to mix with the soil 10-20 tonnes of FYM/ha during the last ploughing.

Sowing/transplanting

When it is grown as a garden crop on the hills, the seeds are sown in March-April. The seedlings are transplanted in May and the crop is ready for harvest in November. In the plains, the seedlings preferably brought from the hills are transplanted in September-October and the crop is ready within three months. In the plains, the seeds are generally sown from the middle of September to October and the transplanting of seedlings, 30 cm apart in rows spaced at 30-45 cm, is carried out in January. It is also grown as a directly seeded crop, on a smaller scale in Amritsar District.

Manures and fertilisers

Before planting the seedlings, about 10-20 tonnes of FYM/ha is added. On medium soils, about 80-200 kg Nitrogen, 33-60 kg Phosphorous and 20-40 kg Potassium/ha is applied to the crop. Half of the Nitrogen, a full dose of Phosphorus and Potassium are applied in rows at the time of planting and the remaining Nitrogen is given as a top-dressing after one month. Sometimes Bo and Mn deficiency in soil can cause black heart and cracked stem, respectively and deficiency of Mg can cause chlorosis. In such soils, 10 kg each of Borax and Magnesium sulphate are applied at the time of land preparation (10 days before planting).

Irrigation and inter-cultivation

The crop requires 10-12 irrigations at an interval of 8-7 days during the crop period. Once established, the crop requires 3 to 4 hoeings to keep down the weeds. Application of chemical weedicides like Basalin (pre-planting) at 2-2.5 lit/ha or Stamp at 2 lit/ha or TOK-E-25 + Lasso @ 2.5 lit./ha or Fluchloralin 0.90 kg/ha and Pendimethalin 0.60 kg/ha are reported to successfully control the weeds.

Pests and Diseases

Pests

Leaf miner (*Liriomyza trifolii*) is found to attack the older outer leaves of the celery plant and cause damage to most of the leaves. The insect can be effectively controlled by spraying a systemic insecticide like Quinolphos (0.1%). Carrot rustfly, tarnished plant bug, celery leaf tier, celery fly, aphids, caterpillars, flea beetle, mole cricket, spring trails, army worms, wire worms, red spider mite, etc., are the other insect pests which cause losses in some areas. Suitable control measures may be taken up as and when the pests are noticed.

Trichoplusia sp. and *Spodoptera exigua* are two major insect pests of celery in California.

Diseases

Cercospora apii is reported to cause early blight; *Septoria petrosulnii*, late blight and *Phyllosticta apii* cause leaf-spot diseases to celery crop. These diseases can be controlled by spraying Thiophanate methyl (0.5%) + Manzeb (0.1%) or Fentin hydroxide (0.2%) at an interval of two weeks. Petiole-rot caused by *Sclerotium* sp. and *Fusarium* sp. and yellow mosaic are the other diseases reported on this crop.

Harvesting and yield

In cooler climates and on the hills, celery is a perennial plant and produces seeds only in the second year. It takes about 4-5 months from the time of sowing to seeding. In the plains, the crop matures in about 3 months after transplanting. The crop is harvested when about 80% of the buds begin to turn light brown. Since seed shattering is common, harvesting it in the early morning hours is advisable. The harvested crop is stacked in the field for a few days, and then threshed to obtain the seeds. Care is taken to avoid prolonged exposure to the sun. The seeds are cleaned by winnowing. They are graded through sieving and stored in gunny bags in a cool dry place.

The average yield of celery is about 1000-1500 kg/ha. Celery seeds yield 2-3% of pale yellow volatile oil with a persistent odour. The volatile or essential oil contained in the seed is isolated by steam distillation. The seed should be crushed and immediately put through distillation to avoid the loss of oil by evaporation. It is important that the seeds be evenly spread on the perforated grids with which a still serving for seed distillation should be equipped. The distillation of one batch lasts for 10-12 hours. The distillation wastes are usually redistilled.

Value-added products

Celery seed oil, celery herb (leaf) oil, celery chaff oil, celery seed oleoresin, celery salt, celery pepper, dehydrated celery leaves and freeze-dried celery are some value-added products.

Chilli

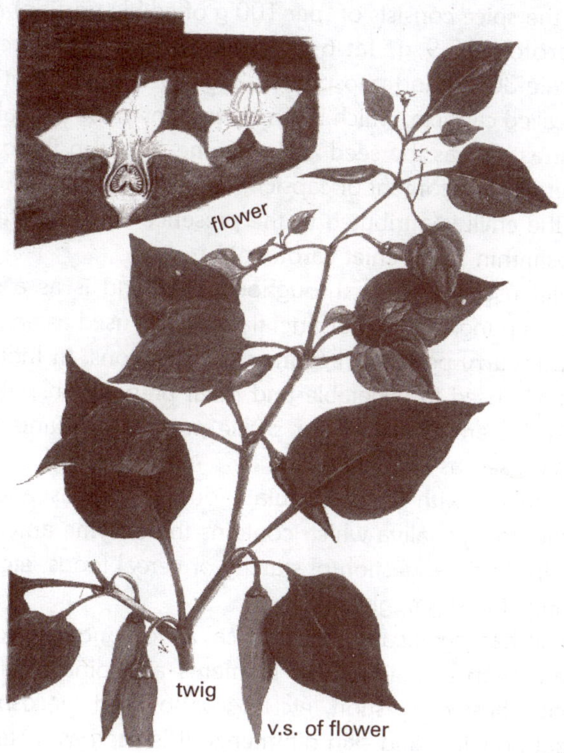

Plant Profile

Family	: *Solanaceae*
English name	: Chilli, Red pepper, Hot pepper, Hot chilli
Indian name	: *Lal Mirch* (Hindi), *Menasina Kayi* (Kannada), *Mulaku* (Malayalam), *Milagai* (Tamil), *Mirapakayi* (Telugu).
Species	: *Capsicum annum* L. *C. baccatum* L. *C. fruitescens* L. *C. chinense* L. *C. pubescence* L.
Distribution	: South Central America, Chile, Mexico, India, Africa, Japan, Turkey, USA.
Uses	: Spice, culinary purposes.

Chilli or red pepper or hot pepper or hot chilli is the dried ripe fruits of *Capsicum annum* Linn. of the family *Solanaceae* and the leading vegetable-cum-spice of India. The dried fruits which constitute the spice consists of (per 100 g of edible portion) moisture-10.0 g, protein-15.9 g, fat-6.2 g, minerals-6.1 g, fibre-30.2 g, carbohydrate-30.6 g and phosphorus-270 mg. The pungent principle of chilli is called capsicin, which is mainly present in the central placenta of the fruit as well as the seed and to some extent in the ovary wall also. Generally, the content of capsicin varies from 0.2-0.4%. The red colour of the chilli is attributed to the presence of 20 carotenoids, of which capsanthin is the chief carotenoid.

The chief use of chillies throughout the world is as a spice, on account of its pungency and distinct flavour. It is used as an important ingredient in curry powder in culinary preparations. In India, chillies are essentially used in vegetable and meat preparations, either fresh or dry. Chillies are also used for pickle preparations and the sweet varieties are used as vegetables.

Chillies taken with food stimulate our taste-buds and thereby increase the flow of saliva which contains the enzyme amylase. This, in turn, helps in the digestion of starchy or cereal foods, etc., into the easily assimilable sugar, glucose.

Capsaicin has significant physiological action and is used in many pharmaceutical preparations like liniments and ointments for cold, sore throat, chest congestion, etc. It is also used in cosmetics like prickly heat powders and skin ointments. It is used as a carminative, tonic and stimulant. The extract of chillies is used in the preparation of ginger beer and other beverages. Oleoresin, known as oleoresin capsicum, is used in pain balms, vapours, liniments, etc.

Origin and distribution

Chilli is said to be a native of South Central America and it was domesticated in Mexico from ancient times. The introduction of chillies into India is attributed to the Portuguese. Chillies are cultivated mainly in tropical and subtropical countries, particularly Africa, India, Japan, Mexico, Turkey, the USA, etc.

India is a major producer, exporter and consumer of chillies. Chillies produced in India are generally highly pungent. In India, chillies are grown in almost all states. Andhra Pradesh has been the largest chilli growing state followed by Karnataka and Maharashtra. The productivity of chillies is the highest in Andhra Pradesh followed by Arunachal Pradesh and Punjab. The productivity of chillies in Andhra Pradesh and Punjab is higher since the crop is raised under irrigated conditions as compared to Maharashtra and Karnataka. Indian chillies reach over 90 countries in the world. Bangladesh, Bahrain, Canada, Italy, Israel, Japan, Malaysia, Netherlands, Philippines, Singapore, Spain, Sri Lanka, Saudi Arabia, USA and UAE are the leading importing countries. During 2000-2001, 61000 MT of chillies valued at Rs. 195.25 crores was exported from India, earning a foreign exchange of 13% from spices.

Description of the plant

Though 20 wild species of chillies have been reported, only 5 species are cultivated, viz., *C.annum, C.baccatum, C.fruitescens, C.chinense* and *C.pubescence*. Almost all the varieties of low and medium pungency that are cultivated on a field-scale in India belong to *Capsicum annum*. It is a herbaceous or semi-woody annual. The plants are erect, profusely branched and 0.5 to 1.5 m high with taproots that are restricted after transplanting. The leaves are variable in size and simple, the lamina is broadly lanceolate to ovate, the tip is acuminate and the base is acute. The flowers are usually borne singly and are terminal. The flowers are bisexual and their colour varies from white to blue. The fruits are borne singly at nodes, variable in size, shape, colour and degree of pungency, linear, conical or globose. The unripe fruit is green or purplish, ripening to red or orange, yellow, brown, cream or purplish and the seed is pale yellow.

Capsicum fruitescens is a perennial sub-shrub similar in structure to *C.annum*. Plants usually bear two or more erect berries at the fruiting nodes and the fruits are extremely pungent.

Climate

Chillies are grown in both tropical and subtropical areas, ranging from sea level to 2000 m altitudes in Indian conditions. However, the pungent varieties or varieties suited for spices cannot be grown in higher altitudes. A warm humid climate favours growth while warm and dry weather enhances fruit maturity. A temperature range of 20-30°C is ideal. It can be grown successfully as a rain-fed crop in areas receiving an annual rainfall of 75-100 cm. Excessive rainfall leads to poor fruit set and this, in association with high humidity, leads to rotting of the fruits. Pungent chillies are susceptible to frost.

Soil

A soil temperature of 10°C retards development, while 17°C is good for normal development. The tops continue to develop with higher temperatures, but the root development is retarded at soil temperatures of 30°C and above. In capsicum, the fruit development was found to be adversely affected at temperatures of 37.8°C or more. High temperature, associated with low relative humidity, at the time of flowering increases transpiration resulting in the abscission of buds, flowers and small fruits. An average high night temperature was found to be responsible for the high capsaicin content. A temperature ranging from 20 to 25°C is ideal for chillies. A day length of 9-10 hours light stimulates plant growth and increases the productivity by 21-24%, besides improving the quality.

Chillies are found to grow in a variety of soils provided they are deep, well drained, well aerated and fertile. In poorly drained soils, the plants shed their leaves and turn sickly. The fruit drop takes place due to water-logged conditions. It can be grown successfully in sandy loam soil, provided adequate irrigation and manuring are carried out. Black soils are also suitable for rain-fed crops. An ideal soil for chilli is light loamy or sandy loam, rich in lime and inorganic matter and having a pH range of 6.5-7.5. It is grown on soils of pH 5 in coastal areas, to pH 9 in vertisols in India. In acidic soils, liming is beneficial. Strong acid and alkaline soils are not suitable for chilli growing.

Varieties

There are several varieties of chilli under cultivation differing in the size, shape, colour and pungency of the fruits (pods), and there are two main groups based on this. One is called the *Samba* type in which the fruits are long and the other is the *Gundu* type in which the fruits are round in shape and short. The *gundu* types are generally more pungent than the *samba* types and they are better adapted to the rain-fed culture than the *samba* types. For dried chillies, the varieties having thin pericarp, low seed content and strong spikes are suitable. The important varieties grown in India are given in the following table.

Important varieties and hybrids of chilli

Variety	**Important characters**
Andhra Jyothi* (G_5)	Pods, 3-4 cm and 4-5 cm in girth
Aparna*	Pods, 8.3 cm long, cylindrical with semi-cup shaped calyx and blunt tip. Turn yellow on ripening. Retains colour in storage. High pungency (0.7%). Average yield (dry)-11.7 q/ha. Moderately tolerant to major pests and diseases under field conditions.

(*continues...*)

(...*contd.*)

Variety	Important characters
Arka Lohit*	Dual purpose, highly pungent chilli with an average yield of 3.5 t/ha (dry).
Bhagyalakshmi* (G4)	Immature pods, olive-green turning dark red on ripening, seed content-38-40%. Fairly tolerant to diseases and insects. Average yield (dry)-11.7 q/ha under rain-fed and 50 q/ha under irrigation. Pods 7-8 cm long and 3 cm in girth.
CH-1	F_1 hybrid, yield 249 q/ha (fresh chilli).
CO-1*	Elongated and stout fruits with high pungency. Average yield (dry)-21 q/ha.
CO-2	Pods are small, 2.4 cm long with dark red colour having 7-8 seeds/pod. Crop duration of 200 days. Average yield (dry)-22.50 q/ha.
CO-3* (CA 586)	Suitable for high-density planting. Dual-purpose chilli, with an yield potential of 15-18 t/ha (green chilli) or 35 q/ha (dry chilli).
G3	Pods are 6-7 cm long and 3 cm in girth. Immature green and mature red fruits remain smooth after drying, suitable for rain-fed and irrigated conditions, seed content-44%, pungency high, average yield (dry)-10.6 q/ha.
Hissar Vijay (HC 28)	It possesses a high degree of resistance to viral and fungal diseases. Yield-17.85 t/ha (red ripe). Suitable for ratooning.
Hissar Shakti (HC 44)	Another multiple resistant variety. Yield-18.6 t/ha (wet). Also suitable for ratooning.
Jawahar 218*	Pods are erect, thin, wrinkled, light green and turn rosy-red at maturity, early and high-yielding, average pod yield (dry)-18-22 q/ha. Suited for green and red chilli production. Tolerant to leaf-curl and fruit-rot.
Jwala Muskhi*	A hybrid evolved through recombination breeding of a cross between Vellanotchi and Pusa jwala. Dwarf, low pungency, suitable as vegetable chilli. Yield-22.5 t/ha (green) in 137 days. Tolerant to little leaf, leaf-spot disease and partial shade.
Jwala Sakhi*	A cross between Vellanotchi x Pusa jwala. Yield-19.6 t/ha of green chilli. Tolerant to little leaf and leaf-spot.

(*continues...*)

(...*contd.*)

Variety	Important characters
K1*	Pods are 6.6 cm long with 75-80 seeds, red in colour, smooth surface with pointed tip. Pericarp-0.07 cm thick. Capsaicin content-0.42% and dry yield-17 q/ha.
K2*	Pods, 7.3 cm long, pericarp thickness-0.19 mm, red fruit colour, 80-85 seeds with blunt tips. Capsaicin content-0.42% and dry yield-17 q/ha.
MDU-1*	Pods are 8.0 cm long, red, 50-60 seeds/pod, crop duration-210 days, high capsaicin. Average yield (dry)-19 q/ha.
Musalwadi	Fruits are small, dark green with black patches. Dual-purpose chilli. It is tolerant to die-back and powdery mildew.
NP 46A	Pods are 8-9 cm long, wrinkled, light green, turn light red on ripening, tolerant to thrips, susceptible to viral diseases. Yield-70-80 q/ha (fresh).
Pant C1*	Fruits are erect, small-sized, highly pungent. Yield (dry) is 12 q/ha. Tolerant to mosaic and leaf-curl virus.
Pant C2*	Fruits are pendant, average yield (dry)-14 q/ha. Tolerant to mosaic and leaf-curl virus.
PKM-1*	A hybrid derivative of the cross between AC No.1997 x CO-1, selected in F_4 generation and fixed by selfing. Yield-30-32 q/ha (dry) in 180 days.
Punjab Lal*	Pods are erect, medium-sized (4.25 cm × 0.79 cm), dark green, turn dark red at maturity. Rich in capsaicin (0.7%), oleoresin, dry matter and red pigment. Resistant to TMV, CMV and leaf-curl virus. Moderately resistant to fruit-rot and die-back. Average fruit yield (fresh)-100-120 q/ha.
Pusa jwala*	This is a hybrid derivative of NP 46 A x Puri Red. Pods are long, thin, red in colour and usually curved at the tip. Suited for use as green chilli.
Pusa Sadabahar*	Pods are erect, 8 cm long and 6-14/cluster. Ripe pods are bright red, pod yield (dry)-15-20 q/ha. Resistant to CMV, TMV, leaf-curl viruses.
Sankeshwar 32	Pods are long (25-30 cm), thin wiry, highly wrinkled. Immature pods are light green, turn red on maturity. Retain red colour in storage. Mostly grown as a dry chilli.

(*continues...*)

(...*contd.*)

Variety	Important characters
Sindhur	Fruits are short and stout with a conical shape
Ujjwala*	Chilli with erect fruits borne in clusters. Resistant to bacterial wilt.
X-235	Flowers with yellow anthers serve as a marker gene. Pods are thin, 5-6 cm long. Calyx is deeply cup-shaped, fruit tip is pointed. Seed content is 45%, highly pungent. Maximum yield (dry) potential upto 75 q/ha.

* Gundu types

The other released varieties include RHRC Pendant cluster, RHRC Erect cluster, DH-76.6, PRL1 (Palur 1). A number of F_1 hybrids have also been released from private companies. Of the hybrids, 'Agni' and 'ARCH 228' are the high yielding ones.

Cultivation

Propagation

Chillies are propagated by raising seedlings in seed beds. Only seeds of the recommended variety procured from a reliable source should be used.

Raising seedlings

The time of sowing, and transplanting depends upon temperature, rainfall and availability of irrigation facilities. Generally, chilli is raised by transplanting but, in some areas, direct seeding is also practised for production of dry ripe chillies. The very light seeds of chilli remain viable for 2 years. Extracting seeds one week before planting is beneficial. About 1.5 kg seed is enough for planting one hectare. Nursery beds should be prepared well, near partially shaded areas. The seeds are treated with 2.5 g Bavistin to prevent any incidence of seed-borne diseases and sown thinly in lines spaced 5 cm apart. Sand and well-decomposed compost are sprinkled over the seeds. Mulching with paddy straw soon after sowing conserves moisture. The mulch should be removed as soon as the seeds start germinating. Heavy watering should be avoided in the nursery. Drenching with copper oxychloride at fortnightly intervals keeps damping-off under control. On the 20th day of sowing, 300 g of Furadon 3 G granules has to be applied in between the seedling lines across the bed and stirred with the soil. A foliar spray of 1% Urea is helpful to induce quick growth. The seeds germinate in 6-10 days. The seedlings are ready for transplanting 40-45 days after sowing, when they are about 12-16 cm in height.

Land preparation

The main field should be ploughed 3-4 times to get a fine tilth. FYM @ 25-30 t/ha is incorporated during the last ploughing. For irrigated crops during the summer season, ridges and furrows are made at the recommended spacing. In summer and in the rainy reason planting should be done on the ridges. Adequate shade should be provided during the summer season for a period of one week.

Planting

In Maharashtra and Karnataka, a spacing of 75 cm × 75 cm or 90 cm × 90 cm is generally practised. Whereas, in Andhra Pradesh and Tamil Nadu, a closer spacing of 45 cm × 45 cm or even closer than this is followed. For the tall chillies, Sunkeshwar and Badyagi grown extensively in Maharashtra and Karnataka, wider spacing is adopted. The short-statured chilli Jwala and NP 46 A require a spacing of 60 cm × 45 cm. Generally, closer spacing is ideal in light soils. Heavy soils like black cotton soil require a wider spacing for enabling inter-cultivation. In Tamil Nadu, closer spacings of 30 cm × 30 cm and 30 cm × 20 cm give higher yields if N is applied @ 120 kg/ha. For rain-fed chilli cultivation, a closer spacing of 90 cm × 20 cm is recommended. Single healthy seedlings are transplanted and the field is irrigated immediately for better establishment in the soil.

Manures and fertilisers

Chilli responds well to the application of fertilisers both in irrigated and rain-fed areas. Since chilli has a long growing season, it needs judicious application of manures and fertilisers. Heavy application of organic manure is followed in irrigated areas. The heavy application of nitrogenous fertilisers may increase vegetative growth and delay maturity. The fruit yield (fresh and dry weight) increases up to 180 kg N/ha. Under rainfall conditions, the full dose of P and K and a half of N should be applied two weeks after planting, and the remaining N should be top-dressed one month after the first application. In irrigated areas, N should be split into 3 equal doses. If N is applied at more than 100 kg/ha it should be split into 2-3 doses, but when the dose is less than 50 kg/ha, it may be applied in a single dose. A single application of N at the commencement of flowering (3-4 weeks after transplanting) gives a higher yield, rather than 2-3 split applications of N. However, 10 tonnes of organic manure should be applied as spot manuring. With the application of 120 kg N, 60 kg P_2O_5 and 50 kg K_2O, a yield of 24.8 q/ha may be obtained, while the application of 50 kg K_2O alone gives the lowest yield (3.8 q/ha).

The use of a weak starter solution at the time of transplanting is beneficial. The foliar application of fertilisers is beneficial is adverse soil conditions. The application of Urea + Single superphsophate + Muriate of potash (2:1:1) in liquid form, applied at 300 ml/hill soon after transplanting, gives the highest yield of chilli.

Use of bio-fertilisers

Azatobactor and *Azospirillium* are recommended in chilli. Of the two, *Azosporillum* is more effective. It can be applied as a seed treatment, seedling treatment and as a direct application through the soil. For treating the seeds, 200 g of *Azospirillum* should be mixed with 200 ml of boiled and cooled rice water. About 500 g seeds can be soaked in it and dried in the shade for 30 minutes before sowing. For seedling treatment, a slurry can be prepared by dissolving 40 g of *Azospirillium* in 2 litres of water. The roots should be dipped in the slurry for 15 minutes before transplanting. For soil application, *Azospirillium* 2 kg/ha should be mixed with 20 kg of FYM or compost.

Irrigation

During the first month of transplanting, a light irrigation is required. In summer, irrigation on alternate days is essential in light soils. In Andhra Pradesh and Tamil Nadu, chilli is mainly grown as an irrigated crop. Gap-filling is done during the second irrigation 10 days after transplanting. The spraying of anti-transpirants is advocated in arid and semi-arid areas to prevent water loss from the plant surface.

Interculture

Earthing-up should be done 30 days after transplanting. Mulching helps conserve moisture and nutrients, and prevents weed growth. Dried leaves, paddy straw, sawdust or aluminium foil can be used for mulching. Aluminium foil and sawdust also act as insect repellents. On the 20th , 40th, 60th and 80th days of transplanting, Triacontanol 1.25 ppm (2.5 ml of Vipul dissolved in 10 lit. of water) may be sprayed on the foliage of the plants using a hand sprayer. Triacontanol increases photosynthetic efficiency and thereby the yield upto the tune of 500 kg of dry pods per hectare. To improve the fruit set, as well as to reduce flower and fruit drop, 50 ppm of Naphthalene Acetic Acid (equal to 1 ml of Planofix mixed with 4.5 lit. of water) has to be sprayed.

Weed control

Manual weeding is a common practice for controlling weeds in chilli plantations. However, the non-availability of labour or continuous rains, often do not allow the timely control of weeds at the critical periods of crop growth. Therefore, spraying of Tok E 25 @ 2 lit/ha with one hand weeding helps to check weed growth to a considerable extent. The application of Pendimethalin at 1.25 kg supplemented with one hoeing 25 days after transplanting, followed by the same herbicide without hoeing, is beneficial. Oxyflourfen (0.2 kg) in combination with one hoeing, is effective in checking weed growth. Treflan application before transplanting @ 0.6-1.1 kg/ha is helpful in controlling the annual weeds.

Crop rotation

Under rainfed conditions, chilli is successfully rotated with jawar, ragi, cotton, groundnut and castor. As an irrigated crop it is grown in rotation with sugarcane, turmeric, beans, ragi, maize or with vegetables (except brinjal and potato).

Pests and diseases

Pests

Thrips (Scritothrips dorsalis)

The female thrip inserts about 40-80 eggs into the veins of leaves. The nymph and adults infect the tender leaves and feed on the sap, which causes curling of the leaf. Severe infestation leads to heavy curling of the leaves, stunting of the crop and damage to buds and flower. Thrips also causes transmission of the leaf-curl virus. Thrips population can be minimized by spraying Dimethoate at 0.25 or 0.05% Monocrotophos or Achook (1.5%) or neem oil (5%, 3%) at fortnightly intervals, commencing from three weeks after transplanting. This should be followed by Azorin at 0.05% spray on the 60th, 75th and 90th days after transplanting; this controls the thrips effectively. Parastioids like *Orius maxidentex, Chrysoperla cornea, Amblyseius longispinosus, Hauptamannia* Sp. and *Cheiracantium* Sp. are the key enemies of the thrips.

Aphids (Aphis gossypii) and green peach aphid (Myzus persicae)

They suck the cell sap from the leaves and petioles and cause considerable loss to chilli crop. The green peach aphid can be controlled effectively by the foliar application of Ethiopencarb at 0.05% and Methamidophos 60 E.C. at 0.05 or 0.06%. Aphids can be successfully controlled by the application of Demeton methyl (0.05-0.1%), or Fenvalerate (0.02%) spray. The parasite *Aphidius platensis* has been reported to reduce *Myzus persicae* population by 93.4%. *Verticillium lecani,* a fungal pathogen @ 16 and 20 spores/ml, is also found to be effective against *M. persicae*. A syrphid predator and *Araneus* sp have also been recorded on *A. gossypii*.

Mites (*Hemitarsonemus latus*)

The tiny spider-like mites are found with fine webs on the underside of the leaves. The nymphs and adults suck the cell-sap and devitalise the plants. Wettable sulphur (0.3%) or Kelthane (0.1%) may be sprayed at regular intervals for the effective control of the pest.

The muranai mite *Polyphagotarsonemus latus* (Banks) also sucks the sap from the base of the leaf leading to cupping and crinkling. The potential parasitoids like *Chrysopera cornea, Oligota oviformis, Stethorus* sp., *Amblyseius ovalis, A. longispinosus* and *Cheiracantium* sp. and two fungal pathogens, *Aspergillus flavus* and *Cladosporium* sp. could be employed for biological control of this mite.

Diseases

Damping-off (*Pythium aphanidermatum* and *Rhizoctonia solani*)

It is a common disease during the nursery stage. The plants or the seedlings may be killed before their emergence from the soil; young seedlings are also attacked. They show water soaking and shriveling of the stems which fall over and die. The disease is most damaging on moist soils with poor drainage conditions due to dampness.

The fungus is mainly soil-borne, hence it can be checked by steam sterilisation of the nursery soil or by burning trash on the surface of the soil. Sterilisation of soil with Formaldehyde, drenching it 10-16 cm deep, diluted 50 times with water or with some strong copper fungicide is also effective. The seeds should be treated with some fungicide like Cereson or Agrosan. The seedlings may be protected by spraying 0.5 to 1% Bordeaux mixture or any other effective Copper oxychloride like Fytolan. The seed beds should be raised, to avoid water stagnation.

Anthracnose and Ripe fruit-rot (*Collectotrichum capsici*)

This is a serious fungal disease of chilli found in severe form in all the Southern and Northeastern states of India, appearing mostly on the ripened fruit during the rainy season. The disease is also called ripe fruit-rot and is characterised by circular and sunken spots with black margins. The spotted fruits drop off prematurely resulting in a heavy loss of yield. In severe cases, it may also attack the fruit stalk and spread along the stem causing die-back symptoms. The fungus is seed-borne and spreads with wind-blown rains. The fungus *Aureobasidium pullulans* is also reported to cause fruit-rot of chillies.

Seed-borne infections can be controlled by treating the seed with Bavistin and Agrosan GN at 0.1% with an increase in yield by 13.5%. Difolatan 80% W.P. should be sprayed @ 0.2% per month after transplanting against anthracnose and Dithane M-45 to fight ripe fruit-rot. Plant C-1 exhibits some degree of resistance to anthracnose and fruit-rot. Variety K-2 is also tolerant to fruit-rot.

Fusarium wilt (Fusarium annuum)

This disease is characterised by the wilting of the plant with upward and inward rolling of the leaves. The leaves eventually turn yellow and die. The fungus can live in the soil for several years, so the control measures consist of soil treatment or use of wilt-resistant varieties. Drenching with 1% Bordeaux mixture or Fytolan may control the disease. The disease can also be controlled by seed treatment with Bavistin or Vitavax.

Frog eye leaf-spot (Cercospora capsici)

The common symptoms are choloretic lesions, angular to irregular in shape, later turning grayish-brown with sporulation at the centre of the spots on the leaves. Severely infected leaves drop off prematurely resulting in poor yield. Spraying with Bavistin 0.1% 5-6 times at fortnightly intervals effectively controls the leaf-spot disease.

Powdery mildew (Oidiopsis taurica Syn, Leoeillula taurica)

The symptoms of this disease are the development of a powdery mass on the leaves and rapid defoliation of the plants. Spraying with Karathane (0.1%) at 10-day intervals, 3-4 times or Aureofungin 50 ppm effectively controls this disease. Dinocap @ 0.05% or wettable sulphur @ 0.3% sprayed thrice at fortnightly intervals effectively controls the disease. The varieties S-32, Dondaicha and K-red show resistance to *L. taurica*.

Bacterial leaf-spot (Xanthomonas vesicatoria)

It causes serious damage to the chilli leaves and fruit. The symptoms are yellowish-green spots appearing on the younger leaves and dark and water-soaked lesions developing on the older leaves. Small, blister-like spots occur on the fruit, later becoming warty in appearance. The seeds can be treated with corrosive sublimate, as the bacteria is seed borne. Suitable crop rotation is essential as the bacteria is also soil-borne. The seedlings in the nursery and the plants in main fields may be sprayed with a suitable copper fungicide, such as Bordeaux mixture. The varieties like G-4 and X-206 are resistant, while CA-960, CAP-63, CO-1, G-5 and X-197 are highly resistant to bacterial leaf-spot.

Leaf curl of chilli

Leaf curl of chilli is caused by the Tobacco virus 16 or *Nicotiana virus* 10, which also causes leaf curl in tomato, tobacco and papaya. Curling of leaves, shrinking of size, shortened internodes and general dwarfing of the plant, which assumes a witch's broom appearance, are the main symptoms of the disease. The flower buds abscise before attaining their full size and the anthers do not contain pollen grains. The fruits are small and deformed. The spread of the disease can be reduced to some extent by controlling the insect-vector (*Bemesia tabaci*). The soil application of Carbofuran or Dialfolton brings down the incidence of leaf curl to a great extent and an increased yield of 2 to 3 times after control can be obtained. Furadan (Carbofuran) at 0.03% gives the next best results. The varieties Pant C-1, Sel 16-1, IC-18253, IC-18885, ICA-196, Cross 218, EC-124490 and Karanja are tolerant to the leaf curl disease.

Mosaic of chilli

The symptoms of chilli mosaic are the appearance of yellow and dark green areas on the leaf surface and shrunken or raised (puckering) veins. Sometimes the leaves are greatly reduced in size and become filamentous. The diseased plants produce less flowers and fruits. The fruits are usually deformed and rough. Mosaic incidence in chilli can be reduced by spraying 0.05% Dimethoate or Monocrotophos six times at fortnightly intervals commencing from 21 days after transplanting, or by application of Carbofuran or Phorate granules @ 1.5 kg/ha twice, at 15 and 60 days after transplanting. Varieties like puri red, puri orange, G2 and kondiverum are resistant to chilli mosaic virus.

The other minor diseases of chilli are stem-rot caused by *Sclerotium rolsii*, fruit-rot caused by *Phomopsis* sp. and fruit- and stem-rot caused by *Glamerella* sp. Blossom-end rot, sun scald and salt injury are some of the non-parasitic disorders of chilli that cause considerable loss in the yield and the quality of fruits.

Harvesting and yield

The first picking of chillies should be done at the green stage to stimulate a further flush of flowers and the fruit set. Flowering begins 1-2 months after transplanting and it takes another month for the green fruits to develop. The chilli meant for vegetable purposes is generally harvested at the fully-grown green stage, whereas for dried chilli, the fruits are harvested at the red, ripe stage. For pickles, chillies can be picked either at the green or ripe stage. The ripe fruits are ready for picking at 1-2 weeks intervals after the first harvest. The harvesting continues over a period of 3 months, depending on the cultivar, season and cultural practices. Since dried chilli is an export-earning commodity, its timely harvesting, curing and proper grading are very important. The ripe chillies are generally dried under the sun for a period of 8-15 days. Commercially, it is dried at a temperature of about 54.5°C for 2-3 days.

Normally, 2-2.5 tonnes of dry chillies and 7.5-10 tonnes of green chillies are obtained from one hectare. The yield of dry chillies in rainfed areas is lesser (0.5-1.0 tonnes/ha) than the irrigated crop (1.5-2.5 t/ha). About 100 kg of fresh ripe fruits yield 25-40 kg dry chillies, depending upon the cultivars and thickness of the inner walls.

Storage

Dried chillies can be stored for a longer period compared to green chillies. At 0°C and 95-98% relative humidity, green chillies can be stored for about 40 days. Green chillies in 200 gauge polythene covers can be stored for 5-6 days under normal conditions and up to 15 days under refrigerated conditions. The weight loss and damage of fresh red chillies can be minimised by packing it in perforated plastic bags of 1 kg capacity and further storing it in bamboo baskets or cartons. Dried chillies can be stored for months together in dry places, if they are well protected from insect pests.

Processing

The primary processing of chilli essentially consists of drying and de-spiking. The traditional sun-drying approach results in poor quality of the finished product, due to the breakage of the pods and the loss of seeds. Depending on the weather, it takes 3-10 days for sun-drying to lower the moisture content from 80-10%. Better retention of colour and a higher yield of finished products can be obtained by dipping the fresh chillies in Dipsol for 5 minutes and then drying them on racks having multi-tier wire-net trays. Dipsol can be prepared by mixing Potassium carbonate (625 g), purified groundnut (250 g), gum (25 g) and Butylated Hydroxy Anisole. (BHA 0.25 g). About 100 kg of ripe chillies require around 25 litres of Dipsol solvent. Dried chillies generally contain about 6% stalks, 40% pericarp and 54% seeds. In the market, mainly 3 types of dry chillies—with spike, without spike and calyx and without spike with calyx are available.

Grading

Commercially, there are various grades, such as 'First sort', 'Second sort' and 'Mixtures', etc., or 'Special', 'Medium' and 'Fair' grades as prescribed under Agmark.

Seed production

Chilli is a self-pollinated crop but cross-pollination also takes place to some extent. Therefore, to obtain pure seeds, the two cultivars may be kept 250-400 m apart to avoid any cross-pollination. True to type, healthy and well-developed fruit should be selected for producing seeds.

The seeds can be extracted by a simple mechanical seed extractor. A curry powder grinder, after suitably adjusting the grinder plate can be advantageously put to use to break the pod into pieces, while the seeds are completely shed. A winnower or sieve shaker fitted with a BSS 6 wire mesh is used for the separation of the seeds. By this method, seeds are safer and the extraction costs are also less in comparison to the traditional methods.

Chilli oleoresin

The chilli oleoresin offers many added advantages over the bulky, whole or powdered spices, i.e., elimination of microbial contamination, uniformity of colour, flavour, economical transport and storage. Oleoresins offer a whole profile of the colouring pigment and pungency of the spice. Many advanced countries are shifting towards the spice oleoresin rather than using whole or powdered forms of the spice.

Extraction procedure

The laboratory procedure for extraction of chilli oleoresin is as follows: 40 gram of finely-ground chilli powder should be extracted repeatedly for five hours in a Soxhlet extraction apparatus at 50°C, using 200 ml acetone as solvent. After extraction is over, the solvent is evaporated from the flask by slow heating and the weight of the oleoresin should be recorded. The oleoresin percentage can be calculated using this formula:

Oleoresin (%) = (Weight of the oleoresin/Weight of the chilli powder) × 100

The pungent principle capsaicin and the colouring matter of oleorsin can be determined using the methods suggested by Bajaj and Kaur (1979) and Rosebrook *et al.* (1968), respectively.

Cinnamon

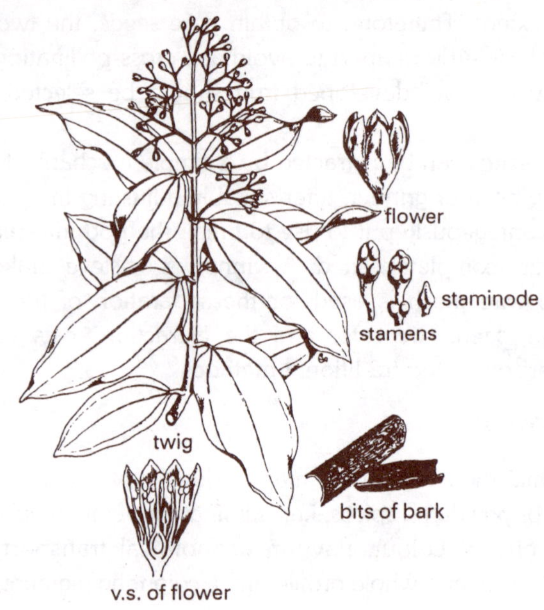

Plant Profile

Family	:	*Lauraceae*
English name	:	Cinnamon, Sweet wood, Ceylon cinnamon, True cinnamon
Indian name	:	*Dalchini*, *Darchini* (Hindi),*Chekke* (Kannada), *Lavangapattai* (Malayalam), *Darushila* (Sanskrit), *Sanna-lavangapattai* (Tamil).
Species	:	*Cinnamomum verum* Prel. Syn. *C. zeylanicum* Blume *C. tamala* *C. camphora* C. cassia C. burmani C. lourneirii C. oliveri *C. malabaricum*
Distribution	:	South Central America, Chile, Mexico, India, Africa, Japan, Turkey, USA.
Uses	:	Spice, preservative, flavouring agent, confectionery culinary purposes.

innamon (*Cinnamomum verum* Presl. Syn *C.zeylanicum* Bheme) or sweet wood or true cinnamon (*Dalchini* or *dorchini*) or Ceylon cinnamon had been prized for many centuries in the Orient and is one of the earliest known tree spices in India. Cinnamon is famous for its bark and leaves which are strongly aromatic. The bark has a sweet, agreeable taste and contains moisture-9.9%, protein-4.6%, fat-2.2%, fibre-20.3%, carbohydrates-59.5%, total ash-2.5%, calcium-1.6%, phosphorus-0.05%, iron-0.004%, sodium-0.01%, potassium-0.4%, vitamins-(mg/100 g), B_1 0.14, B_2 0.21, C 39.8, A 175 IU, niacin-1.9, calorific value-355/100 g. The cinnamon bark contains oil, which is light yellow in colour when freshly distilled. On storage, it becomes reddish coloured. The bark oil contains mainly cinnamic aldehyde (60-75%), eugenol (10%), etc. The leaf oil has a slight camphoraceous odour similar to clove oil due to the presence of 70-95% eugenol. The bark, either as small pieces or powder, is extensively used as spice or condiment. It is aromatic, astringent, stimulant and carminative. It possesses the property of checking nausea and acts as an anti-emetic. Powdered cinnamon is a constituent in chocolate preparation in Spain. It is also used for flavouring cakes and sweets, candy, gum, curry powders, incenses, dentifrices and perfumes. The leaf oil is used in the manufacture of some cheaper types of perfumes, soaps, toothpastes, hair oils and face creams. It is used commercially as an agent for flavouring liquor, and also in the synthesis of vanillin. In the flavouring industry, it is used as a modifer. The cinnamon buds are as good for flavouring and spicing like the bark itself. Cinnamon wood provides a soft timber for use as a low-grade board wood. The root bark also yields 3% oil, which differs from both the stem-bark and leaf oil. It is a colourless liquid with a camphoraceous odour. It contains camphor, pinene, cineole, dipentene, phellandrene, eugenol, safrole, caryophyllene, borneol and cinnamic and benzoic aldehydes.

The seeds contain 3% fixed oil, used for making candles. The seed oil is also called 'cinnamon suet' and is obtained by boiling the crushed ripe fruits, suspended in water. The oleaginous matter rises to the surface and solidifies on cooling.

The bark oil obtained from the matured stems is very little and expensive in comparison to the cost of its extraction. It is used in costly perfumes, for flavouring confectionery, in liquors and in pharmaceuticals, especially to mask the unpleasant taste of certain preparations and in soap manufacture. In addition, the true cinnamon of commerce cassia is obtained from various sources like *C. cassia,* (True cassia or Chinese cassia) *C. burmanni,* (Indonesian cassia) and

C. lourneirii (Saigon cassia) and *C. tamala* (Indian cassia). The other economically important species include *C. camphora, C. oliveri* and *C. malabaricum*.

Origin and distribution

Cinnamon is a native crop of Sri Lanka and the Malabar coast of India. Besides India and Sri Lanka, the other major producer of cinnamon is the Seychelles Island, from where it has spread to Java and other places. It is also cultivated in Burma, Vietnam, China, Malayan Peninsula and Brazil. The cinnamon growing areas in India are the Naga hills in Assam, the coastal districts of South Karnataka and the Western Ghats. It is estimated that the area under cinnamon in India is about 350 ha. The cultivated type *C. zeylanicum* is confined to the lower elevation of the Western Ghats in Cannanore, Calicut and Kottayam districts of Kerala and the lower Nilgiris of Tamil Nadu. The biggest cinnamon estate in the world is the Randathara estate of Auyarakandy near Telicherry in Kerala with an area of 85 hectares and the cinnamon produced here is considered better in quality than that produced in Sri Lanka. The cinnamon produced in India is being exported to about 15 countries, of which Japan, Taiwan and France are the major importers.

Description of the plant

Cinnamon is a bushy, evergreen tree belonging to the family *Lauraceae*. It may reach a height of about 6-15 m, but in cultivation it is generally coppiced or cut back periodically. Its bark and leaves are strongly aromatic. The leaves are stiff, evergreen, with the petiole 1-2 cm long, grooved on the upper surface, the lamina is ovate or elliptic, 5-17 × 3-10 cm in area. It is strongly 3-veined from near the base, the lateral veins are 3/4 or more in length, it is reddish when young, and turns dark green above with paler veins and pale glaucous green beneath. The flowers are axillary borne in terminal panicles at the ends of twigs, the peduncle is creamy, white and 5-7 cm long. The individual flowers are very small, about 3 mm in diameter, with a foetid smell, each subtended by a small ovate hairy bract. It has 6, 9 sepals, companulate, pubescent, the stamens occur in 3 whorls with glands at the base, with hairy filaments and 4-celled anthers. The fruit is a fleshy berry, dark purple, 1-seeded, ovoid, 1.5-2 cm long with an enlarged calyx at the base. In Karnataka, the tree flowers in January and the fruits ripen 6 months later. The somatic chromosome number of *C. verum* is 2n=24.

Soil and climate

Wild cinnamon trees are confined to tropical evergreen rain forests upto 1800 m MSL. Cinnamon is a hardy plant, tolerant to a wide range of soils and climate conditions. In the West coast of India, the crop is confined to the laterite and sandy patches with poor nutrient status. Sandy loam soils with a mixture of humus or vegetative mould is the best for the sweet and fragrant bark. Cinnamon grows well in places from sea level upto an elevation of about 1000 m above MSL. Since it is mostly raised as a rain-fed crop, an annual rainfall of 150-250 cm is considered ideal. A hot and moist climate with an average temperature of 27°C is suited for the cultivation of cinnamon. Prolonged spells of dry weather are not conducive for its successful growth. However, the best cultivated cinnamon is grown at low altitudes in Sri Lanka, with an average temperature of 30°C and 200-250 cm rainfall per annum. Proximity to the sea, humid conditions and brackish water are good for the crop.

Varieties

The National Conservatory of Cassia Germplasm consists of 262 accessions. In India there are some varieties like *pat* or *mat-pat* and *kurunchi* which are considered the best for cultivation.

In Sri Lanka there are two popular varieties locally known as 'Sweet Cinnamon' and 'Honey Cinnamon'. Camphor cinnamon, Snake cinnamon, Astringent cinnamon, etc., are some of the others. Konkan Tej, and Yercaud-1 (YCD-1) have been released for commercial cultivation in the Konkan region and Karnataka and Tamil Nadu, respectively. Among the other improved varieties 'Navashree' and 'Nityashree' have also been released by the Indian Institute of Spice Research, Calicut. RRL (B) C-6 is another improved variety.

Navashree

Navashree is a superior selection. It has a high and stable regeneration capacity (6-7 shoots/ year), high yield (average yield 56 kg/ha in the first 4 years), high bark recovery (40.6%), and excellent bark characteristics (bark oil 2.7% with a very good cinnamaldehyde content of 37%, bark oleoresin 8%, leaf oil 2.8%). Its very young flushes are purple, which turn green in 7-10 days. It is recommended for cultivation both in the plains and high altitudes.

Nityasree

Nityashree is also a good stable selection, which produces 4-5 shoots per year and the average yield is 54 kg/ha in the first four harvests.

The bark contains 2.7% oil, 10% oleoresin and 3% leaf oil with a very good eugenol content of 78%. The young flushes are purple in colour and turn green in just 2 days. This variety is also recommended for both the hills and plains.

YCD-1

This is a selection made at AICRPS, Yercaud (TNAU). The variety is ready for harvest after 3 years of planting and its economic life-span is 20 years. The bark yield is high (359.8 kg/ha; 3800 kg leaf yield/ ha) with a recovery of 35.3%. The quills and the leaves contain 2.8 and 3.0% essential oil, respectively. The plants have a good regeneration capacity. The quills are sweetish-bitter in taste. This variety is recommended for places that receive 100-200 cm of rainfall and have about 500-1000 m elevation.

Cultivation

Propagation

The most common method of propagation of cinnamon is through the seeds. Cinnamon flowers in December-January and the seed ripens in May-June. The fruits are picked when the outer pulp begins to turn black and the seeds are gathered, dried and heaped up in a shady place till the pulp rots and turns black. To free the pulp, the whole mass of seeds is trampled and then washed. After separation of the pulpy mass, the seeds are separated. They are dried in the air taking care not to expose them to the sun. If they are exposed, the seed coat cracks and causes damage to the seeds. The seeds soon lose their viability and should be sown immediately after the removal of the pulp, in raised nursery beds or pots or polyethylene bags (30 × 15 cm) filled with a mixture of sand, cattle manure and soil (3:3:1) during July-August. Frequent irrigation is required for maintaining adequate humidity. The seeds germinate within 15-20 days of sowing. The seedlings raised in nursery beds are transplanted into polybags or baskets 4 months after germination of the seed. Artificial shade is required for the seedlings till they are about 6 months old. According to local conditions and the growth rate of the seedlings, 10-12 month old seedlings are ready for planting into the main field.

The plants can also be raised vegetatively by cuttings off young, three-leaved shoots and by layering the shoots and division of old rootstocks. It can also be propagated by air-layering and ground-layering and tongue and ring methods, of which the ring method of layering is more successful. The growth regulators, Indole acetic acid and Indole butyric acid (50 ppm) have been reported to promote rooting in air-layers. Micro-propagation protocols for the rapid multiplication of cinnamon, synseed technology and the somatic embryogenesis and regeneration procedures have been standardised.

Preparation of land and planting

The area selected for planting cinnamon is cleared and 50 cm^3 pits are dug at 2 m x 2 m or 3 m × 3 m spacing. They are then filled with compost and top soil before planting. As in the case of other tree spices, cinnamon is also planted during June-July, to take advantage of the monsoon for the establishment of the seedlings. In each pit, 5 seedlings can be planted. In some cases, the seeds are directly dibbled in the pits that are filled with compost and soil. Partial shade in the initial years is advantageous for rapid growth and development of healthy cinnamon plants. Care should be taken to provide a drainage channel from the pit along the slope, since water stagnation is detrimental to the young seedlings.

Manures and fertilisers

Cinnamon responds well to fertilisers. Apart from 20 kg of FYM, a dose of 20 g N, 20 g P_2O_5 and 25 g K_2O per seedling is recommended in the first year. The dosage is gradually increased to 50 kg FYM/ compost, 200 g N, 180 g P_2O_5 and 200 g K_2O, for grown up plants of 10 years and above. The fertilisers should be applied in two split doses, in May-June and September-November.

Irrigation

Cinnamon is raised mostly as an unirrigated crop. However, in the initial 2-3 years, watering is done twice a week during summer months. The quantity of water depends on the soil moisture level and the growth of the plants.

Inter-culture

Inter-cultivation is confined to 2 weedings during June-July and October-November in a year. This is done by digging the soil around the bushes, once, during August-September. Mulching the soil around the plant with the weed and grasses or with spent cinnamon leaves after the distillation for oil, and then covering them with the soil is very effective in conserving soil moisture in the root zone.

Pruning/Cropping

In about 2 years, the seedlings grow upto a height of 2 m and form healthy bushes. The young trees are cut close to the ground to a height of 15 cm from the stump to encourage side shoots. This process is called coppicing. This is done until the whole tree assumes the shape of a low bush with the side shoots springing forth in profusion.

Pests and diseases

Pests

Cinnamon Butterfly (Chilasa clytic)

The caterpillars of the pest feed voraciously on the young leaves and defoliate the plants. They are pale yellow with dark stripes on the sides of the body. The pest can be controlled by spraying 0.05% Quinalphos. Infestation by the leaf miner (*Melanagromyza* sp.), which is sometimes serious on tender leaves of seedlings in the nursery, can be controlled by spraying 0.05% Monocrotophos. Shoot and leaf webber, leaf miner (*Conopomorpha civica*), tussock caterpillar and leaf psyllid are a few other pests attacking cinnamon. They can be controlled by spraying Dimethoate (0.03%) or Demacron (0.4%). Pink shoot borer, red ants, termites, loopers (*Fascellina castanea* and *Hyposidra infixaria*), *Cleora alienaria* sub sp. *rasanaria, Lopharcha* sp, beetles (*Dercitina* sp) and *Coptosoma variegata* are also reported to cause damage to cinnamon plants.

Diseases

Pink disease

This is caused by the fungus *Corticum javanicum*, during the rainy season. It develops a pale pinkish-white crust on the stems of the plants, destroying the corky layer and causing the death of the shoot or twig. The disease can be controlled by spraying 1% Bordeaux mixture.

Seedling Blight

In nurseries, blighting of the stem with light brown patches is noticed. This results often in the death of seedlings. *Diploida* sp. has been isolated from such seedlings. Spraying the plant with 1% Bordeaux mixture and 0.3% Copper oxychloride controls the disease.

Nursery leaf-spot and die-back disease (Colletotrichum gloeosporoides)

On young nursery seedlings, small brown specks appear which gradually enlarge resulting in the drying of the leaf. From the leaves, the infestation spreads to the stem, resulting in necrosis from the apex downwards. On old seedlings and mature trees, light and dark brown concentric zonations occur on the stem. The central portion of the necrotic tissue causing shot-hole symptoms also appear in certain cases in seedlings. Spraying 1% Bordeaux mixture during the rainy season controls the disease.

Leaf-spot

Dark brown leaf-spots, which appear brittle and sunken, are noticed in some trees. *Gleosporium* sp. is usually associated with these diseases. Algal leaf-spot, caused by *Cephaleuros* sp., is also quite common. The other diseases of cinnamon include grey blight caused by *Pestalotiopsis palmarum* and sooty mould (*Phargmocapnius betle*). Foliar spray with 1% Bordeaux or 0.3% Copper oxychloride controls the diseases.

Harvesting and processing of cinnamon

For the preparation of the quills, the plants are harvested 3 years after planting. The crop is harvested in the months of May and November, by cutting the shoots for the extraction of the bark. As soon as the rain stops, the cutting of shoots for peeling of the bark is commenced. At this time the new growth of bushes is stopped and the trees have mature leaves. This stage is also indicative of the free flow of sap between the bark and the wood. This is the appropriate stage to obtain the best quality bark. After cutting, young shoots spring up from the stump which will be ready for removal in the subsequent seasons within 18 months. The bulk of the bark is obtained from the shoots that are 1 to 2 years old. The shoots selected for cutting are usually 1.5-2 m long and 2 to 2.5 cm thick. To judge the suitability for peeling, the peeler makes an oblique cut and lifts the bark to see whether it separates easily with a free flow of sap that facilitates easy peeling. If there is any difficulty in peeling, the peeler rejects the shoots. The shoots that are ready for peeling are removed from the stumps from the terminal ends of shoots.

Peeling

Peeling is done with a small knife with a round edge at the end. The cut stems or sticks are given longitudinal slits from one end to the other. By working the knife both ways, the two halves of the entire bark are removed. If the bark does not peel easily, the sticks are rubbed in between hard pieces of wood which enables the easy detachment of the bark.

Rolling

The barks are packed together and placed one above the other and pressed well. The length of the bark slip is reduced to 20 cm, which are then piled up in small enclosures made by sticks. They are then covered with dry leaves or mats to preserve the moisture for the next day's operation and also to aid in the slight fermentation. Preservation of moisture is very important for the subsequent operation, known as piping.

Piping

After peeling and rolling, the slips (quills) are bundled and taken to the piping yard for the piping operation. Three sticks are driven into the ground in such a way that they cross each other at a height of 30 cm from the ground level. This serves as a support to keep the fourth stick resting on the knot. The operator sits down and places the slips one by one on the fourth stick to scrape off the outer skin with a small curved knife. The scraped slips are sorted into different grades according to their thickness.

The graded slips are rolled to form pipes by fitting them over the outer edges of the slips. Soon after piping, they are allowed to dry. The bark-free ones of finest, smoothest quality are graded as "00000", the coarsest being grade "O" and the remaining ones are graded as 'chips', 'pieces', 'quillings' (broken pieces) and 'graded featherings'. The outer bark possesses a slightly acidic flavour and its removal enhances the delicate aroma.

Good quality cinnamon should not be thicker than a thick paper. It should be light brown with wavy lines and produce a fractured sound when broken. When chewed, it should become soft, melt in the mouth and sweeten the breath. The bark of a large shoot is coarse, whereas the tender ones are very thin and straw coloured. The shoots which are exposed to the sun are said to be more spicy than those grown under the shade. The best quality cinnamon is always obtained from the thin bark from the shoots in the centre of the bush and from the middle portion of the shoot.

Commercial bark should not be more than 0.5 cm thick and the thinner the bark the better the grade. Broken quills are exported as quillings and the inner bark of twigs and twisted shoots as feathering. They are used mainly for grinding or for the distillation of cinnamon oil.

Yield

Under favourable conditions, when the plants reach a height of 1-2 m. after two or three years of transplanting, the first cutting is made. About 65-125 kg of quills can be obtained per hectare from the first crop. When the trees are 10 or 11 years old, 225-300 kg of quills per hectare can be obtained normally. Barks which cannot be taken out like tubes are called quillings and the scraped pieces are called 'feathering'. In addition, about 70 kg of quillings and featherings are obtained from one hectare of cinnamon.

Leaf and bark oil from cinnamon

The leaf and bark oil of cinnamon can be obtained by distilling the dried cinnamon leaves and bark, respectively. The dried cinnamon leaves are steam distilled in special distilleries. About 4 kg of bark oil can be obtained from a hectare of cinnamon plantation.

Cinnamon bark oil contains cinnamic aldehyde which is the principal aromatic substance and eugenol. The clips yield 0.5-1.0% oil. The higher the cinnamic aldehyde content, the higher is the price. The leaves yield 0.5-0.7% oil, and the oil contains eugenol and cinnamic aldehyde.

Value-added products

Whole and ground cinnamon, the essential oils of the leaf and bark, seed oil, root-bark oil and oleoresin are the value-added products of cinnamon.

Clove

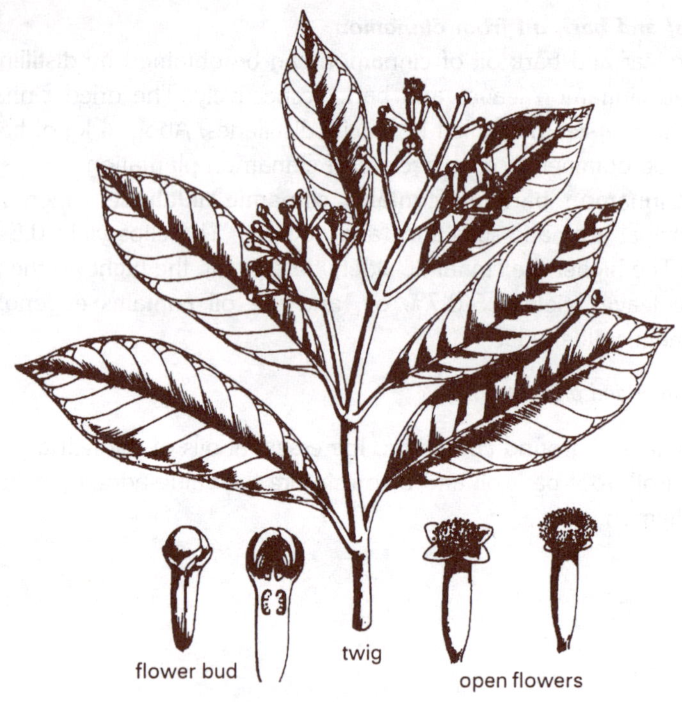

Plant Profile

Family	:	*Myrtaceae*
English name	:	Clove
Indian name	:	*Lung, Lavang* (Hindi), *Lavanga* (Kannada), *Grambur, Grambu* (Malayalam), *Lavanga* (Sanskrit), *Kirambu, Lavangam* (Tamil), *Lavangalu* (Telugu).
Species	:	*Syzygium aromaticum* (Merrill *et* Perry) Syn. *Eugenia caryophyllus* (Spregal) Bullock *et* Harrison
Distribution	:	Indonesia, Zanzibar, Mauritius, Pemba, Madagascar, Malaysia, Sri Lanka, Haiti and India.
Uses	:	Flavouring curries, domestic culinary spice, baked foods, medicines.

ommercial clove is the fully grown but unopened, aromatic, dried flower bud of an evergreen tree, *Syzygium aromaticum* Merrill and Perry. [Syn. *Eugenia carryophyllus* (Spregal) Bullock *et* Harrison], belonging to the family *Myrtaceae*. The term 'clove' is derived from the French word 'clov' and the English word 'clout', both meaning 'nail'. The word may have been based on the likeness of the flower bud of the clove tree to a broad-headed nail. Clove is one of the most ancient and valuable spices of the Orient, known as far back as the 1st century BC. Clove is valued as a spice and for its essential oil. The composition of clove varies with the agro-climatic conditions and post-harvest processes. The dried clove bud has the following composition; moisture-5.4%, protein-6.3%, volatile oil-13.2%, non-volatile ether extract (fat)-15.5%, crude fibre-11.1%, carbohydrates-57.7%, mineral matter-5.0%, ash insoluble in HCl-0.24%, calcium-0.7%, phosphorus-0.11%, iron-0.01%, sodium-0.25%, potassium-1.2%, vitamins-(mg/100 g), B_1-0.11, B_2-0.04, C-80.9, A-175 I.U, niacin-1.5, calorific value (food energy)-430 (calories/100 g). The volatile oil obtained from the clove bud contains mainly eugenol-(80–90%) and caryophyllene (4-8%). The major use of cloves, is for domestic culinary purposes, as a spice to flavour both sweet and savoury dishes and in the preparation of pickles and sauces. Cloves, both whole and ground, are used in baked goods, cakes, confectionery, chocolates, puddings, desserts, sweets, syrups, preserves, etc. Clove is used for flavouring curries, gravies, ketchup and spice mixtures. The inferior culls are used for the production of clove oil, which is used as a flavourant for all kinds of food products, in soap preparation, perfumery and the synthetic preparation of vanillin. Owing to the antiseptic property of eugenol, clove oil is invariably an ingredient in chewing gums, toothpastes and mouthwashes. In dentistry, eugenol is used in combination with zinc oxide for the temporary filling of cavities. It is also used for flavouring of 'Kretek' cigarettes in Indonesia. It is reported to aid in digestion and is also used as an antispasmodic and counter irritant.

Cloves are acrid, bitter, aromatic, refrigerant, ophthalmic, a digestive, carminative, stomachic, stimulant, antispasmodic, antibacterial, rubefacient, aphrodisiac, appetising, expectorant, emolient, anthelmintic, sialogogue, rejuvenating, galacto-purifier, diuretic, febrifuge and tonic. They are useful in halitosis, odontalgia, ophthalmology, flatulence, colic, gastropathy, anorexia, cough, asthma, vitiated conditions of *kapha* and *pitta*, burning sensation, skin diseases, helminthiasis, agalactia, impurity of breast milk, strangury, fever, cephalagia, neuralgia, lumbago, dental caries, hyperacidity, vomiting, dyspepsia, hepatopathy, general debility and tuberculosis. The oil is also useful in catarrh, cough, bronchitis, vitiated conditions of *vata*, gastrohelcosis, odontalgia and ceplialalgia. Externally, the oil is used as a rubefacient and counter-irritant.

Origin and distribution

Clove is indigenous to the Molluccas Island of Eastern Indonesia. Later, it was introduced to Mauritius and later on established in the islands of Zanzibar and Pemba. The important clove-producing countries in the world are Tanzania (Zanzibar), Pemba, Madagascar and Indonesia. Clove is also grown in Malaysia, Sri Lanka, Haiti and India. France, USA, India and West Germany are the major importers of clove. Tanzania accounts for roughly 75% of the world output. The world's annual production of clove is 40-50,000 tonnes, out of which nearly 15,000 tonnes are used for the production of 'Kretek' cigarettes.

In India, clove was introduced in 1800 AD by the East India company and is now cultivated in Tamil Nadu, Kerala and Karnataka. The production of clove in India has not kept pace with its increasing demand and most of the clove requirement in India is met through import from Tanzania, Singapore, Malagasy and Indonesia. However, there is scope for extending the cultivation of this spice along the Western Ghats.

Description of the plant

The clove is an evergreen tree growing to a height of 7-15 m. The trunk is conical when young, later becoming roughly cylindrical. It begins to fork near the base, into two or three main erect branches. The smaller branches are semi-erect, the twigs are brittle, smooth and greyish-white in colour, and the whole head is bushy and dense. The leaves are simple, opposite, exstipulate, glabrous and aromatic owing to plenty of oil glands on the lower surface. The lamina is lanceolate or narrowly elliptic, bluntly acuminate at the apex, cuneate and bare. The new leaves appear in flushes and are bright pink. Later, the upper surface becomes glossy and dark green and wavy with recurved leaf margins. The hermaphrodite flowers are borne on a terminal, corymbose, trichotomous panicle. The inflorescence is shortly pedunculate, branched from the base and shorter than the leaves. The bracts and bracteoles are narrow, acute and fall quickly. The number of flowers varies from 3 to 50; a fleshy hypanthium is present, surrounded by the sepals. The hypanthium is green in the young bud, flushed pink at anthesis and turns deep red after the stamens fall.

Above the sepals, there are four dome-shaped whitish petals. After fertilisation, the stamens and styles invariably fall. The lower part of the flower along with the calyx develops into a fleshy, dark, one-seeded drupe. The sepals are reduced to triangular projections and this is popularly known as the 'mother of clove'. The stamens are numerous, the anthers are pale yellow, with a small, pale brown, inconspicuous connective gland. The style is very stout, swollen at the base, pale green and dotted with glands. The two-celled, multi-ovate, inferior ovary is embedded at the top of the hypanthium. The fruit (mother-of-clove) is usually a single-seeded drupe, but occasionally contains two seeds also.

Climate

Clove grows best in the insular, maritime climates of the tropics, with an annual rainfall ranging from 150-300 cm and a relative humidity of 70% and above. It grows well from sea level to an altitude of 700-900 m and a mean temperature range of 20-32°C. Alternating periods of dry and wet weather are suitable as against a very moist climate, which affects flowering. The clove cannot withstand high winds, therefore, windy locations should be avoided. Providing shade during the early period of growth is necessary for its good growth.

Soil

Deep red or acidic loam and laterite soils rich in humus, having a pH between 4.0 and 5.6, are found best for clove cultivation. Sandy soils and water-logged conditions are unsuitable for clove cultivation. It can be grown well in both sloping and level ground.

Varieties

Clove plantations in India are reported to have originated from a few seedlings obtained originally from Mauritius. The germplasm collections (215) made within the country and from abroad have not yielded any appreciable variability, mainly due to the self-fertilising nature of clove. Although there are no named cultivars as such, in trade the differences of the spice are recognised on the basis of the places of their cultivation. The large, plump and bright reddish 'Penang Cloves' are considered the best in appearance, followed by the Zanzibar and Madagascar types. From India, Burliar-1 is one of the two high yielding progenies selected in Tamil Nadu, the other is from Odetham estate. One of the cultivars is also recognised by the name 'Amboyan clove'.

Cultivation

Propagation

Vegetative propagation is difficult in clove. Several attempts have been made in different clove-growing areas, using terminal leafy softwood cuttings and by budding, but have been met only with occasional success. However, the approach method of grafting using its own rootstock has been successful in clove. Generally, clove is only propagated through seeds.

Seed propagation

To raise the seedlings, the seeds are extracted from the ripe fruits (mother-of-clove) obtained from regular bearing trees. The fruits for seed collection are allowed to ripen on the tree itself and drop down naturally. The seeds are then soaked in water overnight prior to sowing, in order to dehusk them. Afterwards, only fully-developed, uniform-sized seeds, which show signs of germination by the presence of pink pedicle, are used for sowing and the remaining are discarded. Since the viability of the seeds is very short, they should be sown as soon as possible in raised nursery beds of 90-100 cm width, 15 cm height and of convenient length, at 2 cm depth at a spacing of 12-20 cm both ways. The seeds sprout in about two weeks depending on the individual vigour of the seeds and germination is completed in 40 days. If fresh seeds are sown, seed germination to the extent of over 90% may be obtained. The seedlings are very slow in their initial growth and when they are 50-60 cm tall they start branching and are transferred to polythene bags (30 cm × 15 cm) containing a mixture of good soil, well decomposed cowdung and sand (in the ratio of about 3:3:1). After they are one to two years old, they are ready for transplanting into the main field. While the seedlings are in the nursery, sufficient shade and irrigation should be provided. To avoid damage by crickets, Carbaryl dust may be applied in the nursery.

Planting

The area selected for raising the clove plantation should be cleared of wild growth before monsoons. For planting, pits of 60-75 cm^3 are dug 6-7 m apart in rows spaced 6 m apart, about a month or two prior to planting. If planted as an intercrop, the spacing should be adjusted based on the spacings of the major crop. The pits are filled with a mixture of top soil, burnt earth and FYM or compost. Transplanting should preferably be done during June-August, and in low-lying areas towards the end of the monsoon in September-October. Clove prefers partial shade. Under Indian conditions, it is best suited for mixed-cropping in older coconut or arecanut plantations or in coffee estates.

In order to provide a cool, humid micro-climate, intercropping with banana is found to be very good. In the vast majority of cases, clove is planted in garden lands together with various other crop plants. Under conditions in Kerala, such gardens will contain coconut, banana, jackfruit, mango and miscellaneous crop plants. Cloves can be fitted into this system without much of a problem. The sites should be selected based on the availability of light.

Manures and fertilisers

The organic manure along with a half dose of fertilisers may be applied during May-June and the remaining quantity of fertilisers may be given during September-October, as a top dressing. The plants may be applied with manures and fertilisers as given below.

Recommended doses	Per plant	Per hectare
I. Organic Manure		
FYM *or* compost		
Before planting	15 kg	3.5 tonnes
After planting	15 kg	3.5 tonnes
II. Fertilisers (NPK)		
First year	20:18:50 g	4.64:4.17:11.6 kg
After 2nd year	40:36:100 g	9.28:8.34:23.2 kg
After 5 years	100:90:250 g	28.20:20.85:58.0 kg
After 10 years	200:180:150 g	46.40:41.70:116.0 kg
After 15 years	300:250:750 g	69.60:58.12:174.0 kg

The application of coconut meal, bonemeal or fish meal at 2-5 kg/ plant is beneficial. The manures can be applied in shallow trenches dug around the plant, normally about 1-1½ m away from the base.

Irrigation

In the first 3-4 years, extreme care should be taken especially during the summer months. Plant-based water application has to be very scrupulously followed.

Interculture

Immediately after transplanting, shade is provided by planting banana, Cassava, *Glyricidia* sp., *Peltophorum* sp, *Acacia* sp., *Albizzia* sp., *Erythrina* sp., Subabool, etc., in between the rows and the plants are irrigated till their establishment. Besides, protection from heavy rains and winds are required during the early stages of growth. Clove is generally grown along with orange and mangosteen.

The plots should be kept weed free by regular weeding. Usually, the young plants are ring weeded. The pruning of diseased branches should be done regularly. It also helps in preventing overcrowding of the branches. The basins of the trees are mulched with dry leaves to conserve the soil moisture.

Inter-cropping

Clove can be inter-cropped in coconut, arecanut, nutmeg and banana plantations of the midlands. In higher elevations, it can be mixed cropped with pepper or coffee.

Pests and diseases

Pests

Shoot borer (Sahyadrassus malabaricus)

The adult female lays eggs on the branches, sometimes near the growing tips. The larvae, on hatching, bore into the branches making long galleries, and may reach the main trunk. In severe cases, it leads to die-back. To control the pest, the plants should be sprayed with 0.05% dimethoate or Methyl dimethon or Quinalphos (0.1%) should be injected into the bore hole. Termites, leaf eating caterpillar, black scale (*Saissetia eugeniae*), masked scale (*Mycetaspis personata*), wax scale (*Ceroplastes floridensis*), *Kilifia accuminata*, *Attacus atlas*, crickets, mealy bugs, grasshoppers and cockchafer grubs are the other common pests of clove. Termites can be controlled by drenching the soil with 0.05% Chlorphyriphos EC. Spraying of 0.05% Monocrotophos controls the leaf eating caterpillar and scale insects. *Hindola striata* and *H.fulva* are reported to be the vectors of the Sumatra disease.

Diseases

Sudden death of cloves (Valsa eugeniae)

This disease is characterised by heavy leaf-fall, coupled with wilting symptoms, usually during prolonged drought periods. This wilting can be arrested by providing copious irrigation to the trees.

Die-back (Cryptosporella eugeniae)

This disease is commonly observed in neglected clove plantations and is associated with low soil fertility, weed competition and algal infection. The symptoms include dark red spots on both sides of the leaves, which gradually spread and increase in size. The leaves turn chlorotic and ultimately fall off. The young leaves are more prone to this disease.

These diseases can be controlled by adopting proper management practices along with spraying the trees with 1% Bordeaux mixture.

Seedling wilt (*Cylindrocladium* Sp, *Fusarium* sp, *Colletotrichum* sp, *Rhizoctonia* sp., *Phythophora* sp.), leaf-spot, leaf rot (*Cylindrocladium quinqueseptatum*), twig blight [*Colletotrichum gloeosporiodes* (*Glomeralla cingulata*)], grey leaf blight (*Pestalotiopsis versicolor*), Sumatra disease (*Pseudomonas syzygii*) and flower bud shedding are some of the other fungal diseases which are found to attack clove trees. To control these diseases, the plants could be sprayed with Bavistin (0.2%) or 1% Bordeaux or 0.30% Copper oxychloride. The infected plants promote further spread of the seedling wilt disease, hence they should be removed and the remaining seedlings should be drenched with any of the copper fungicides.

Problems

Apart from the diseases, some other problems connected with clove cultivation are: very low percentage and slow germination of seeds and slow seedling growth, difficulty in harvesting due to the inaccessible heights of the plant which results in nearly 30 to 35% loss of the crop and which deserves attention. There is also the need to develop short-statured plants with early yielding habits.

Harvesting and yield

Though clove trees flower from 4-6 years of their planting, the trees generally start bearing an economic yield 18-20 years from the time of planting and the production continues for 80 years or more. The bearing between the years varies quite a lot and a bumper crop can only be expected about once every 4 years, being influenced by the weather and the previous crop load. The flowering season varies from September-October in the plains to December-January at high altitudes.

Cloves, which are the unopened flower buds, are produced on the terminal shoots of the twigs. The buds are collected when they are dull red or pink in colour and less than 2 cm long. The inflorescence is harvested without damaging the branches when the buds have reached their full size, but before they open, so that the petals together with the stamens inside form the head of the dried clove. Delayed picking, i.e., after the opening of the buds, will devalue the spice.

The yield of cloves is found to vary from year to year. The average yield at Burliar is 2 kg per tree (500 kg/ha) per year. But, there are individual trees which are reported to give 8-10 kg in some years. In comparison, the average yield in Zanzibar from a well-grown bearing tree is reported to be very high (40 kg/year). Yields upto 80 kg/tree/year have also been recorded. About 11,000-15,000 dried cloves weigh one kilogram.

Cloves are normally packed in double jet sacks of 50-60 kg capacity each.

Processing

The appearance, size, content and aromatic characteristics of its volatile oil are the factors which decide the quality of the dried spice. Also, they should be free of mustiness and mould. The best prices are obtained for whole dried cloves of a good bold size with a bright, uniform, reddish-brown colour. The features mentioned above are, in turn, influenced to a great extent by the care taken in the harvesting, drying, cleaning and sorting operations, and the storage conditions.

Prior to drying, the buds are removed from the stems and then piled in separate heaps for later individual drying, during which, over-ripe cloves and gleanings of fallen flowers are sorted out. Drying is undertaken as soon as possible, after the buds have separated from the clusters. If the buds are left too long in heaps they will ferment and the dried spice will have a whitish, shrivelled appearance (*Khoker* cloves). In sunny weather, drying may take four to five days to produce a brightly-coloured dried spice of attractive appearance. The correct stage of drying is reached when the base of the bud is dark brown, and the rest of the bud lighter brown in colour. On drying, the cloves retain about two-thirds of their original fresh green weight. Then another sorting is done to separate 'mother-of-cloves' and '*khoker* cloves'. A final thorough cleaning, sorting and grading is carried out by the exporting firms prior to packing.

The stems remaining after the separation of the buds from the freshly-harvested clusters are dried similarly and are used to distil clove oil by the steam distillation method.

The duration of distillation ranges from 8-25 hours depending upon the size of the still, the nature and volume of steam and the condition of the cloves. The leaves and small twigs yield clove-leaf oil. Clove-stem oil is obtained from the stems attached to the buds and flowers, and bud oil from the buds. The essential oil yield is 17-19% from clove buds, 6% from the clove stems and 2-3% from the leaves.

Grading

Whole cloves are graded as special (Hand-picked), Grade-2, Grade-3, Ground (powdered) cloves, while the defective cloves are named as *Khoker cloves,* Headless cloves, Mother cloves, Extraneous matter, etc.

Value-added products

Clove oil, ground clove, oleoresins, clove-stem oil, clove-leaf oil, oil of mother of cloves and clove-root oil are some of the value-added products of clove.

Adulterants

Cloves are sometimes adulterated with headless cloves and clove stems. They may also be adulterated with *Khoker* cloves or mother cloves and other extraneous matter like dust, dirt, stones, clay particles and pieces of wood.

The adulterants of clove-bud oil are stem oil, and clove-leaf oil. Another form of adulteration is the addition of clove terpenes, synthetic terpineol, dibenzyl or dibenzylether and acetins in clove-bud oil.

Coriander

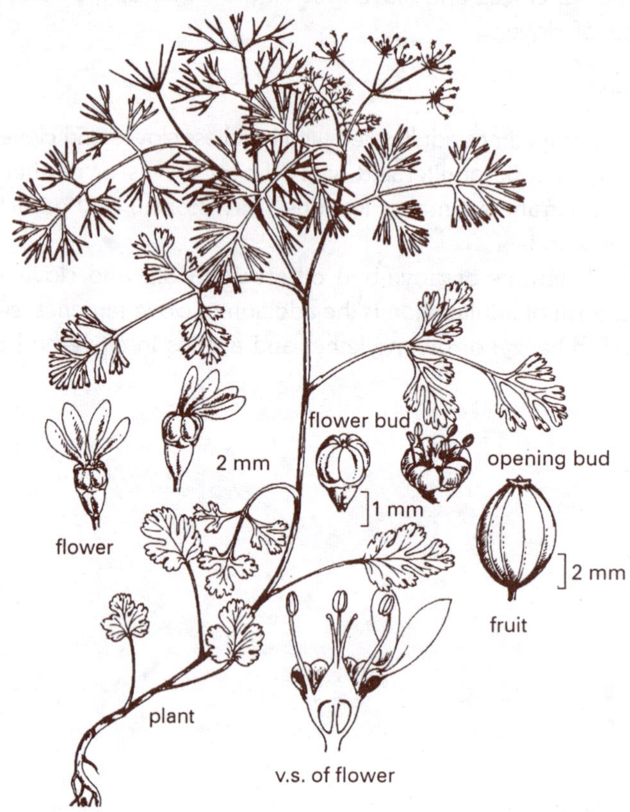

Plant Profile

Family	:	*Umbelliferae / Apiaceae*
English name	:	Coriander
Indian name	:	*Dhania, Dhanya* (Hindi), *Kothambri, Kothamiri bija* (Kannada), *Kothumpalari bija* (Malayalam), *Dhanyaka* (Sanskrit), *Kothamalli* (Tamil), *Dhaniyalu* (Telugu).
Species	:	*Coriandrum sativum* L.
Distribution	:	Southern Europe, India, USA, Pakistan, England, Turkey.
Uses	:	Spice and flavouring agent, culinary purposes, medicine.

Coriander is the dried fruit of *Coriandrum sativum* L., an aromatic spice crop of the family *Umbelliferae* or *Apiaceae*. It is a very old flavouring substance and its usage both for its leaves, stems as well as fruits has been mentioned in Egyptian, Herbrew and Roman literature as early as 5000 BC. The composition of coriander seeds is moisture-6.3%, protein-1.3%, carbohydrate-24.0%, volatile oil-0.3%, phosphorus-0.44%, potassium-1.2%, calcium-0.8%, sodium-0.02%, iron-0.006%, vitamins - A-175 IU (mg/100 g), B_1-0.26, B_2-0.23, niacin-3.2, vitamin C-12.0, while the green leaves contain moisture-87.9%, protein-3.3%, fat-0.6%, carbohydrate-6.5%, calcium-0.14%, phosphorus-0.06%, iron-0.01%, etc. Coriander seed oil contains mainly α-pinene-0.96-7.97%, p-cymene and phellandrene-3.69-9.91%, linalool-59.55-72.61%, borneol and decyl aldehyde-5.31-7.43%, myrcene-5.44%; terpinene-5.44%, camphor-2.46%, etc. The essential oil content of the seeds is about 0.1-0.7%. The seeds also contain 19-21% fatty oil, which solidifies upon keeping and is used in the manufacture of sodium soaps and has a pleasant odour. The residue left after extraction of volatile oil is used in cattle feed. All parts of the plant are indispensable food adjuncts in Indian cookery. The young plants are used as a spice in the preparation of chutneys, sausages, curries and soups for flavouring. The dried fruits are an important ingredient of curry powder, sausages and pickling spices. In the USA and Europe it is used for flavouring spirits and gin. In food industries, coriander is one of the important ingredients in the manufacture of bakery products, imitation flavours, pork, meat, fish and salads, soda and syrup, gelatin, dessert, candy, preserves, chocolates and liquors. The essential oil of coriander seeds is a valuable ingredient in perfumery industries.

Coriander is also used as a carminative and flavouring agent and to correct the griping qualities of other medicines. A herbal tea of coriander with milk cures bleeding of piles. The fruit is antispasmodic, diuretic, aphrodisiac, refrigerant and stimulant. A decoction of the dried fruit is given for treating flatulence and colic. It is also reported to lessen the intoxicating effects of spirituous liquors.

Origin and distribution

The genus *Coriandrum* consists of about six species of annual aromatic herbs of which *Coriandrum sativum* is widely cultivated in India for its spicy fruits. The plant is considered to be a native of the Eastern Mediterranean region or Southern Europe. Precisely, Italy is presumed to be the native place of coriander. It is extensively cultivated in Morocco, Romania, India, France, Spain, Italy, the former USSR,

Holland, Yugoslavia, Pakistan, Turkey, Mexico, Hungary, Poland, Argentina, Guatemala and to some extent in England and the USA.

Area and production

In India, it is cultivated practically in all the states with a total production of 2,90,000 tonnes of seeds over an area of 5,46,500 hectares. Rajasthan stands first both in area (1,94,700 ha) as well as in production (12,03,000 quintals), followed by Andhra Pradesh with a production of 2,08,000 quintals (78,800 ha), Tamil Nadu with 1,98,000 quintals (49,000 ha) and Madhya Pradesh with 1,61,000 quintals (48,700 ha). Karnataka stands fifth with a production of 15,000 quintals (13,800 ha). It is also cultivated in Gujarat, Maharashtra, West Bengal and Bihar.

Description of the plant

The plant is a smooth, erect annual herb, 30-90 cm high, with conspicuously enlarged nodes and hollow internodes. The stems are vertically ridged. The leaves are pinnately compound and often decompound. The lower leaves are broad with serrately-lobed margins. The upper leaves are finely cut with linear lobes. The petiole is often swollen, even, hollow at the base and sheathing the stem. The leaf arrangement is alternate. The plant comes to the flowering stage in about 45-60 days after sowing. The flowers are small, white or pinkish in compound terminal umbels. There are 5 sepals, 5 petals, 5 stamens and 2 carpels which are free with an epigynous ovary. The fruit is a schizocarp, globular, yellow in colour with brown ribs. The size of the seed is about 3 mm in diameter and ripe seeds are aromatic. At dehiscence, the 2 carpels called mericarps separate, each containing a single seed with a copious endosperm and a minute embryo. Anatomically, in a transverse section, on the mericarp, vittae (oil glands) and vascular bundles can be seen. Coriander is a diploid with 2n=22. Similarities in chromosome number and morphology are reported in variants. However, variants having somatic nucleus with 2n=15, 23, 24, 25-27 chromosomes are also prevalent. The structural alterations of chromosomes is thought to have contributed to the evolution of the different strains within the species.

A good deal of variability is seen for most biometric characters. Environmental factors also play a significant role in the expression of yield and related characteristics.

Climate

Coriander is a tropical crop and can be successfully cultivated in the *rabi* season in areas that are free from severe frost during the flowering and seed setting stages. Dry and cold weather favours higher seed production. Cloudy weather during the flowering and fruiting stages increases pest and disease incidence. The time of sowing however varies in different locations in India. In West Bengal and Uttar Pradesh, the seeds are sown during the *rabi* season, in Maharashtra during the *kharif*, in Tamil Nadu in the *kharif* and *rabi* and in Karnataka from May to August and from October to January.

Soil

Coriander is cultivated both as an irrigated and rain-fed crop. As an irrigated crop, coriander can be cultivated on almost all types of soils, provided sufficient organic manure is applied. A rain-fed crop may be raised only on the heavier types of soil which have better water-holding capacity. The crop is best cultivated on heavy black, clayey cotton soils and rich silt loams which are well drained with distributed moisture. Saline, alkaline and sandy soils are not suitable for this crop.

Varieties

The number of germplasm accessions of coriander being evaluated and conserved by AICRPS in India is 1467. A number of cultivars, distinguished by the name of the locality, are under cultivation. The yield varies depending on variety and location. The varieties are grouped as the bold and small seeded types. The important cultivars of coriander are described below.

RCr-41

This was developed by the Rajasthan Agricultural University through recurrent half-sib selection from a local collection of Kota and released in 1988 for irrigated conditions. The plants are tall, erect with small-sized (9.3g/1,000) grains. The cultivar is highly resistant to stem gall and wilt but only moderately tolerant to powdery mildew. It matures in 130-140 days and produces an average yield of 9.2 q/ha.

Rcr-20

This was developed by the Rajasthan Agricultural University through recurrent half-sib selection from a local collection of Jaipur and released in 1997 for limited moisture conditions and the heavier soils of southern Rajasthan. The plants are bushy, spreading, with a medium height. It produces oval grains of large size (18.0g/1,000). It is moderately tolerant to powdery mildew, wilt as well as stem gall. The cultivar matures in 100-110 days and produces an average yield of 10.0q/ha.

RCr-435

This has been developed by the Rajasthan Agricultural University through half-sib selection from a local collection of Jalore district and identified for release in 1995 for irrigated conditions. The plants are bushy with a quick, early growth and medium-sized grains (14.0 g/ 1,000). It matures in 110-130 days and produces an average yield of 10.5 q/ha.

RCr-436

This was developed by the Rajasthan Agricultural University through half-sib recurrent selection from a local collection of Kota and identified for release in 1995 for limited moisture conditions. The plants are bushy with quick early growth and bold grains (16.0 g/1,000). It matures in 90-100 days and produces an average yield of 11.09 q/ha under limited moisture conditions.

RCr-446

This was developed by the Rajasthan Agricultural University through half-sib selection from a local collection of Jaipur district and identified for release in 1977 for irrigated conditions. The plants are leafy and erect with higher number of grains per umbel. The seeds of this variety are medium in size (13.5 g/1000). It matures in 110-130 days and produces an average yield of 12 q/ha.

GC-1

This was developed by the Gujarat Agricultural University through selection in local material and released in 1974 for early sowing conditions. The plant is erect with medium sized (13.2 g/1,000 grains) and round, yellow-coloured grains. It is moderately tolerant to wilt and powdery mildew. It matures in 112 days and produces an average yield of 11.0 q/ha.

GC-2

This was developed by the Gujarat Agriculture University through mass selections from Co-1 and released in 1985, for early sowing. The plants are of a semi-spreading habit, and have dense foliage with dark green leaves and bold grains (14.8g/1,000 grains). It is moderately tolerant to wilt and powdery mildew. It matures in 110 days and produces an average yield of 14.5 q/ha.

CS-4 (Sadhna)

This was developed by the Andhra Pradesh Agricultural University through mass selection from the Nandyal collection and released in 1989 for rain-fed and late-sown conditions. The plants are semi-erect with medium-sized (16.8g/1,000 grains), oval, straw-coloured grains. The cultivar is tolerant to white fly and mites. It matures in 100 days and produces an average yield of 10.3 q/ha.

CS-6 (Swathi)

This was developed by the Andhra Pradesh Agricultural University through mass selection from the Nandyal collection and released in 1989 for rain-fed and late-sown conditions. The plants are semi-erect with medium-sized (16.8g/1,000 grains), oval, brownish-yellow grains. The cultivar is tolerant to white fly, grain mould and wilt, and produces an average yield of 8.8 q/ha.

CO-1

This was developed by the Tamil Nadu Agricultural University through selection and released in 1972 for the rain-fed conditions of the southern districts of Tamil Nadu. The plants are small with globular, small (14.3 g/1,000), dusty brown grains. It matures in 100-120 days and produces an average yield of 4.0 q/ha.

CO-2

This was developed by the Tamil Nadu Agricultural University through selection from P_2 cultivar of Gujarat and released in 1985. It is a dual-purpose cultivar suited for use as greens and for the production of grains. It can be grown in water-logged, drought, saline and alkaline conditions. The plants are erect, with oblong, medium-sized (15.0 g/ 1,000 grains), dull yellowish-brown grains. It matures in 90-110 days and gives green leaves in 40 days and produces an average grain yield of 5.2 q/ha and green yield of 100 q/ha.

CO-3

This was developed by the Tamil Nadu Agricultural University through selection from ACC 695 of IARI and released in 1991 for irrigated and rain-fed conditions. The plants are dwarf with medium-sized (16.2 g/1,000), oblong, brownish-yellow grains. It is tolerant to wilt, powdery mildew and grain mould. It matures in 86-104 days and produces an average yield of 6.5 q/ha.

CS-287

This was developed by the Tamil Nadu Agricultural University through reselection from CS-6 and released in 1991 to irrigated and rain-fed conditions. The plants are early maturing with medium-sized (14.9 g/ 1000 grains), oblong, straw-yellow coloured grains. It is tolerant to wilt and grain mould. It matures in 78-97 days and produces an average yield of 6.0 q/ha.

RD-44 (Rajendra Swathi)

This was developed by the Rajendra Agricultural University, Bihar, through selection from the Muzaffarpur Local and released in 1987. The plants are medium-sized with fine, round, aromatic grains (12.5 g/ 1000 grains). It is resistant to stem gall and moderately resistant to wilt, aphid and weevil. It matures in 100 days and produces an average yield of 13 q/ha.

DH-5

This was developed by the Harayana Agricultural University, Hissar, through selection from a local collection and identified in 1993, for irrigated conditions. The plants are medium-tall, bushy with round attractive grains of medium-size (13.8 g/1000). It matures in 120-130 days and produces an average yield of 18-20 q/ha under good management.

UD–20

This variety is a recurrent selection from the Jaipur local developed at SKN College of Agriculture, Rajasthan Agricultural University, Jobner, during 1983 for rain-fed conditions. The plants are tall, erect with oblong, bold grains (13.5 g/1000). It matures in 110 days and produces an average yield of 12 q/ha. The variety is resistant to wilt and stem gall diseases.

CIMPO, S-33

This is an improved strain developed by intensive selection at the Central Institute of Medicinal and Aromatic Crops, Bangalore from the Bulgarian introduction. The variety is a tall, late-maturing type with small seeds and can yield 21 q/ha of seeds (three times more) compared to 7.5 q/ha from the local cultivars. It has an oil content of 1.3% as against 0.18% of the local, giving 27.5 2 kg/ha of oil which is seven times more compared with the local variety.

A Russian cultivar 'Lucs' is reported to give the highest yield and essential oil. The other important cultivars are CS-15, CS-362, CS-52, CS-7, CS-11, CS-358, Rcr-648, Hissar Anand, Pant Haritima, Azad Dhania, DWD-3, Gwalior 5365, V1, V2, Naranul Selection, NP (D) 95, NP (D) 172, NP (K) 24, Amber and Pusa selection 360.

Cultivation

Propagation

Coriander is propagated by seed. Before sowing the fruits are rubbed until the two mesocarps are separated and then sown.

Land preparation and sowing

The land should be well prepared by ploughing 3 to 4 times and brought to a fine tilth. For an irrigated crop, beds and channels of convenient size (3 x 2 m) are formed. The seed rate requirement in coriander varies from 10-15 kg for irrigated and 25-30 kg per hectare for rain-fed crops. The seeds that are stored for 15-30 days exhibit an early and better germination. At the time of sowing, the seeds are trampled or crushed into halves with the foot or by rubbing with hands. The seeds are then treated with Agrosan-GN or any other mercurial fungicide or Carbendazim at 2 g per kg of seeds. Sowing is done either by broadcasting or by drills in rows at 20-30 × 10 cm spacing. The seeding depth should not exceed 3.0 cm. The broadcasted seeds are raked up to mix them with the soil. Soaking the seeds in cold water for 12-24 hours can also be done to hasten the germination process before sowing. Soaking seeds in a solution of 50 ppm Gibberellic Acid is beneficial. Seeds soaked in 1% Potassium dihydrogen phosphate solution or in 2% leaf extract of *Calotropis* and *Prosopis* registered an improved germination and seedling growth especially in the rain-fed crop. The seed germination is rather slow and may take 10-15 days. The thinning of plants is done after 30 days of sowing leaving only two plants per hill.

Manures and fertilisers

A crop of coriander yielding 63.1 kg/ha of green coriander requires 12 kg N, 4.5 kg P_2O_5 and 11 kg K_2O. About 10-20 tonnes of FYM per hectare is applied to the field well before planting. Fertilisers @ 40:40:40 or 30:40:20 kg or 15:40:20 or 30.:20:0 or 20:40:20 or 60:40:20 NPK per hectare are applied depending on the location at the time of sowing. The recommendation is based on the research conducted at different locations. Nitrogen is applied in 2 split doses. The first dose is given at the time of sowing and the next one is applied after 6 weeks of sowing. For a rain-fed crop, an application of 20 kg Nitrogen per hectare is sufficient. The application of neem cake (60 kg/ha) along with the recommended NPK dose (60:30:90 kg/ha N in 3 equal splits) has been found ideal for irrigated conditions in Rajasthan. A foliar application of Cu, Zn and Mo is reported to increase the seed yield.

Irrigation

The crop is irrigated immediately after sowing in order to ensure even germination, and later irrigation is provided at 7-10 days' intervals, depending upon the soil and climatic conditions. The crop requires in all 4-5 irrigations, 30-35, 60-70, 80-90, 100-105 and 110-150 days after sowing.

Inter-culture

The crop requires two or three weedings. The first weeding, hoeing and earthing-up should be done at 40-45 days after sowing, when the seedlings are well above the ground. Weeds can be checked by the application of Pendimethalin or Fluchloralin @ 1.5 kg or 5-6 kg Propanil per hectare alone or in combination as pre-emergent and post-emergent weedicides, which can bring about the optimum control of weeds with a considerable increase in seed yield. Hoeing is done to conserve the soil moisture for better yields. A spray of 250 ppm Cycocel (CCC) one month after sowing is beneficial for the rain-fed crop. The companion cropping of coriander plus mustard has been found to be economical.

Use of growth regulators

A spray of GA_3 100 ppm at the five-leaf stage increases the male flowers, while Ethrel and CCC reduce the ratio of bisexual and male flowers. GA_3 from 5 ppm upto 50 ppm hastens and improves flowering, when sprayed at the five-leaf stage; BA counters the effect of GA. Soaking the seeds in 50 ppm IAA improves flowering, while ascorbic acid increases the frequency of bisexual flowers. The application of Magnesium chlorate at 8 kg/ha at the waxy stage of ripening makes harvesting easier and improves the essential oil content.

Pests and diseases

Pests

Aphids (Hyadaphis coriandri; Myzus persicae)

The attack of aphids usually occurs at the time of flowering. This insect is easily controlled by a number of insecticides (0.15% Dimethoate at 10 days interval). However, the use of these insecticides after the crop has started flowering kills the bee population, which affects pollination in the crop. The varieties, RC-15 and Rumanian exotic EC-69 are found to be resistant to aphid attack. Whiteflies are also reported to cause damage to the crop.

Red mites (*Petrobia latens*), weevil, stink bugs (*Nazara viridula* and *Agonoscelis nobilis*), leaf eating caterpillars (*Spodoptera litura* and *S.exigua*), white fly (*Bemisia tabaci*), etc., are found on the plants. A chalcid fly, *Systole albipennis*, has been reported to damage the fruits. A foliar application of 0.03% Phosphomidon or Dimethoate or Methyl dematon (0.05%) controls the leaf eating caterpillars, while red mites could be controlled by a foliar spray of 0.3% Dicofol or sulfex, 2-3 times at 10 day intervals. Other pests attacking coriander are thrips *Scirtothrips dorsalis, Thrips tabaci* and cut worms. The other stored product pests include *Stegobium panicum, Lusiderma serricornai, Rhizopertha dominica, Cribbium psoylloids, Trigoderma granarium, Attagenenus fasciatus* and *Tribolium casteneum*. Keeping the moisture level in the grain below 11% is a prerequisite for discouraging these storage pests. Exposing the grains to temperatures of 50, 60, 70, 80 and 100°C for periods of 120, 60, 45, 25 and 15 minutes, respectively, reduces the incidence of storage pests. Fumigation with sulphur dioxide before storage is effective in arresting the storage pests. Other fumigants like Phosphine, Ethylene Dibromide or Methyl Bromide are also effective when used at the rate of 1.0-1.38 mg/litre.

Diseases

Wilt (Fusarium oxysporum f. Sp. coriandrii)

This disease inflicts heavy losses in all coriander growing areas. Wilt infection occurs in the roots and drooping, yellowing, wilting and drying symptoms are expressed in the aerial parts. Though no direct control measures are available against it, preventive measures like the use of disease-free seeds, seed treatment, use of resistant varieties, crop rotation and deep summer ploughing may help in reducing the wilt incidence. Seed pelleting with *Trichoderma viride* (formulation with CFU 10^6) @ 4 g/kg of seed and neem cake application @ 150 kg/ha followed by a foliar spray of Hexaconazole (0.05%) helps. Seed dressing as well as oil drenching with Carbendazin (0.2%) or Thiophenate methyl on the 25th, 40th and 55th day after sowing are recommended. *T. harzianum* and *Pseudomonas fluorescens* could also be used for biocontrol. The wilt tolerant cultivars, Rcr-41, Rcr 4, GC-1 and 2 and Sindhu, should be grown in wilt-affected plots.

Powdery mildew (Erysiphae polygoni)

This disease is characterised by the appearance of a whitish, powdery mass on the leaves and stem. It is severe in the initial stages of the crop, arresting seed formation. But in case of its occurrence at later stages, the seeds produced will be of poor quality and the total quantity is also reduced. This can be effectively controlled by repeated spraying of Wettable sulphur at 0.3% concentration or 0.1% Karathane once at the time of disease infestation and 10 days later with Carbendazim

(0.1%) or Dinocop (250 ml/ha) or dusting of Sulphur at 20-25 kg per hectare. Eco-friendly techniques of using 5% onion leaf extract as a foliar spray thrice also keeps the disease under check. The disease-tolerant lines like Sindhu, Swathi, CO-3, GC-1 and GC-2 could be used.

Blight

This disease is caused by the fungus *Alternaria poonensis*. The symptoms include the appearance of dark-brown spots on the stems and leaves. Spraying with 0.2% Carbendazim or Mancozeb 20 days after the grain set effectively controls the disease.

Stem gall (Protomyces macrosporus)

It is difficult to control stem gall by any of the fungicides. The Karan (UD 041) variety is found to be highly resistant to this disease under field conditions. Crop rotation is recommended to reduce the incidence of the pest. However, to keep the disease under check, seed treatment with Thiram 2.5 g/kg of seed or Carbendazim 1 g/kg of seed before sowing is followed. Spraying of Bayleton or Systhane (0.4 ml/l of water) and a repeat spray after 20 days is recommended.

Stem-rot caused by *Sclerotinia sclerotiorum*, root-rot (*Rhizoctonia solani*) and grain mould (*Alternaria, curvularia* and *Helminthosporium* sp.) have also been reported. The grain mould infestation can be reduced by spraying 0.1% Carbendazim twice at grain set and after 15 days. Cilantro dwarf is a new disease reported from Brazil and the disease is caused by the nematode *Rotylenchulus reniformis*. The root knot nematode *Meloidogyne incognita* is also reported to cause damage to the crop.

Harvesting

The coriander plant matures in 90-120 days for grains and 40 days for greens. Leaf plucking, to the extent of 50%, when the crop is 60-75 days old gives economic returns, under irrigated conditions. The crop should not be allowed to turn over ripe as this will spoil the quality, particularly the colour of the produce. The *kharif* season crop matures earlier than the *rabi* season crop. Harvesting has to be done when the fruits or 50% of seeds are fully ripe and start changing from green to brown colour. High yields could be obtained when the plants are harvested at the stage where 100% of the fruits turn yellow. For harvesting, plants can either be uprooted or cut back using sickles. The plants are then tied in small bundles and stacked for drying.

When the bundles are dried, the grains are separated by beating them gently against the threshing floor. The grains are winnowed and stored in a moisture-free godown, till they are sold. The moisture content should be moderate as the quality will decrease with excess moisture.

Yield

On an average, a seed yield of 400 to 500 kg/ha under rain-fed conditions and 1000 to 1250 kg/ha under irrigated conditions may be obtained. Under favourable conditions, yields of 1700 kg to 2240 kg/ha have been recorded.

Storage of seeds

The whole, dried seeds are usually packed into sacks and stored in a cool, dry room. At the time of storage the seeds should not contain more than 0.5 to 9% moisture. It is reported that no more than 5% volatile oil would be lost during two year storage period. It has been recommended that the spice should be placed in hermetically-sealed cans immediately after drying, in order to ensure that quality deterioration during storage is minimal. The crushed or ground spice should be stored in air-tight containers.

The major types of the spice

The spice entering international trade varies considerably in its physical and chemical characteristics according to the geographical source, and consequently, some users express preferences for certain types in particular applications.

In commerce, coriander is broadly divided into two types according to the size of the fruit, which is an indication of its volatile-oil content and suitability for particular end-uses, as the var. *vulgare* (diameter 3-5 mm) or the var. *microcarpum* (diameter 1.5-3 mm). The large fruited types supplied mainly by tropical and subtropical producing countries, e.g., Morocco and India, contain a low volatile-oil content (0.1-0.35%) and are used extensively for grinding and blending purposes. The smaller-fruited types are produced in temperate regions and usually have a volatile-oil content of more than 0.4%. The very small-fruited types grown in the USSR and some other countries of Eastern and Central Europe contain between 0.8 and 1.8% volatile oil and are highly valued as a raw material for the preparation of essential oil.

Extraction of essential oil and adulterants

Commercial coriander oil is prepared by steam distillation of the mature, dried fruits. The yield of the essential oil obtained varies from 0.88 to 0.92%.

The commercial volatile oil is extensively adulterated with sweet orange oil, turpentine and anethole or aniseed oil or cedar wood oil.

Value-added products

The oil of the coriander grains is a valuable ingredient in perfumes. Its soft, pleasant and slightly spicy note blends into scents of oriental character. Decydenyde (0.1% volatile oil) is also used in perfumery.

Good quality oleoresin is extracted from coriander seeds. The oleoresin is used for flavouring beverages, pickles and sweets. Soluble coriander is prepared by properly blending and dispensing a minimum of 3% of the total extractives of coriander on a soluble, dry, edible carrier.

Cumin

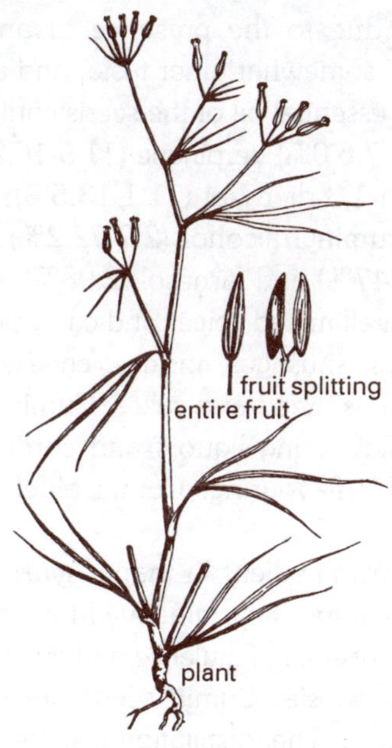

Plant Profile

Family	:	*Apiaceae/ Umbelliferac*
English name	:	Cumin
Indian name	:	*Jira*, *Jeera*, *Zira* or *Safaid Jeera* or *Zeera* (Hindi), *Safaid Jira* or *Zeera* (Bengali and Punjabi), *Jiru* or *Jeeru* (Gujarati), *Jeerige* (Kannada), *Zyur* (Kashmiri), *Jeerakam* (Malayalam), *Jeregire* (Marathi), *Jira*, *Jeera* (Oriya), *Jiraka*, *Jira* (Sanskrit), *Zero*(Sindhi), *Ziragam* or *Jeeragam* (Tamil), *Jidakara*, *Jikaka* (Telugu)
Species	:	*Cuminum cyminum* L.
Distribution	:	Egypt, Syria, India, Argentina, Cyprus, Denmark, Mexico, Russia, Pakistan, Syria, Iran, Morocco, Turkey, China, Indonesia, Japan
Uses	:	Ayurvedic medicines, Veterinary drugs, Spice, Culinary purposes.

Cumin or *Safaid jeera* (*Cuminum cyminum* L.) is one of the oldest spices known to mankind. It is the ripe fruit of a slender herb. Cumin seeds are yellowish to greyish-brown and have an aromatic odour due to the presence of an aromatic alcohol, cuminol, and a spicy somewhat bitter taste, and are extensively used as a condiment. The essential oil of the seeds contains α-pinine (14.3-19.7%), p-cymene (2.7-6.0%), terpinene (11.5-16.2%), cuminaldehyde (20-22.4%), p-menth-1,3-dien7-al (11.1-13.5%), p-menth-1, 4 dien-7-al (23.6-24.3%), cuminyl alcohol (2.7-72.2%), sabinene (3.38%), benzyl cinnamate (4.47%) and farnesol (2.98%). Cumin is used as an essential ingredient in all mixed spices and curry powder for flavouring soups, pastries, stews, sausages, pickles, cheese and for seasoning bakery products such as bread and cakes. Cumin oil is used in soaps, perfumery and for flavouring liquors and cordials. The absolute is superior to the oil for flavouring. Cumin aldehyde is also used in perfumery.

It is also used as a ingredient in many *Ayurvedic* medicines. It is prescribed as a stimulant and carminative in indigenous medicines. It is an astringent and is useful in flatulence and griping, and in conditions of diarrhoea and dyspepsia. Cumin seeds are now chiefly used in many veterinary drugs. The distillation-residue water of the oil is given to children as carminative and is useful in flatulence and griping. The residue left over after the extraction of volatile oil contains 17.2% protein and 30% fat. It can be used as a cattle feed. The fixed oil could also find use in the oil, fat and soap industries.

Origin and distribution

Cumin is a native of Egypt, Syria, Turkey and the Eastern Mediterranean region. It is an important cash crop of India, Argentina, Cyprus, Denmark, Mexico, USSR, Pakistan, Syria, Iran, Morocco, Turkey, China, Southern Russia, Indonesia and Japan. Iran is the major world exporter of the cumin seed known as 'green cumin'. In India, it is cultivated on a commercial scale in Rajasthan, Gujarat, Tamil Nadu and to some extent in Uttar Pradesh, Punjab and Madhya Pradesh. Gujarat is the leading cumin seed producing state in India accounting for about 52% of the production. The area under cumin in India was about 2,96,990 hectares and production about 1,38,220 tonnes during 2001-02. There are several types of cumin well-known in the international trade. These are Iranian cumin, Indian cumin, Egyptian cumin, Turkish cumin and cumin from other regions.

Description of the plant

The cumin plant belongs to the family *Apiaceae* (*Umbelliferae*) or the carrot family. It is a small, slender, annual herb of about 30-45 cm height, with a much-branched angular or striated stem bearing 2 or 3 linear leaves, bluish green in colour and having a sheathing base. The flowers are white or rose coloured, borne in compound umbels. The fruits are greyish, bristly, about 0.6 cm long, tapering towards both base and apex and compressed latterly with ridges covered by papillose hairs. The hairs may be absent in some forms.

Climate

Cumin thrives well in tropical and sub-tropical climates. It can be grown from sea level upto an elevation of 1800 to 3000 m MSL. It flourishes well in a mild cool climate rather than the hot plains. It is grown as a *rabi* crop during October-November. In areas where the atmospheric humidity is low during the months of February and March, which coincides with the flowering and fruiting periods, it induces the development of diseases like powdery mildew and blight and insect pests to which the crop is highly susceptible.

Soil

Cumin can be cultivated in all types of soils but a well-drained, medium to heavy textured, medium to highly fertile soil with a good water-holding capacity is ideally suited for its cultivation. Sandy loam or loamy soil is supposed to be the best for its successful production. A soil pH range between 7.0 to 9.5 is reported to be optimum. The crop can tolerate salinity where other crops fail to grow. The incidence of wilt disease is higher in light-textured soils.

Varieties

The germplasm holding of cumin at the AICRPS centres in India is 495. There are only a few varieties available at present. The important varieties grown in India are given below.

RS-1

This is an early maturing variety, with 10 to 12% higher yields than the local varieties. The seeds are bold, hairy, sweet and aromatic.

MC-43 and Vijapur-5

These varieties bear bold seeds, rich in pleasant aroma and give a good yield. MC-43 gives an average yield of 580 kg/ha. It has the capacity to survive under conditions like water stress and disease incidence like *Fusarium* wilt, *Alternaria* blight and powdery mildew. Its duration is 113 days.

UG-19

This has good attractive seeds and shows better resistance to wilt.

UC-220

This variety produces about 208 kg seeds having a volatile oil content of 3.53% in seeds and yielding about 7.34 kg of oil.

UC-217

The seed yield in this variety is 218 kg and the oil recovery is 7.42 kg/ha. The volatile oil content of the seeds is 3.4%.

JC-147

This variety can yield 269 kg seeds from which 9.16 kg of volatile oil could be recovered. The seeds contain 3.4% volatile oil.

Gujarat cumin-1

This variety gives 23% higher yield than MC-43. It has the capacity to yield 700 kg/ha and has better resistance to wilt and powdery mildew diseases. This variety is also free from lodging and seed shattering. The crop duration is 105 days.

Gujarat cumin-2 (MC-43-73)

This is a pure line selection from M2 of γ-irridiated seeds of MC-48. The plants are bushy with a good branching habit. It gives an average yield of 700 kg/ha. It is moderately resistant to *Fusarium* wilt, *Alternaria* blight and powdery mildew diseases. The variety matures in 100 days.

Gujarat cumin-3

This variety (culture sel. 84-1) is derived by recurrent selection from an introduction from West Germany (EC 232689) received through NBPGR, New Delhi. It is the first wilt-resistant variety of medium maturity (98 days). The plants are a bushy type (21.8 cm). The variety produces 663 kg/ha and possesses 3.3% essential oil. The seeds are pungent with a good aroma and are dark grey in colour.

S-404

This is an old selection from local germplasm made at the Spices Research Station, Jaudan, Gujarat Agricultural University, Gujarat. It gives an average yield of 350 kg/ha and is moderately resistant to powdery mildew disease. The crop duration is 105 days.

RZ-19 (Rajasthan zeera-19)

This is a recurrent selection from UC-19 made at the SNJ College of Agriculture, Jobener, Rajasthan. This is a late variety, maturing in 125 days and is well adopted to late sown conditions. The variety can yield 500 kg/ha and is tolerant to *Fusarium wilt* and *Alternaria* blight diseases.

UG-198

This is highly tolerant to wilt and is very rich in volatile oil (5-6% compared to 3% in other varieties), but its grains are some what fragile and are not suitable for marketing.

UG-223

This variety can yield 301 kg of seeds and the oil recovery is 10.23 kg/ha. The seeds contain 3.40% of essential oil.

Sel.7-3

This variety has been released from the Indian Institute of Agricultural Research, New Delhi. It is a high yielder and is resistant to wilt disease.

CJS-182

This variety was released from the Indian Agricultural Research Institute, New Delhi. This is a high yielder preferred for commercial cultivation.

UC-52, UC-91 NP (D)-1, NP (J) 126, NP (J) 149, Rcr 1 g are the other recommended varieties of cumin. The exotic cultivars EC-232684, EC-243373, EC-243375 and EC-109635 are found resistant to *Fusarium* wilt disease.

Cultivation

Propagation

Cumin is propagated through seeds and the seed rate varies with the method of sowing. The broadcast method requires more and the line sowing requires less quantity of seeds.

Land preparation

The root system of the crop is confined to the upper six inches and, hence, shallow cultivation is advised. The porousness and friableness of the soil plays a very vital role in germination and crop growth. The land is prepared well by ploughing and planking. Plots of 2 × 2.5 m are made after the final preparation of land, just before sowing.

Sowing

The best time for sowing is the first fortnight of November. If the soil moisture is not adequate, a pre-sowing light irrigation should be given prior to the tillage operations. However, sowing can be done up to 15th December. The late sowing of the crop results in reduced in yield and increased infestation of insect pests and diseases. The sowing of seeds can be done either by broadcasting or line sowing. In case of the broadcasting method of sowing, generally 10-20 kg seeds/ha is required, whereas in line sowing 9-12 kg seeds/ha is needed. In order to hasten germination, soaking the seeds with Potassium nitrate (100 ppm) for 24 hours is best. Seed treatment before sowing is a must, which can be done by using Agrosan GN or Difoltan at 3 g per kg of seed. Line sowing is done at 22.5 to 30 cm × 15 cm spacing and is preferred over broadcasting. The depth of sowing should be 1.5 to 2 cm. Normally, if sowing is done properly, the seeds germinate in 5-7 days.

Manures and fertilisers

A crop yielding 545 kg seeds/ha is reported to draw 18 kg N, 4.7 kg P_2O_5, 10 kg K_2O, 5.5 kg Ca, 1.2 kg Mg, 79 g Fe, 16 g Mn, 26 g Zn and 9 g Cu per hectare. Hence, it is recommended to apply well rotten FYM @10-15 t/ha to the field at the time of land preparation. In addition, an application of 25:20:20 or 50:50:80 kg or 30-40 kg each of NPK per hectare is considered to be optimum for the satisfactory growth of the crop. Half the dose of N, the whole of P_2O_5 and K_2O may be applied as a basal dose followed by the remaining half of N at 30 days after planting, as a top dressing.

Irrigation

First, a light irrigation has to be given just after sowing and the second after 7-8 days of sowing. The crop has to be irrigated at IW/CPE ratio of 0.7. The irrigation interval between two irrigations can be kept at 12-15 days, taking into consideration the climate and the soil type. The crop requires totally 4-5 irrigations. It is essential to provide irrigation at the time of flowering and seed formation. Irrigation should be withheld at the time when the crop is maturing as this may adversely affect the seed quality.

Interculture

For its proper growth, the first weeding should be done 30-40 days after sowing. At the same time, the plants are also thinned to a spacing of 12-15 cm within the rows. Another 1-2 weedings will help in better crop growth. The herbicides, Terbutyrin (0.5 kg/ha) or Oxidiazone (0.5 kg/ha) or Pendimethalin (1.0 kg/ha) are recommended for the control of weeds.

Pests

Cumin is often attacked by aphids (*Myzus persicae, Hyadaphis coriandari*). They can be effectively controlled by spraying a 0.3% solution of Phosphomidon (0.03%), or Dimethoate (0.3%) or Methyl parathion (0.2%) or Metasystox (0.3%) or Endosulfan (0.07%) or Monocrotophos (0.01%) or tobacco decoction. Weevils are also reported to cause damage to the crop.

Diseases

Cumin wilt

This is caused by *Fusarium oxysporum, F. cumini* and *F. equisetti*. The organism can infest the plant at any stage of growth. The infected plants will wilt and die. It has been estimated that crop losses due to this disease vary from 70% to 80%.

Although control of the disease is rather difficult, the severity of the disease can be minimised by the use of healthy and disease-free seeds, treating the seed with a mixture of Captan and Bavistin @ 4 g/kg of seed before sowing, fallowing, etc. A three-year crop rotation sequence to accommodate cumin once in three years and deep summer ploughing and soil solarisation during summer months is effective. Organic amendments like neem cake are also useful in checking wilt. The exotic lines UC-198 from Egypt and UC-199 from Libya are found to be more resistant to wilt.

Cumin blight

This disease is caused by *Alternaria burnsii*. The blight-infected plants show dark brown lesions on the leaves, and the stem and tops tend to bend downward. The infection spreads rapidly if there is cloudy weather and it becomes very difficult to protect the crop in the advanced stages of infection. The plants are susceptible to this disease at the flowering stage. In order to control this disease, seed dressing with Bavistin or Captan (4 g/kg) or Dithane M-45 (2 g/kg) and a foliar application of Dithane Z-78 at the concentration of 0.2% or Blitox-50 at fortnightly intervals is recommended.

Powdery mildew

This disease is caused by *Erysiphe polygoni*. The whole plant becomes almost white with powder, and does not exhibit proper seed formation.

In the initial stages of the disease, the crop should be sprayed with a 0.3% solution of Wettable sulphur/Dinocop (0.25%), Karathane, or 0.2% Calixin at 15 days' intervals commencing from 45 days after germination, or dusting the crop with 20 to 25 kg of Sulphur dust per hectare would also effectively control the disease.

Witches broom, caused by phytoplasma, has also been reported to affect cumin.

Harvesting and yield

The crop will be ready for harvest in about 80 to 120 days after sowing. At this stage, the leaves of the plant become yellow. The crop is harvested before the fruits shatter, by uprooting the whole plant in the morning. The uprooted plants are stacked for 2-3 days in the sun for drying. The seeds are separated by rubbing the plants manually or beating them with bamboo sticks. The clean and dried seeds are filled in gunny bags lined internally with a polyethylene sheet and stored in a cool dry place till they are marketed.

A disease-free field receiving the above package of practice would easily produce 8 to 15 quintals of cumin seeds per hectare.

Processing

The dried fruit or seed is crushed and distilled immediately to obtain the essential oil; steam distillation is usually carried out. The oil recovery ranges from 2.5 to 4.5%; older seeds yield lesser oil. On an average, an oil yield of 25-30 kg/ha is obtained. The volatile oil is colourless or pale yellow, turning dark on storage. The oil could also be extracted by the Super-Critical Fluid Extraction (SCFE) process.

Cumin oil is often adulterated with synthetic cumin aldehyde, the presence of which in small quantities cannot be detected by routine analysis, and higher percentages affect the optical rotation. In addition to volatile oil, the seed also contains about 10% of fixed greenish-brown oil with a strong aromatic flavour. It is a semi-drying oil with an iodine value of 92. Cumin oleoresin is also obtained from the seeds.

Curry Leaf

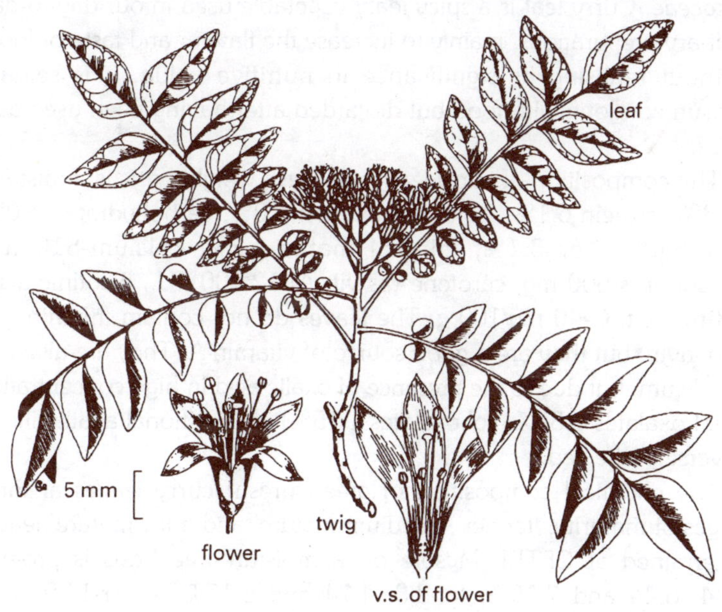

Plant Profile

Family	:	*Rutaceae*
English name	:	Curry leaf
Indian name	:	*Kathnim, Mitha neem, Curry or Kari patta, Gandhela, Barsanga* (Hindi), *Karibevu* (Kannada), *Karriveppiler, Kariveppilei* (Malayalam), *Karivampu, Karuveppilei* (Tamil), *Karepaku* (Telugu).
Species	:	*Murraya koenigii* (Linn.) Sprengel *Syn. Bergera koenigii* Koen., *Chaleas koenigii* Carz ex. Swingle
Distribution	:	India, Sri Lanka, Myanmar, China, Pakistan, Australia, Pacific Islands.
Uses	:	Spicy leafy vegetable, medicine.

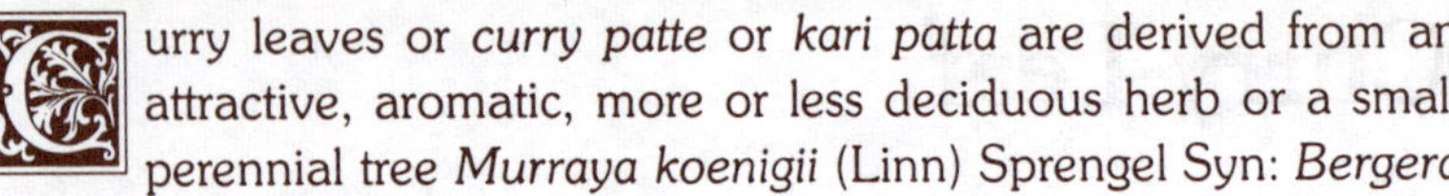

Curry leaves or *curry patte* or *kari patta* are derived from an attractive, aromatic, more or less deciduous herb or a small perennial tree *Murraya koenigii* (Linn) Sprengel Syn: *Bergera koenigii* Koen., *Chaleas koenigii* Curz ex. Swingle, of the family *Rutaceae*. Curry leaf is a spicy leafy vegetable used in our day-to-day culinary preparations, mainly to increase the flavour and taste of food. Without realising its significance in nutritive value, it is seldom consumed along with food but discarded after having been used as a spice.

The composition of curry leaves on fresh weight basis is: moisture-66.3%, protein-6.1%, fat (ether extract)-1.0%, carbohydrate-16.0%, fibre-6.4%, iron-3.1%, mineral matter-4.2%, calcium-810 mg, phosphorus-600 mg, carotene (as vit. A)-12,600 I.U., nicotinic acid-2.3 mg, vit. C-40 mg/100 g. The leaves do not contain thiamin and riboflavin, but they are a good source of vitamin-A. They are also rich in calcium, but due to the presence of oxalic acid in high concentration (total oxalates 1.35%; sol.oxalates 1.15%) its nutritional availability is adversely affected.

The chemical composition of green (fresh) curry leaves at three stages of maturity, 'tender', 'medium mature' and 'fully mature' leaves determined at CFTRI, Mysore on a moisture-free basis is protein-5.44, 6.44 and 7.19%; fat-3.3, 4.74 and 6.15%; sugar-14.9, 17.9 and 18.9; starch-11.4, 14.2 and 14.6%; crude fibre-5.8, 6.2 and 6.2%, volatile oil-0.82, 0.55 and 0.48%; acetone extractive (oleoresin)-1.6, 1.4 and 1.3%; total ash (mineral matter)-12.7 and 13.1% and acid insoluble ash-1.2, 1.3 and 1.35%.

It is seen that with advancing maturity, there was a gradual decrease in volatile oil and oleoresin (acetone extractive), while there was a progressive increase in all the other items of analysis. Vaccum-shelf drying of curry leaves gave a better product of greenish colour than leaves dried by other methods.

The free amino acids present in the leaves are: asparagine, serine, aspartic acid, glutamic acid, threonine, alanine, proline, tyrosine, tryptophan, α-aminobutyric acid, phenylalanine, leucine, isoleucine, traces of lysine, arginine and histidine. The leaves, fruits, bark and roots contain a crystalline glucoside, *koenigin*, and a resin. The twigs and leaves contain 0.8% potash on a dry matter basis. The flowers contain a glucoside *murrayin*.

The fresh leaves, on steam distillation under pressure (90 Ib/sq.in), yield 2.6% of a volatile oil (curry leaf oil) which may find use as a fixative for heavy type of soap perfume. Distillation at ordinary pressure gives very poor yields of oil, while distillation with superheated steam (temperature 220°C) yields a dark coloured, foul-smelling oil. Rectified curry-leaf oil is deep yellow in colour with a strong spicy odour and pungent clove-like taste. The essential oil obtained by simple water distillation has the following characteristics: yield-1%, colour-light yellow, smell-aromatic with a spicy note, Sp.gr.at 25°C-0.8589, Opt.rot-36°14, acid value-1.17, Sap. value-8.18, Sap. vaue after acetylation-52.81. The major chemical constituents of the oil are–1-sabinene-34%, 1-α-pinene-27.3%, dipentene-15.9%, 1-terpenol-7.7%, 1-caryophyllene-6-7%, 1-cadinene-5.2% and unidentified residue-3.2%. Using modern techniques, CFTRI scientists have reported the presence of γ-pinene in curry-leaf oil for the first time.

The leaves of this plant have slightly pungent, bitter and feebly acidic taste, and they retain their flavour and other qualities even after drying and have been used for centuries in South India as a natural flavouring agent in various curries, *sambar, rasam* and *chutneys*. Ground curry leaf with mature coconut kernel and spices form an excellent preserve.

The leaves, bark and the root of the plant are used in indigenous medicine as a tonic, stomachic, stimulant and carminative. An infusion of the roasted leaves is used to stop vomiting. Externally, they are also used to cure eruptions and the bites of poisonous animals. The green tender leaves are eaten raw as a cure for dysenty. A decoction of leaves is sometimes given with bitters as a febrifuge. It has also been used as an antiperiodic, and many a time, the powdered dry leaf, mixed with honey and juice of betelnut is recommended in the *Ayurvedic* system of medicine. The aqueous extracts of the leaves, when administered parenterally to female guinea pigs, not only raise the phagocytic index but also mobilise a greater number of leucocytes to take part in phagocytosis; the effect does not last long though. The juice of the root is taken to relieve pain associated with the kidney.

The fruit is edible. It yields 0.76% of a yellow volatile oil with a neroli-like odour and peppery taste, accompanied by an agreeable sensation of coolness on the tongue.

The wood (43-50 lb/cft) is grayish-white, hard, even, close-grained, and durable. It is used for agricultural implements.

Origin and distribution

Curry leaf is native to the *Terai* region of Uttar Pradesh in India and is found almost throughout India and the Andaman Islands upto an altitude of 1500 m. Two related species of curry leaf are *M. paniculata* indigenous to Myanmar and *M. exotica,* an ornamental shrub indigenous to India, Sri Lanka and Myanmar. Curry leaf is cultivated in Myanmar, Sri Lanka, China, Pakistan, Australia and the Pacific Islands. It is commonly found in forests, often as gregarious undergrowth along the foothills of Himalayas from the River Ravi to Sikkim and Assam. It is also found in Bengal, Madhya Pradesh and South and the Southwestern States. On a commercial scale, it is cultivated in large areas in Tamil Nadu, Andhra Pradesh and Karnataka. In Tamil Nadu, the cultivation is concentrated in Coimbatore, Salem, Trichy and pockets of Ramanathapuram districts. In Coimbatore district, its cultivation is mainly concentrated in places like Karamadai, Mettupalayam, Thayanur, Kanvanur and in Thangacchimadam (Ramanathapuram Dist.). In Karnataka, it is being cultivated on plantation scale in Dharwad, Belgaum and the Uttar Kannada districts.

Description of the plant

Curry leaf is an attractive, aromatic, more or less deciduous shrub (0.9 m) or a small tree that grows upto 6 m height (in the Himalayan regions) and 15-40 cm in diameter. The bark is dark brown or almost black; the leaves are imparipinnate, with 9-25 leaflets that are ovate, lanceolate or somewhat rhomboid, irregularly crenate-dentate, acuminate, obtuse or acute. The base is usually oblique, almost glabrous above, pubescent beneath, gland dotted and strongly aromatic. The flowers are in terminal corymbose cymes, white and fragrant. The berries are sub-globose or ellipsoid, purple-black when ripe and 2-seeded. The roots are woody, widely spread and produce many suckers. The tree bears flowers from February-May and the flowers are self-pollinated, hence its variability is very limited. The diploid chromosome number is 2n=18.

Soil and climate

Though curry leaf can be cultivated in a wide range of soil conditions, red sandy loam with good drainage is ideal for its normal and fleshy growth which results in a better leaf yield. To a certain extent, also it can tolerate drought. Heavy clay with poor drainage is not suitable for its cultivation. It can tolerate a maximum temperature ranging from 26° to 37°C. But when the temperature falls below 16°C, the vegetative buds become dormant arresting the new growth of the plant. However, this varies from variety to variety. It can be grown in places upto an elevation of 1500 m MSL.

Varieties

'Senkaampu' is a local cultivar grown in different parts of Tamil Nadu, especially in the Karamadai tract of Coimbatore district. The petiole in this variety is purplish-red in colour. The leaves have a good aroma and flavour due to its high oil content.

There are two types of curry leaf, viz., broad-leaved and small-leaved types. The small-leaved types are more fragrant and hence used for extraction of essential oil.

At the Department of Horticulture, University of Agricultural Sciences, Dharwad, two genetically distinct cultivars, viz., DWR-1 and DWR-2, have been identified and are being multiplied.

DWR-1 (Suvasini)

This is a clone obtained from root suckers that have dark green, highly aromatic, shining leaves. It is sensitive to low temperature in the winter season and hence the bud burst is poor during winter. The leaves are dark green (0.1629 mg of chlorophyll content per gram of fresh leaf), shiny and highly aromatic, and have an oil content of 5.22%. It can be dehydrated at 50°C without any loss of quality, and made into dry powder. The plant could be trained to grow as a low bush (up to 90-120 cm height). This variety is resistant to *Cylindrosporium* leaf-spot.

DWR-2

This is an open-pollinated seed progeny with pale green leaves with less aroma (4.09%). It is not very sensitive to low winter temperature and is much superior in the number of bud burst and internodal length. It is also 8 times higher in the growth of shoot length and weight of new shoots than DWR-1. Due to its winter tolerance, it can give extra income to planters.

Cultivation

Propagation

Curry leaf is propagated through seeds. Raising seedlings in the nursery and then planting them in the main field is the normal practice that is being followed. For raising seedlings, good and well-ripened fruits should be harvested from high yielding mother trees. Though there is stray flowering and fruit set throughout the year, the main season of availability of curry leaf fruits is July-August. Within 3-4 days of the collection of fruits, the seeds should be pulped and sown in nursery beds or polybags filled with soil and FYM. If seed sowing is delayed there will be a drastic deterioration in the germination of the seeds. Each fruit contains about 2-3 seeds.

The nursery beds should be prepared in a well-drained area since water stagnation will severely affect germination. Beds of 1 m × 1 m size with 30 cm height are formed. Adequate FYM should be applied before preparing the beds. Lines are drawn at a spacing of 10 cm and seeds are sown in these lines. The seeds will germinate in about three weeks' time. The seedlings can also be raised in polybags. One-year-old seedlings are best suited for planting. Even young seedlings of about 3 months of age (20 cm in height) are planted in 15 cm^3 pits.

A protocol has been developed for the *in vitro* propagation of curry leaf.

Land preparation and planting

The field is ploughed 3-4 times to get a fine tilth. Before the last ploughing, well decomposed FYM is applied @ 20-26 tonnes/ha. Pits of 30 cm^3 or 60 cm^3 size are dug, at a spacing of 1.2-1.5 m or 3-4 or 3 × 1.5 m (222 plants/ha) one to two months before planting the seedlings. In soils of very low fertility, a spacing of 90 × 90 cm will suffice. Before planting, the pits are filled with a mixture of soil, sand and FYM. One seedling is planted at the centre of the pit. The seedlings can be planted during any season of the year. However, June-July planting is the best.

Manures and fertilisers

The fertilisers recommendation (g/plant/annum) is as follows:

	1st year	**2nd year**	**3rd years and above**
Nitrogen	50	150	300
Phosphorus	25	37	50
Potassium	25	37	50

Research conducted at the Tamil Nadu Agricultural University, Coimbatore, has indicated the manurial schedule of 90:10:10 kg NPK/ha.

The fertilisers should be applied in four divided doses (25% each application) at tri-monthly intervals after each harvest of leaves. After each harvest, 20 kg of FYM per plant is also applied and mixed with the soil.

Irrigation

Immediately after planting, the pits are irrigated. On the third day the second irrigation is given; thereafter, irrigation is given once in a week. Channels can be made along the line of the pits and flow irrigation can be provided.

Weeding

The inter-space is kept free of weeds by periodical hoeing in the first year and one inter-crop like pulses can be taken. The weeds can also be kept under check by spraying Paroquat (3 ml/lit). After each harvest, the field should be thoroughly weeded.

Intercropping

Leguminous crops like cowpea, black gram, green gram or soybean could be raised as inter-crops during the first year.

Pruning and training

Initially, the terminal bud is allowed to grow to a height of 1 m. After attaining this height, it is cut off to encourage basal branching. This will maintain the plant in a bushy state so as to facilitate the harvesting and other operations. In total, 5-6 branches are maintained per bush. Branches which grow tall in a vertical fashion have to be removed so that it will not become a tree. Ten to twelve months after planting, the first harvest begins.

Pests and diseases

Pests

Among the pests attacking the curry leaf, the caterpillar of citrus butterfly (*Papilio demoleus*) is very important. Early instar larvae are dark with white patches resembling the droppings of birds. When grown up, they turn deep green in colour, stout and cylindrical in shape. Hand picking and destruction of the larvae can be done wherever possible. If the population is heavy, an insecticide like Malathion (0.1%), which has less mammalian toxicity, can be sprayed.

The branches affected by citrus psylla (*Diaphorina citri*), aphids and scale insect (*Anoidiella orientalis*) have to be removed immediately. If necessary a systemic insecticide like Dimethoate (Rogar) can be sprayed at 0.1% concentration. But the leaves should not be harvested for atleast 20-25 days. The stem (bark) borer (*Indorbela tetraonis*) is reported to cause drying of the branches occasionally and mites are also found to infest the leaves, for which no control measures have been suggested.

The nymphs and adults of *Dasynus antennatus* are reported to infest the growing tips of the branches and cause the drying of young leaves, adversely affecting the growth of the plants.

The other pests of curry leaf include black fly (*Aleurocanthus woglumi*), leaf roller (*Tonica zizyphi*), tartoise beetle (*Silana farinosa*), citrus leaf miner (*Phyllocnistis citrella*), leaf weevils (*Myllocerus discolor, M.viridanus*), hairy caterpillars, coreid bugs and termites.

Diseases

Leaf-spot (*Phyllosticta murrayae* and *Cylindrosporium* sp.) is a serious disease of curry leaf which causes severe defoliation. Carbendazim (1%) or Mancozeb (2%) could be sprayed 15 days prior to harvest to control this disease. Sap-rot (*Fomes pectinatus*) and collar-rot of seedlings (*Rhizoctonia solani*) are the other diseases, but the economic loss due to these diseases are minimal.

The spraying of Sulphur compounds should be avoided since it results in leaf shedding.

Harvesting and yield

In about a year after planting, the plants grow 1 to 1.5 m in height, and they should be harvested at that stage. Though the crop is harvested at the end of the first year, economic yields are obtained from the 2nd year onwards. While harvesting, the new shoots produced above the final pruning cut are harvested with sharp knives at the dense green leaf stage, at tri-monthly intervals, four times a year. The yield of leaves per hectare is as follows;

At the end of first year 250-400 kg of leaves/ha can be harvested.

II year	:	1800 kg/ha once in 4 months, which works out to 5400 kg/ha per year
III year	:	Around 5400 kg/year.
IV year	:	2500 kg/ha once in 3 months, which works out to 10,000 kg/ha/year
V year onwards	:	5000 kg/ha once in 3 months, which works out to 20,000 kg/ha

The plants can be maintained for 25 years.

Post harvest management

The young shoots with tender leaves are harvested, packed in gunny bags or tied in bundles and transported. Water is sprinkled on the bags/bundles. No research has been done on the post-harvest management of curry leaf. The leaves are dried and ground into powder and used as curry-leaf powder.

Value-added products

The value-added products of curry leaf are volatile oil and dehydrated curry leaves.

Dill

Plant Profile

Family	:	*Apiaceae/ Umbelliferae*
English name	:	Dill, Garden Dill, Anet
Indian name	:	*Sowa, Soya* (Hindi, Bengali, Punjabi and Urdu); *Surva (*Gujarati), *Sabasige (*Kannada), *Soi* (Kashmiri), *Surva, Shepu (*Marathi), *Satapushpi* (Sanskrit), *Sathakuppi Sompa* (Tamil), *Sabasige* (Telugu)
Species	:	*Anethum graveolens* L.; *A. sowa Roxb*
Distribution	:	Iran, Egypt, Abyssinia, Russia, Africa, Asia, USA, Hungary, Germany, Holland, Sweden, Asia Minor, England
Uses	:	Seasoning soups, salads, Processed meats, Sauces, Pickles, Condiment

Dill belongs to the family *Umbelliferae*. It comes under the genera *Anethum*, in which there are two species under cultivation, namely, The European dill, *A.graveolens* L. and another closely related Indian dill, *A. sowa* Roxb. ex Flem.

The composition of the seeds of the two types of dill are given below.

Composition	Indian dill
Moisture	4.5–7.5%
Volatile oil	1.5-4.0%
Mineral matter	5.89–11.54
Acid insoluble ash	0.55–2.71
Extraneous matter	3.16–12.93

Composition	European dill
Moisture	6.6%
Fat	17.9%
Crude fibre	20.7%
Carbohydrates	35.7%
Mineral matter	6.0%
Calcium	1.6%
Phosphorus	0.21%
Iron	0.012%
Sodium	0.01%
Potassium	1.1%
Vitamin-B_1	0.42
Vitamin B_2	0.28
Niacin	21.8
Vitamin-C	12.0
Vitamin-A	175 IU/100 g
Calorific value	435 calories/100 g

The entire dill plant is aromatic, but most of the volatile oil is in the seed (fruit). The herb and fruit oils possess different flavour constituents. The principle constituents of the herb oil is phellandrene and that of the seed oil is carvone. It is said that the quality of fruits present in harvested material and their state of ripeness determines the oil content of the crop and its flavour. The herb oil is a colourless to brownish-yellow mobile liquid, while the seed oil is light yellow to pale yellow in colour, becoming dark on storage. In addition to phellanderene, the herb oil contains: pinene, β-pinene, 3-carene, terpinene, limonene, α-terpinene, cymene, terpinolene, undecane, cis-3-hexanyl acetate, cis-3-hexanol, trans-2-hexanol, terpinene-4-ol, terpineol, carvone, thymol, carvacrol, myriciticin and apiole. And the seed oil in addition to carvone, contains: apiole, carveol, caryophyllene, dihydro-carveol, dihydrocarvone, dillapiole, D-limonene, D-phelliandrene, eugenol, iso-engenol, L-pinene l-terpinene and myristicin.

The oil content of the fresh herbage varies from 0.60 to 2.84%. The oil content in the leaves rises progressively from the rosette stage (0.6%) to the commencement of flowering (1.5%) upto the mass flowering (1.7%); thereafter it declines. It is 1.2% at the milky ripening of fruits, 0.65% at mid-ripe fruits and 0.4%, when the fruits grow to maturity on the plant. The flowers at the bud stage contain 0.8% oil rising steeply to 4.8% at the mass-flowering stage.

Dill leaves are used for seasoning soups, salads, processed meats, sauces and particularly pickles. The seed is employed as a condiment in India and is also used as a carminative, aromatic stimulant and diuretic in *Ayurvedic and Unani* medicines. The emulsion of the seed oil in water is an important constituent of gripewater, useful in flatulence, colic pain, vomiting, diarrhoea and hiccups.

The European dill seed oil is preferred in Pharmacy over the seed of the Indian dill or 'sowa' due to the presence of a large percentage of dillopiole varying from 15.6 to 39.6% and a lower percentage of carvone in the sowa seed oil. Dillopiole is a toxic substance, not found in European dill seed oil. The dried residue left after the distillation of essential oil from seeds of *A. graveolens* contains fat-16.8% and protein-15.1%. This may be used as cattle feed.

Origin and distribution

Anethum graveolens is a native of the Orient and Mediterranean countries, and is found growing wild in Iran, Egypt, Abyssinia, various parts of Europe, Southern Russia, Africa and Asia. It is cultivated in the United States, Hungary, Germany, Holland, Sweden, Asia Minor, North Africa, India and England. In India, it has been successfully cultivated on an experimental scale in Jammu and Kashmir, Maharashtra, Gujarat, Andhra Pradesh, Madhya Pradesh and Rajasthan. Indian dill is a native of Northern India.

Description of the plant

Dill is a hepaxanthic annual or biennial aromatic herb, growing to a height of about 1 to 1.20 m. The leaves are feathery and stand on sheathing foot-stalks, with linear and pointed leaflets. The leaves are compound and light green in colour. The stem is smooth, shiny and hollow, having an upright growth and bearing flat terminal compound umbels, with numerous yellow flowers, whose small petals are rolled inwards. The flowers are regular, bisexual, pentamerous and dichogamous. The fruits are flat, dry and groomed, commonly but erroneously these so-called seeds are produced in great quantities. They are very pungent and bitter in taste and are very light in weight. It consists of two brown, broadly oval, compressed, indehiscent, one-seeded carpels (mericarps), 3-4 mm long and 2-3 mm broad, with a thickness of 1.0 mm. They have five ridges, the three dorsal ridges being inconspicuous, brown, and the other two being lateral, yellowish and wing-like. The chromosome number is 2n=22. The seeds of the Indian dill are especially bold (large and good looking). They also tend to be longer and narrower in shape and their oil has a somewhat different profile.

Climate

Dill is quite a hardy plant and may thrive in much cooler climates, provided it finds a warm situation. It is a long day, cold weather crop, growing during the *rabi* season in the Indian plains. Strong winds and heavy rains during the flowering period may be injurious and extreme heat during maturity reduces the oil yield to practically nothing.

Soil

Dill grows well in any good garden soil with high fertility, sandy loam soils with good drainage being most suitable. Light sandy or heavy clay should be avoided. It can be successfully grown in soils with a pH upto 8.6.

Varieties

'Haldwani selection' and 'English stock' are the varieties recommended for cultivation in India. Among the foreign varieties, 'Gribovskii' is recommended for early sowing, and high herbage and seed yield. 'Lesnogorodski' has best flavoured leaves and the highest oil content in the leaves and seeds, besides 'K-299', 'K-253', 'Mammoth Long Island' and 'Herkules' which are the other varieties of the crop.

Cultivation

Dill is propagated by seeds. The seeds are usually sown in March-April or in mid-October by drill or broadcast, in well-prepared land supplied with FYM @ 19 t/ha. The recommended seed rate is 3 kg/ha, sown in rows 60 cm apart and later thinned to a 30 cm spacing after about 30 days from sowing when they are 13 cm high. The seeds have a viability of about three years and take 7-9 days for germination, when sown during the month of October, with 96 to 98% germination.

If the crop is sown late, the vegetative phase in reduced, which seriously lowers the yield of the seed crop. A spacing of 45 × 20 cm and sowing in lines 30 or 45 cm apart have also been found suitable for this crop.

Manures and fertilisers

The crop responds well to the application of nitrogenous and phosphatic fertilisers. A dose of 40 kg each of N, P and K/ha is recommended for obtaining economical yields in dill.

Weeding and irrigation

The plant requires frequent cultivation. The plots should be kept free from weeds. Weedicides, Influtalin and Ethafluralin (1.1 kg/ha), Sethoxydim (4.5 kg/ha), Linuron (1 kg/ha), Chlorbromuron (4.5 kg/ha), Thiobencarb (6-8 kg/ha), etc., have been recommended for chemical weed control. Since it is a herbaceous crop, regular irrigation at an interval of about 8 to 12 days, depending on the soil and climatic conditions, is necessary. Generally, 8-9 irrigations are quite adequate for the maturity of the crop.

Intercropping

Dill can be intercropped with *Rauwolfia serpentina* or *Mentha arvensis* or *Salvia sclara*.

Use of growth regulators

Gibberellic acid (50 ppm) sprayed 15 days after emergence and again 10 days later, increases the yield and improves the flavour. Ascorbic acid (50 ppm) treatment is reported to yield oil with a high concentration of carvone and dihydroxycarvone and the lowest amount of dillapiole.

Pests and diseases

Pests

Aphids

They occasionally cause damage and are generally controlled by spraying Malathion (0.5%) 2 to 3 times during infestation.

Diseases

Powdery mildew (*Erysiphe umbelliferum*)

It is the major disease noticed at the flowering and early seedling stages. It can be controlled by spraying Wettable sulphur (0.2%), 2-3 times at weekly intervals. *Fusarium* sp. causes root-rot disease, and this can be controlled by phytosanitory measures, seed treatment with

Agrosan @ 3 g/kg of seed and by foliar spray of Bavistin (0.1%). *Systola albipensis* has been reported to infect the seeds. The plants are also reported to exhibit mottling, leaf necrosis, dwarfing and malformation due to viral infections.

Harvesting

Harvesting should be properly timed, since the quality of oil depends upon the state of maturity of the herb and the fruits. The plants sown in mid-October will be ready for harvest by the end of April; i.e., the plants take about five and a half to six months for maturity. The main axis of the stem ends in a large-sized compound umbel, which is axillary in position, or in branches. The crop continues to produce umbels on the tertiary branches and these are at various stages of maturity, at harvest time for a seed crop. The crop is prone to shedding on maturity and, therefore, should be carefully collected at the right time of maturity.

Generally, for obtaining dill oil, harvesting is done when the crop is about 3½ months old and is in the milky, ripening seed stage and the most advanced fruits are turning brown. After harvest, the plants are allowed to wilt in the field for 1-2 days and then distilled. When the crop is grown for its fruits, it is harvested when most of the fruits are fully developed (ripe), but still green in colour. The fruit at this stage contains maximum oil and carvone content and this procedure also avoids the shedding of seeds. To avoid shedding in very dry weather, harvesting is preferably done early in the morning, when the plants are still damp with overnight dew. The harvested material is dried in the field till the fruits can be easily threshed. The threshed fruits should be spread in a thin layer and frequently turned over until thoroughly dry. The seeds, after drying and cleaning, are packed in gunny bags and kept in a cool and dry place. Dill has a shelf-life of 6-9 months.

Yield

The herbage yield is about 2.5 to 3.0 t/ha, which on distillation gives about 18-20 kg of oil containing 30% of carvone, while the seed yield is about 5 to 7 q/ha.

Distillation

The essential oil is extracted by steam distillation from the herb, including the immature fruits and also from the mature and separated fruits. The two oils, however differ in composition, odour and flavour. The herb or 'weed' oil (isolated from the above-ground parts, including the unripe seeds) is preferred by the food industry, because of its more characteristic dill herb flavour. The seed oil, with its high carvone content, resembles the oil of caraway. During recent years, dill-herb oil has largely replaced the fruit oil for flavouring and seasoning purposes.

Value-added products

The value-added products of dill are dill-seed powder, dill-seed oil, dill weed (dill leaves), oleoresin dill, etc.

Fennel

Plant Profile

Family	:	*Apiaceae*
English name	:	Fennel
Indian name	:	*Saunf*, *Sonp* (Hindi), *Pan*, *Muhiri*, *Mauri* (Bengali), *Variari* (Gujarati), *Badi sompu* (Kannada), *Perum-jeerakam* (Malayalam), *Badishep* (Marathi), *Saunf* (Punjabi), *Madhurika* (Sanskrit), *Shombei* (Tamil), *Sopu*, *Pedda-jilakara* (Telugu)
Species	:	*Foeniculum vulgare* Miller ; subspecies: *capillaceum and dulce*
Distribution	:	India, Africa, Asia, Bulgaria, Southern Europe, France, Germany, Hungary, Italy, Morocco, Romania, Russia, Spain
Uses	:	Culinary purposes, Drugs, Flavouring, Perfumery

ennel (*Foeniculum vulgare* Miller), commonly known as *Saunf* or *Badi saunf,* belongs to the family *Apiaceae*. It is cultivated mainly for its seeds from which the spice oil is obtained. The main constituent of the oil from the fruits of cultivated *F. vulgare* is trans-anethole. Fennel oil of good quality contains 50-60% trans-anethole. The other constituents are: d-2 fenchone, methyl chavicol, 2-pinene, camphene, d-2 phellandrene, dipentenes, β-pinene, terpinene, limonene, 1,8-cineole, α-terpinene, linalool, cis-anethole, anisic aldehyde, anisic ketone and foeniculin. The presence of a toxic ketonic constituent has been recently reported. Indian fennel oil contains over 70% anethole and 6% fenchone. The sweet or Roman fennel fruits contain anethole, phellandrene and d-limonene. The high percentage of anethole (upto 90%) and the absence of fenchone are responsible for its delicate sweet odour and flavour. Anisaldehyde and anesic acid are also reported to be present.

The fruits contain volatile oil. The percentage of oil is lowest in the fruits of Indian origin (0.7-1.2%) and highest in the fruits of Eastern Europe (4-6%). They also contain fixed oil (9-13%), pentosan and pectin. Starch is present in small quantities. Besides this, trigonelline and choline are also present. The percentage of ash does not exceed 12%. The fruits also contain iodine (C:88 g/100 g), vitamin-A (139 IU/100g), thiamine, (36r/100g), riboflavin, (36r/100g), niacin, (038mg/100g) and ascorbic acid (9.31 mg/100g). Traces of aluminium, barium, lithium, copper, manganese, silicon and titanium are also present.

The plant is pleasantly aromatic and is used as a pot herb. The leaves are used in fish sauces and for garnishing. The leaf stalks are used in salads. Medicinally, the leaves are regarded to have diuretic properties while the roots are purgative. The dried fruits are extensively used for flavouring soups, meat dishes and sauces, bread rolls, pastries and confectionery. They are also used for flavouring liquors and pickles. The fruits are aromatic, stimulant and carminative; they are considered useful in treating diseases of the chest, spleen and kidney. They are also employed as a corrective for less pleasant drugs, particularly senna and rhubarb.

Fennel oil is largely used as a flavouring agent in culinary preparations, confectionery, cordials and liquours. It is useful in infantile colic and flatulence. It also checks griping in purgatives and is used as a germicide against hookworms. It is also used in scenting soaps in addition to perfumery. The residue left over after the distillation of the essential oil from the seeds is used as feed for cattle. The seed contains 14-22% proteins and 12-16% fat.

Origin and distribution

The plant is a native of Southern Europe, Asia and Africa. It is widely cultivated throughout the temperate and subtropical regions of the world, particularly in France, Italy, Bulgaria, Morocco and Spain for its aromatic fruits used as a culinary spice, and the extraction of essential oil. In India, it is mainly grown as a garden crop or as a small-scale crop during the winter, mainly in the states of Rajasthan, Maharashtra and Gujharat and also in some parts of Punjab, Uttar Pradesh, Madhya Pradesh, Haryana and Andhra Pradesh. In Karnataka, it is grown on a small scale in Belgaum and Dharwad districts. The area under this crop is reported to be about 20,610 hectare and production around 30,460 tonnes. Inspite of having a number of industrial and medicinal uses, and a good foreign exchange earner, at present the crop is not being exploited commercially.

The major types of fennel in the world market are Indian fennel, Egyptian fennel and Chinese fennel.

Description of the plant

Foeniculum is a small genus of annuals, biennials and perennial herbs. The biennials and perennials have firm and round, greenish-blue, finely straighted stem, 150-200 cm high. There are 2-4 leaves which are pinnate, the segment is always very narrow and often filiform. The flowers in large numbers are found in the second year of growth; these are small and yellow. The fruits at first are bluish, then become brownish-gray. All the parts of the plant are aromatic. Different races of fennel exist with fruits that are sweet or bitter or acidic in taste. The flowers are small, yellow and borne in compound terminal umbels. The fruits are oblong, cylindrical, 6-7 mm in length, slightly curved, greenish or yellowish-brown in colour, 5-ridged, with prominent furrows with oil glands.

Climate and season

It is cultivated throughout India at altitudes upto 2000 m. It requires a fairly mild climate and is cultivated as a cold weather crop in parts of North India. It does not succeed so well in South India, except at high elevations.

The time of sowing is during October-November in the plains and March-April on hills. Under Bangalore conditions, the crop can be grown as a winter season crop during the months of September, October and November.

Soil

Fennel grows in any good soil, but thrives best in rich, well-drained loam or black sandy soil containing sufficient lime. Any soil which has a high salt content in it and is prone to water-logging is unsuitable for its cultivation.

Species/varieties

There are a large number of varieties and races differing in size, odour and taste of the fruits, and they are hardly distinguishable from one another. They are regarded as races, varieties or subspecies of *F. vulgare*. The varieties which yield commercially important volatile oils are reduced generally to the subspecies *Capillaceum* and placed under the distinct varieties: *vulgare* giving bitter fennel and var. *Dulce* producing sweet fennel oil. The variety *dulce* is mainly cultivated in India, while *vulgare* is cultivated in Russia, Romania, Hungary, Germany, France, Italy and in South Europe. The genetic resources holding in this crop in India is 420.

PF-35, S-7-9, Guj. Fennel-1, Guj. Fennel-2, RF-125, RF-101, Rajendra Sourabha, Azad Sauf–1, Pant Madhurika and CO-1 are some of the varieties of fennel released through the All India Coordinated Research Project on Spices. The varieties S-7-9, PF-35 and Guj. Fennel-1 are of long duration and can be grown as a transplanted or directly seeded crop and give 20% more yield than the local variety. The local variety, UF–32 (Lucknow type) is identified as best for edible purposes. The yield potential of the released varieties is as follows.

Variety	Grain yield (kg/ha)	States of cultivation
S-7-9	1000	Gujarat
PF-35	1280	Gujarat
Guj. Fennel-1	1700	Gujarat
Guj. Fennel-2	1940	Gujarat
CO-1	570	Tamil Nadu

Cultivation

Propagation

It is propagated readily by seeds, but can also be grown by root or corm divisions. The seeds are broadcasted or line-sown by shallow drills. Soaking seeds prior to sowing improves germination. The seeds can be sown in a nursery bed and later the seedlings are transplanted to the field when they attain 8-10 cm height. For transplanted crop, seeds are sown in the nursery during the months of May-June and seedling of 1½ – 2 months are transplanted in the main field in August-September.

Land preparation and sowing

The field is brought to a fine tilth by continuous ploughing and laid out into beds of convenient size along with the irrigation channels. It is manured with 25 tonnes of FYM along with the basal dose of nitrogen, phosphorus and potash. The seeds are either broadcasted or drilled in rows at a 45-60 cm spacing for rain-fed and 40-80 cm apart for irrigated crops. About 8-10 kg seeds per hectare are required for drilling and 3-4 kg/ha for transplanted crop. When the seedlings are 8-10 cm in height they are thinned in rows at 10 cm in the rain-fed and 40-45 cm in irrigated crop.

Manures and fertilisers

A fennel crop yielding 1113 kg/ha is reported to require 27 kg N, 11.2 kg P_2O_5, 21 kg K_2O, 18.9 kg Ca, 6.1 kg Mg, 290 g Fe, 60 g Mn, 67 g Zn and 21 g Cu per hectare. The application of 30-100 kg of N, 25-60 kg P_2O_5 and 25-90 kg of K_2O results in good seed and oil yield. Of this, half of the nitrogen and the entire quantity of phosphate and potash are applied as a basal dose while the remaining half of the nitrogen is given as a top-dressing when the crop is about a month old. The application of micronutrient zinc (0.3%) and boron (0.1%) as foliar spray has been recommended.

Irrigation

The field is irrigated immediately after seeding if there are no rains. Initially, the irrigation upto one month is given at 3-4 day intervals and thereafter at weekly or 15-20 day intervals has been found to improve the yield considerably.

Irrigation at IW/CPE ratio of 0.9 from the day of sowing and thereafter at 15 days' intervals (9 irrigations) is economical in Gujarat.

Inter-culture

As the crop is delicate, frequent weedings at initial stages of growth is essential. Subsequent weedings may vary with the situation. About 3-4 weedings may be enough for the entire life cycle of the crop. A pre-emergence application of Pendimethalin @ 1.0 kg/ha with one hand weeding 50 days after sowing controls the weeds effectively.

Inter-cropping

Intercropping fennel plus radish is found to be remunerative. Fennel is also grown as a mixed or inter-crop with chilli.

Pests and Diseases

Pests

Thrips (*Hercothrips indicus*) are reported to cause considerable damage to the crop, which can be controlled by spraying 0.15% Dimethoate. Semilooper (*Tysanoplusia orichalcea*) and *Dolycoris indicus* are the other pests of the fennel crop.

Diseases

Not many diseases and pests are reported to attack this crop. The leaf blight (*Ramularia foeniculi* and *Alternaria tenius)* and mildew (*Erysiphae polygoni)* are among the other diseases that attack this crop. They can be controlled by spraying 0.2% Diathane M-45 and 0.3% Wettable sulphur, respectively. Collar-rot is a soil-borne disease, and the spread of the disease can be checked by drenching the soil with Bordeaux mixture (1%) or Copper oxychloride (0.2 or 0.3%). The fungi *Aspergillus* and *Penicillium* are reported to damage the seeds during storage. Sugary disease is caused by *Sclerotinia sclerotiorum*; the disease is spread by aphids. For vector control, spraying with Phosphomedon (0.2%) or Dimethoate (0.03%) or Quinolphos/DPVPC (0.05%) commencing from the 50% emergence of flowers and also at the seed development stage is recommended.

Physiological disorder

Fennel is most vulnerable to frost damage at the flowering and early seed-formation stages. The frost damage can be minimised by spraying 0.1% Sulphuric acid solution, irrigating the crop prior to the incidence of frost, using wind-breaks and creating smoke cover in the early morning.

Harvesting and yield

The crop is harvested after 5-6 months of sowing, before the fruits are fully ripe, to avoid shattering. The umbels are harvested at 'half length size of grain yet green stage'. For the production of the chewing type, the crop can be harvested when the grains are at half length size (around 30 days after anthesis). While harvesting, the stems are cut with a sickle and spread out in loose bundles to dry in the sun. The dried fruits (after 4-5 days) are threshed and cleaned by winnowing. The per hectare seed yield in fennel ranges from 20-25 quintals which, in turn, may yield about 35-40 kg of essential oil under irrigated conditions. The dried and cleaned seeds should be stored in jute bags in damp-free, aerated stores. The seeds are cleaned with the help of a vacuum gravity separator and stored in an aerated store.

The fennel fruits (seeds as they are known in commerce) are classified for trade purpose according to their place of origin. Some of the well-known types in India are Bombay, Bihar and Uttar Pradesh. The fennel seeds from Lucknow are considered to be the best for culinary purposes and are higher priced than those from other areas.

Distillation of Oil

To obtain the maximum oil, the fruits are crushed before distillation. The oil obtained from fruits by steam distillation is a pale yellow-coloured liquid with a characteristic taste and odour. Two types of oil are recognised in commerce: (i) Sweet fennel oil from the fruits of var., *dulce,* and (ii) Bitter fennel oil from the fruits of var., *vulgare.*

Fenugreek

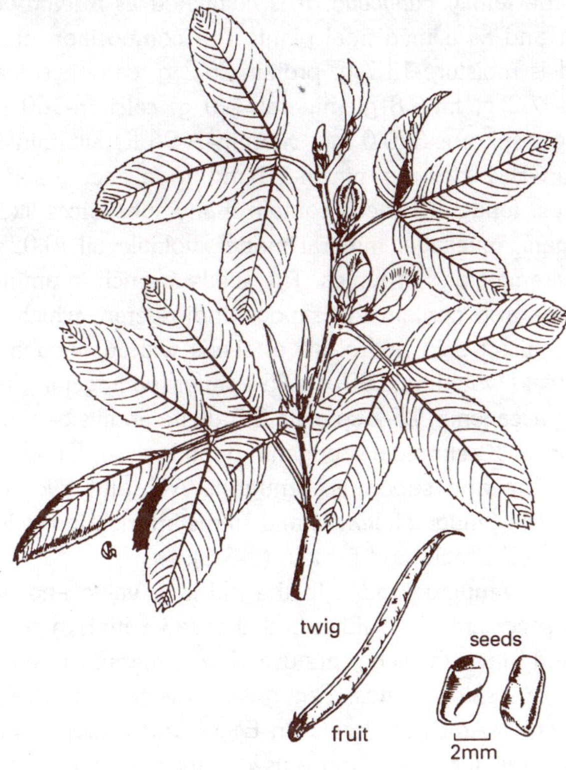

Plant Profile

Family	: *Fabaceae*
English name	: Fenugreek, Greekhayes
Indian name	: *Methi, Methe* (Hindi, Sanskrit), *Menthya* (Kannada), *Ventayan, Uluva* (Malayalam), *Vendayam, Venthiyam* (Tamil), *Mentulu, Menthulu* (Telugu).
Species	: *Trigonella foenum-graecum* L., *T. corniculata*.
Distribution	: Mediterranean region, Central Asia, South- Eastern Europe, Ethiopia, China, Pakistan.
Uses	: Seed spice, culinary purposes, ayurvedic medicines.

Fenugreek, commonly called Greekhayes or *Methi,* is the dried ripe fruit of the pulse *Trigonella foenum-graecum* L. (2n=16) of the family *Fabaceae*. It is cultivated as a leafy vegetable, condiment and as a medicinal plant. The composition of 100 g of dried seed is moisture-13.7 g, protein-26.2 g, carbohydrate-44.1 g, crude fibre-7.2 g, fat-5.8 g, minerals-3.0 g, calcium-160 mg, iron-14.1 mg, phosphorus-370.0 mg, vitamin–A-96 IU, vitamin–B_1-10.34 mg, vitamin–B_2-20.29 mg, niacin-1.1 mg, food energy-333 calories. In a nutshell, fenugreek seeds contain many substances like protein, starch, sugars, mucilage, mineral matter, volatile oil (0.02%), fixed oil, (7%) vitamins and enzymes. The seeds are rich in amino acids.

The seeds also contain the sapogenindiosgenin, which could be isolated by direct acid hydrolysis of seeds without incubation. The diosgenin yield would be 0.9%, its 25-epimer (yamogenin), tigogenin, gitogenin, yuccagenin, 25-2-spirosta-3.5-diene and its b-epimer. They also contain a C_{27}–steroidal sapogenin peptide ester, fenugreekine, 4-hydroxyleucine and saponins, fenugrins A-E and two furostanol glycosides. The fenugreek leaves and stems are rich in calcium, iron, carotene, ascorbic acid and protein (3-5%).

As a spice, fenugreek adds to the nutritive value and flavour of foods. The seeds are eaten either boiled or raw with honey in Greece. In the United States the seeds are used in the manufacture of clustrey and in various spice blends. Fenugreek is a popular ingredient of bread known to Arabs as '*hubla*' in Egypt and Ethiopia. In Gujarat and Maharashtra, fenugreek flour is used in the preparation of *chapathi*.

Medicinally, the leaves are refrigerant and aperient and are given internally for vitiated conditions of *pitta.* The seeds are used for colic, flatulence, dysentery, diarrhoea, dyspepsia with loss of appetite, dropsy, fever, vomiting, anorexia, cough, bronchitis, enlargement of the liver and spleen, rickets, gout, diabetes and colonitis. An infusion of the seeds is a good cool drink for smallpox patients. The seeds are used in India to induce lactation during the post-natal period. The seeds also find application in the synthesis of sex hormones as oral contraceptives. The powdered seeds are used in veterinary medicines. An aqueous extract of the seeds possesses antibacterial property.

The seeds are also used in hair tonic preparations. A mixture of fenugreek and cotton-seed powder fed to milch cattle will increase the flow of milk. The seeds are good for the elimination of bad breath and body odour. The flowers are also used for flavouring. Fenugreek paper paste, developed at the Cardamom Research Institute (Spices Board) is used for the coating of bamboo mats, for drying black pepper, and for yielding better quality and more hygienic paper.

Origin and distribution

Fenugreek is indigenous to countries bordering the eastern shores of the Mediterranean, extending to Central Asia and Southeastern Europe. An independent origin also exists in Ethiopia. The genus *Trigonella* comprises 50 species, mostly of Mediterranean and Oriental origin.

It is grown in India, parts of North Africa, Argentina, Southern France, China, Pakistan, Morocco and Lebanon. Fenugreek is the third largest seed spice in India after coriander and cumin. In India, Rajasthan (84%) and Gujarat (15%) are the major producing states followed by Uttar Pradesh (1%). It is also cultivated in Andhra Pradesh, Madhya Pradesh, Tamil Nadu, Haryana, Maharashtra and Punjab. The total area in India is around 92990 hectares and the production 100780 tonnes per annum. The major international markets for fenugreek seeds are Saudi Arabia, Japan, Sri Lanka, Korea, Singapore, UAE, the Netherlands and the United Kingdom.

Description of the plant

There are two species of the genus *Trigonella,* which are of economic importance, viz., *T. foenum graecum* or the common *methi* and *T. corniculata* or the Kasthuri *methi*. Fenugreek is an annual herb, 30 to 90 cm tall, and has light green leaves which are pinnately trifoliate. The flowers are papilionaceous and white or yellow in colour. Anthesis takes place between 9 a.m. and 6.0 p.m. peak, at 11.30 a.m. and 12.30 p.m. The stigma becomes receptive 12 hours before the flower opening and remains receptive for about 10 hours after the opening of the flowers. The plants flower in about 30-37 days after sowing and the duration of the flowering phase is 7-8 days. The species is typically self-pollinated and destogamous. The pollen fertility ranges from 95-98% in the unopened flower buds and 67-80% in the open flowers. The plant produces slender beaked pods, 10 to 15 cm long, which mature within 60-70 days of sowing. A pod contains 10 to 20 small, hard, yellowish-brown seeds, which are smooth and oblong, about 3 mm long. It is pleasantly bitter in taste and possesses a distinct flavour of its own.

Climate

Fenugreek can be grown in the tropics and in temperate regions. It is grown from sea level upto an altitude of 2000 m. It requires a moderately cool climate for its proper growth and high yield. It is fairly tolerant to frost, but is most vulnerable to frost damage at the flowering and early grain formation stages. Cloudy weather and high humidity, particularly during active grain-filling period, increases the incidence of aphids and powdery mildew, which adversely affect the yield as well as the quality. It can be grown either as a rain-fed or irrigated crop.

Soil

Fertile and well-drained loamy or sandy or clayey loam soils are best suited for fenugreek cultivation. However, it can be grown in all types of soils which are rich in organic matter, with good drainage. It also tolerates more salinity compared to other leguminous crops. However, saline and acidic soils should be avoided. The optimum pH for its best growth and production is 6 to 7.

Varieties

There are two types of fenugreek: (i) non-scented or *desient*-which is quick growing, upright and bears white flowers (ii) scented–a slow-growing plant which is mainly confined to North India. The salient features of the improved varieties of fenugreek are given below.

RMt 1 (Prabha)

This is a pure line selection from the Nagpur local released in 1989. Its plants are semi-erect, tall and moderately branched with bold, typically yellow grains. It is moderately resistant to root-rot and tolerant to powdery mildew. It matures in 140-150 days with an average yield of 14.7 q/ha. The plants flower in 60 days.

RMt 143

The grains of this variety are bold with a typical yellow colour. It is moderately resistant to powdery mildew. It takes 140-150 days to mature with an average yield of 16 q/ha. It is especially recommended for the heavier soils of the Chittor, Bhilwara, Jhalawar and Jodhpur areas.

CO-1

This is a reselection from TG 2336 of IARI, released in 1982. Its plants are short and green with medium-sized, brownish-orange seeds. It is tolerant to root-rot. It matures in 95 days with an average yield of 6.80 q/ha. It is a quick-growing, dual-purpose type.

ML-150

This hybrid was derived from the cross T8 × T36 and released for cultivation in Punjab in 1995. The variety has more pods, matures early and produces 18.9 and 18.2% higher grain and fodder yields, respectively, compared to T-8.

Prabhat (NLM)

This is the most suitable variety for Gujarat and Rajasthan. It normally yields 15-20 q/ha, but under favourable conditions, it yields as high as 25-28 q/ha.

Rajendra Kanti

This is a mass selection (RM 16) from the Reghunathpur germplasm released in 1987. Its plants are tall and bushy green with medium-sized, golden-yellow seeds and is moderately resistant to powdery mildew, caterpillar and aphids and *Cescospora* leaf-spot. It matures in 120 days with an average yield of 12.50 q/ha. The variety takes 45 days for flowering. It is suitable for inter-cropping in both the *kharif* and *rabi* seasons.

Lam selection 1

This is a selection from the germplasm collected from Madhya Pradesh, released in 1992. Its plants are bushy, green, with medium-sized, golden-yellow seeds. It is tolerant to root-rot, powdery mildew, caterpillars and aphids. It matures in 90 days with an average yield of 7.40 q/ha. It is a dual-purpose, early maturing type and gives a high yield of greens. It takes 35 days to flower.

Hissar Sonali

The plants of this variety are bushy, semi-erect with bold, yellow, attractive grains. It is moderately resistant to the leaf-spot and root-rot complex of diseases. It matures in 140-150 days with an average yield of 19.0 q/ha.

HM 103

The plants of this variety are bushy, semi-erect with bold, yellow, attractive seeds. It is moderately resistant to the leaf-spot disease. It matures in 140-150 days with an average yield of 20.1 q/ha.

HM 57

This variety is late maturing and is recommended for cultivation in Haryana. It gives an yield of 16.0 q/ha.

The other high-yielding varieties released under AICRPS are Gujarat Methi-1, RMt 303, CO-2, UD-44, Pusa Early Bunching, Rajendra Khushaba, UM 34, UM 35 and Pant Ragini. UM-34 and UM-35 are found resistant to the root knot nematode and UD-44 is completely free from the stem gall disease. The mutant lines SPS 22, SPS 23 and

SPS 27 have been found to be free from powdery mildew. The genotypes UM-94 and UM-98 are suitable for cultivation in saline conditions. The total fenugreek germplasm accessions collected, evaluated and being conserved is 944.

Cultivation

Propagation

Fenugreek is best grown as an annual crop from seeds which are sown following the broadcast or the line sowing method. The diploid and auto-tetraploid strains of fenugreek are reported to regenerate complete plantlets from meristamatic cells.

Land preparation

The land should be thoroughly prepared by repeated ploughing and harrowing. Beds of convenient size (3 × 2 m) are prepared at the same time. For an irrigated crop, irrigation channels are made along the alternate rows of beds.

Sowing

The ideal sowing time for fenugreek in Northern India is from the last week of October to the first week of November. In the southern parts, it is grown both in the *kharif* and *rabi* seasons. In the *kharif*, the yield is less than in the *rabi* crop. Sowing should be done from the second fortnight of June to July end, in the *kharif* crop and for the *rabi* crop, first fortnight of October is the best sowing time. If the broadcasting method is followed, the broadcasted seeds are raked lightly to cover them. Line sowing is done by drilling the seeds 30 cm apart in rows with a plant-to-plant spacing of 10 cm. A spacing of 20 × 30 cm is followed in Rajasthan. The seed should not be sown deeper than 5 cm. A seed rate of 15-25 kg for the seed crop and 30-35 kg for the leafy vegetable crop per hectare is sufficient. The seeds are soaked in water for 2 days prior to sowing to enhance germination. The seeds may be soaked in a cycocil solution at 50-100 ppm for improving germination and enhancing seedling growth. Soaking the seeds in a solution of 0.6% EMS is recommended for improved germination and survival of seedlings. The seeds should be treated with *Rhizobium* culture before sowing. The seeds germinate within 6-8 days after sowing. The line sown crop is thinned within the row at the 3 to 4-leaf stage to retain 1-2 plants per hill.

Manures and fertilisers

FYM or compost @ 10-15 t/ha should be added to the soil at the time of land preparation. The recommendation and methods/time of application of inorganic fertilisers differs from place to place. The recommended fertiliser doses are 25:25:50 or 40:60:20 or 50:25:40 or 40:40:0 kg/ha of NPK. At the time of sowing, the entire dose of NPK or NP is drilled in the soil, or half the quantity of N along with the entire quantity of P and K are applied basally and the remaining part of N is applied 30 days after sowing as a top dressing. The quantity of N may be reduced if the soil fertility is high. Two foliar sprays of urea (1%) at about 45 and 60 days after sowing gives a better economic green-leaf yield as well as seed yield per unit area. One can take 2 leaf cuttings before leaving the crop for seed harvest. A crop of fenugreek yielding 1 t/ha of seeds will absorb NPK in the ratio of 2:1:2. The uptake of nutrients like Ca, Fe and Zn is also fairly high compared to other elements. In order to support sustainable, high and profitable yields, not only N but Zn should also be included in the fertiliser schedule.

Irrigation

The first irrigation should be given immediately after sowing. The crop requires about 5-7 irrigations in all. It should be irrigated at IW/CPE ratio of 1.0 and 0.40. The number of irrigations required, of course, depends upon the type of the soil and evapo-transpiration potential prevailing during the season. In Rajasthan, a good fenugreek crop needs about 8 irrigations. The number of irrigations may be reduced to 4-5 in heavy soils with good water-retention capacity. The distribution of irrigations during the growing season should be such that the crop does not suffer any water stress during the pod- and seed-development stages. The maintenance of optimal soil moisture is essential to prevent blossom and fruit drop.

Interculture and weeding

Two hoeings and weedings are enough to keep the crop well aerated and weed free. The first hoeing and weeding is done at the time of thinning 25-30 days after sowing, and the crop is earthed-up. The second weeding is done at 50-60 days after sowing. Tipping of the young shoots is done after 10-30 days of sowing. This is done at a height of 4-5 cm from the ground level, if the crop is grown for leaf. A pre-planting application of Fluchloralin or Pendimenthalin @ 0.75 kg/ha supplemented with hand weeding 50 days after sowing and earthing-up of the crop will keep the weeds under check.

Use of growth regulators

A spray of ascorbic acid at 250-450 ppm improves the plant growth, enhances flowering, seed number and size. GA_3 spray @ 10-100 ppm concentration enhances the internodal length, height and number of leaves. It also induces early flowering. A spray of 0.05% Sodium 2, 3–Iso butyrate just before flower initiation decreases the pollen sterility and improves the yield.

Pests and diseases

Pests

Aphid (Aphis craccivora, A. gossypii, Myzus (Nectarosiphon) persicae and Acrythosiphon pisum)

Aphids cause heavy loss to the crop. They can, however, be easily controlled by spraying a 0.07% solution of any of the insecticides like Endosulfan, Dimethoate (0.3%), Malathion (1.0%), Phosphamidon or Monocrotophos (0.1%). The spray should be repeated if insect reapers or cloudy weather prevails, which favours the insects. The varieties BDJ-11, BDJ-86-59 and PLM-98 are highly resistant to *A. craccivora*.

The leaf eating caterpillars (*Spodoptesa litura*) and pod borers also attack the crop sporadically and could be controlled by spraying 0.04% Quinalphos or 0.05% Malathion.

Diseases

Powdery mildew

This disease is caused by *Leveillula taurica, Oidium* sp., *Erysiphe polygonii* and *E. tryphi*. The initial stage of this disease is marked by the appearance of a whitish, powdery mass on the leaves. If not checked at an early stage, the disease intensifies and the whole plant become whitish with the powdery coating. The seed size and the yield are adversely affected.

The disease can be effectively controlled by spraying the crop with 400 to 500 litres/ha of 0.25% solution of Wettable sulphur or a 0.1% solution Kerathane or Dinacop (0.25%) thrice. The spray should be repeated after 15 days starting from 40 days after sowing. Dusting the crop with 300 mesh Sulphur dust at 20 to 25 kg/ha also controls the disease. Dusting may be repeated if the disease reappears. The varieties Prabha, Rajendra Kranti, RMt-1 and Lam Selection 1 are tolerant and the germplasm ILO-2819, 69-2320 and ILO-69-2335 are moderately resistant to this disease.

Downy mildew

It is caused by *Peronospora trigonellae*. The initial stage of the disease is marked by the appearance of a whitish, downy growth on the lower surface of the leaves during February and March. As the disease advances, the leaves become yellow and fall. The plant becomes stunted and in the severe form, it dies. The disease spreads fast if the temperature rises. The disease can effectively be controlled by spraying the crop with 0.3% solution of any of the copper fungicides, Blitox, Cupramar and Fytolan or Diafoltan (0.02%) or Zineb (0.3%) at 15-day intervals. The spray should be repeated if the disease symptoms reappear or the control is not complete.

Root-rot

This disease caused by *Rhizoctonia solani* appears at the late stage of the crop. The disease, beginning with the wilting and drying up of the leaves, may be reduced through adopting a proper crop rotation schedule and deep summer ploughing. The incidence is severe if legume crops, such as *moth* or green gram in *kharif* is followed by fenugreek in the *rabi*. It should therefore, be avoided. The application of neem cake @ one tonne per hectare in combination with seed pelleting of antagonistic fungi like *Trichoderma viride*, *T. harzianum* with talc formulation at 4 g/kg of seed or drenching Carbendazin (0.1%) once at the initial appearance and again at 30-day intervals or Brassicol (0.1%) will keep the disease under check.

Leaf-spot (Alternaria alternata* and *Cercospora traversiana)

This is a minor disease of fenugreek and can be controlled by spraying Bavistin (0.05%), Mancozeb (0.02%) or Zineb (0.3%).

Damping-off (Humicola fuscoatra, Pythium aphanidermatum* and *Rhizoctonia solani)

The disease incidence is minimized by seed treatment with Benomyl or Carbendazim (3 g/kg of seed) or Carbendazim (0.1%) spray and crop rotation. The disease can also be managed by drenching with Carbendazim or Brassicol.

Rust (Uromyces anthyllidis)

The pathogen has been observed in the form of minute, brown pustules on both the surfaces of the leaves. No control measure is suggested, as the disease is of minor importance.

Methi mosaic wilt virus

The disease is characterised by vein clearing, severe mottling, reduction in leaf size, stunting and wilting of the plants due to virus infestation. The MMWV is related to cowpea and broad bean mosaic viruses. The disease incidence can be minimised by controlling the insect vectors.

Some of the fungi associated with seeds of methi are *Alternaria alternata, Curvularia lunata, Fusarium equisetti* and *Fusarium xysporum*.

Harvesting and Yield

The crop takes 50 to 60 days for flowering after sowing and takes about 80 to 90 days for maturity after flowering, or the crop duration varies from 70 to 150 days depending on the variety. The harvesting should be done when the crop has turned yellow and most of the leaves, except the top ones, have fallen off. The grain will shatter if the harvesting is delayed beyond this stage. The harvested plants are stacked in small bundles. After drying them in the sun for 4-6 days, the grains are separated by beating the bundles on the threshing floor or by using a mechanical thresher. The grains are cleaned by winnowing and then stored.

The average grain yield would be around 10-15 q. Yields as high as 30 q/ha can be obtained under favourable conditions and good management.

Storage

The dried and clean seeds are filled in bags and stored in damp-free, aerated stores. On a commercial scale, the seeds are cleaned with the help of a vaccum gravity separator or a spiral gravity separator. To get good prices and easy marketing, the produce should be graded and stored properly. The seed can be stored for 2 years.

For seed purposes, the seeds could be stored in 700-gauge polyethylene bags for higher viability and vigour with 9% moisture content after treatment with Thiram, at an equibrium relative humidity of 40%.

Grading

Fenugreek seeds are graded as 'Special', 'Good' and 'Fair' and the grade specifications (%) are as follows.

	Special	**Good**	**Fair**
Inorganic foreign matter	0.25	0.50	1.00
Organic foreign matter	0.50	1.50	2.50
Damaged, discoloured and weevil infested	0.50	3.00	5.00
Shrivelled and immature green	1.05	3.00	5.00
Moisture	10.00	10.00	10.00

Value-added products

The value-added products of fenugreek are fixed oil (7%), volatile oil (0.02%) and oleoresin. The fixed oil consists of fatty acids like linoleic, oleic and linolenic acids. It has marked drying properties. The dried oil has a disagreeable odour and a bitter taste. It is insoluble in ether and is golden-yellow in colour. The volatile oil is brown in colour and slightly odourous.

French Basil

Plant Profile

Family	:	*Lamiaceae/ Labiatae*
English name	:	French basil, Common basil, Roman basil, Sweet basil
Indian name	:	*Baburi tulsi, Gulal tulsi* (Hindi), *Kali tulsi, Marua* (Bengali), *Damaro, Nasabo, Sabza* (Gujarati), *Kama Kasturi, Sajjagida* (Kannada), *Niazbo* (Kashmiri), *Tirunitu* (Malayalam), *Marva, Sabza* (Marathi), *Dhala tulasi, Kapur Kanti* (Oriya), *Furrunj Mushk, Baburi tulasi or Niyazbo* (Punjabi), *Munjariki, Surasa, Varvara* (Sanskrit), *Trinirupachai* (Tamil), *Bhutulasi, Rudrajada, Vepudupacha* (Telugu)
Species	:	*Ocimum basilicum* Linn.
Distribution	:	India, Africa, South America (Brazil), Asia
Uses	:	Cosmetics, Culinary purposes, Drugs, Flavouring, Perfumery

Of the 150 species of *Ocimum*, the sweet or French or common or Roman Basil (*O.basilicum* Linn.) belonging to the family *Lamiaceae; Labiatae* is the most important species being utilised as a source of essential oil (Sweet Basil Oil). The volatile oil contains d-linalool and methyl chavicol as the major components, with the former upto 55% and the latter about 70%, depending on the sources. Other components include methyl cinnamate, which has been reported to be a major component (28%) of a variety of sweet basil. Some of the others are 1,8-cineole, eugenol, borneol, ocimene, geraniol, anethol, 10-cadinols, β-caryophyllene, terpineol, camphor, 3-octanone, methyleugenol, safrole, sesquithujene and 1-epibicyclosesqui-phellandrene. There are great variations in the concentration of these components in the volatile oil from different sources. The other constituents present in sweet basil include protein (14%), carbohydrates (61%), fixed oil, cellulose, pigment, mineral matter and relatively high concentrations of vitamins A and C.

Sweet basil is used for flavouring numerous foods like soups, meat pies, fish, certain cheeses, tomato cocktails, eggplant, zucchini, cooked cucumber dishes, cooked peas, squash and string beans, liquors, etc. Chopped basil is sprinkled over lamb chops before cooking, *pizza* topping, *spaghetti* sauce or *macaroni* and cheese casseroles.

The herb, leaves and seeds are used medicinally in indigenous systems of medicine and homeopathy. Thus, an infusion of the plant is used for cephalalgia, gouty joints and as a gargle for bad breath. The juice, obtained from the leaves, gives relief in irritation of the throat, earache and ring worm. The herb is aromatic, a diaphoretic, stimulant, carminative and expectorant. It is also considered by some to be anti-emetic, anti-diarrhoetic and antihelmintic. The seeds are used internally in gonorrhea, chronic diarrhoea, dysentery, in constipation and piles.

The oil is extensively employed in several European countries and the USA for flavouring of foodstuffs, confectionery goods, condiments and in toiletry products such as mouthwashes and dental creams. It also finds a prominent place in the flavouring of baked foods, such as spiced meats, sausages, tomato pastes, various kinds of sauces, catsups, fancy vinegars, pickles, ketchups and beverages. In the perfumery industry, the oil is used for compounding certain popular perfumes, notably jasmine blends.

Basil oil possesses insecticidal and insect-repellent properties. It is effective against house flies and mosquitoes. It is also bactericidal.

The seeds of the plant are odourless with an oil which is slightly pungent in taste. They contain mucilage (9.3%) which on hydrolysis yields uronic acid, glucose, xylose and rhamnose. The seeds contain a drying oil with the following fatty acid composition: palmitic-7.0, stearic-0.2, olezeic-11.0, linoleic-60.0 and linolenic acid 21%. The unsaponifiable fraction is reported to contain β-stiostarol, oleonolic acid and ursolic acid. The aqueous extracts of the seeds are active against gram-positive bacteria and mycobacteria.

Origin and distribution

The genus *Ocimum* is well represented in the warmer parts of both the hemispheres from sea level to 1800 m. The maximum number of species are from the tropical rain forests of Africa. The main centres of diversity in this genus are Africa, South America (Brazil) and Asia. In India, the cultivation of French basil is mainly concentrated in Uttar Pradesh in an area of about 6000 hectares.

Types of French Basil

In view of the great diversity, the various species and varieties have been classified, in accordance to their chemical composition and geographical sources, into 4 major types.

1. European or sweet basil

The oil obtained from this type mainly consists of methyl chavicol and linalool, but no camphor. This group comprises the French and American sweet basil oils which are in demand because of their high quality and their finest odour. It is distilled in France, Italy, Bulgaria, Egypt, Hungary, South Africa and occasionally in the United States.

2. Reunion basil

The main constituents of the oil obtained from this type are methyl chavicol and camphor, but not linalool. The oil is of a lower quality. It is distilled in the Reunion Island, Comoros, Malagassy Republic, Thailand and occasionally in the Seychelles.

3. Methyl cinnamate basil

The oil of this type principally consists of methyl chavicol and linalool. Methyl cinnamate is also present in substantial amounts. It is distilled in countries like India, Bulgaria, Sicily, Haiti, Guatemala and a few African countries.

4. Eugenol basil

Eugenol is the principle constituent of the oil of this variety. It is distilled in USSR and some North African countries, including Egypt and Morocco.

Nine species are recorded in Eugenol basil from India, of which three are exotic. The more important of these species are:

a) *Ocimum basilicum* Linn (2n=48) (Eng - French or Sweet Basil) It has the following subspecies and varieties.

1. *O. basilicum* Linn. subsp. *minima* Danert (2n=48) (Syn. *O. minima* Linn).
2. *O. basilicum* Var. *majus* Benth
3. *O. basilicum* Var. *difforme* Benth (Curly - leaved basil)
4. *O. basilicum* Var. *purpurascens* Benth (2n=48) (Violet red basil)
5. *O. basilicum* Var. *glabratum* Benth (Common white basil)
6. *O. basilicum* Var. *pilosum* Benth (Syn. *O.pilosum* Roxb)
7. *O. basilicum* Var. *crispa*
8. *O. basilicum* Var. *thyrisiflora* (2n=49)
9. *O. basilicum* Var. *darkopal*

b) *O. americanum* Linn. (Syn. *O.canum* Sims.) (Hoary Basil)

c) *O. gratissium* Linn. (Shrubby basil)

d) *O. kilimandscharicum* Guerke (camphor basil)

e) *O. sanctum* Linn. (Sacred or Holy Basil)

f) *O. viride* willd (Fever plant of Sierra Leone)

Of the above species, *O. basilicum* Linn. alone is cultivated in India on a commercial scale.

Description of the plant

O. basilicum Linn. occurring in nature is a tetraploid (2n=48). It is a large herbaceous, erect, strongly aromatic annual herb, about 30-90 cm high. The leaves are opposite, ovate-lanceolate, 3.75-5 cm long, entire or deviate, with glabrous surfaces. The petioles are very slender and usually slightly hairy. The flowers are 0.72–1.25 cm long, borne in long terminal racemose inflorescences, simple or much branched. The calyx is 5-toothed, the upper tooth is rounded, shorter than others, with 2-lower teeth ovate, lanceolate with brittle points, and 2 lateral shorter than the lower. The corolla is 0.72 - 1.25 cm long, white, pink or pale-purplish, 2-lipped, with a short tube with the upper lip equally 4-lobed, and four stamens that are protruding. The ovary is bicarpellary, syncarpous, bilocular, with a bifid stigma; the fruit is a nutlet and mucilagenous when wet.

Climate

The plant is susceptible to frost. The crop growth is poor in areas which receive heavy and continuous rainfall. In such areas, the crop could be raised provided they are well established prior to the onset of the monsoon and the rain water does not stagnate in the field.

Soil

Moderately fertile, well-drained loamy or sandy loam soils are considered ideal for cultivation. Clayey, water-logged soils are unsuitable for the cultivation of this crop.

Season

The crop can be grown annually from the middle of February to the end of September and also during the *kharif* season in plains of North or South India. In the hilly areas of North India, the crop can be grown during the *kharif*.

Varieties

RRL-011 is an improved strain isloated from the segregating generation. The oil of this strain is sweet smelling and is a better substitute for French Basil oil. It gives an herbage yield of 50 t/ha and an oil recovery of 320 kg/ha. The green herb contains 0.45% oil with linalool (40%) and methyl chavicol (35%) as its main constituents.

'Vikarsudha' is a hybrid variety released from CIMAP, Lucknow. It was developed through hybridisation between an exotic basil from Australia (EC 331886 - CSIRO No.L 6323) and a local adaptive landrace referred as Badaun local. The variety 'breeds true', grows to 75-90 cm in height and matures at about 75-90 days after transplantation. The spacing recommended for this variety is 45 × 30 cm. The variety is day-neutral (photoinsensitive) and can be cultivated throughout the year, except in areas experiencing severe winter where its growth can be retarded. The leaf to stem ration (t/ha) is 75 and it yields 3.73 t/ha herb. The oil content is 0.7% and the oil yield is 261 kg/ha.

'Kusumohak' is a recently released variety from CIMAP, Lucknow, from the seed-raised progeny of the introduced strain from Argentina. The plant has higher linalool (46%) and moderate chavicol (38%) content. The oil yield is 134 kg/ha.

Cultivation

Propagation

The crop is raised through seeds, but direct sowing in the field is not advised. The seedlings are first raised in the nursery and then transplanted into the field. About 250 g seeds are sufficient to raise seedlings for planting 1 hectare of land.

Nursery raising

The nursery area should be well prepared. Well rotten FYM and leaf mould, each at the rate of 1 kg per sq.m, is applied and mixed well into the soil. Beds of 1 m x 4 m, with irrigation channel systems are laid out and the seeds (10-15 g per bed) mixed with fine sand or wood ash are sown in lines, 6 cm apart or broadcast over the beds. The seeds are then covered with a thin layer of fine soil or FYM. The nursery beds are watered immediately after sowing and regularly thereafter. In the plains of North India, the seeds may be sown in the nursery in the months of April-May or August-September, and in the hilly regions in April. The seeds start germinating 3 days after sowing and the germination will be complete in about 10 days. The nursery should be kept weed free. The seedlings will be ready for transplanting after attaining a height of 6-10 cm.

Land preparation

The land is well prepared with 2 or 3 ploughings until a fine tilth of soil is obtained. FYM, if available, may be applied @ 10-12 t/ha before the 2nd or 3rd ploughing.

Planting

The seedlings may be transplanted 40 cm apart, in rows 60 cm apart, preferably in the evening. Cloudy weather and a fine drizzle are considered ideal conditions for transplanting.

Manures and fertilisers

In order to raise a good crop, it is advisable to apply 20 kg N, 40 kg P_2O_5 and 40 kg K_2O per hectare as a basal dose before planting. Nitrogen @ 40 kg/ha should also be applied as a top-dressing in two equal split doses. In order to get increased herbage (68-84% increase) and almost double the oil yield, it is recommended to apply NPK @ 120:100:100 kg/ha. In a study conducted at UAS, GKVK, Bangalore, the maximum herbage and oil yields were obtained by the application of 160 kg N per hectare in two equal split doses along with 80 kg P_2O_5. This crop is tolerant to higher concentrations of copper and zinc and is susceptible to cobalt and nickel.

Irrigation

When it is raised as a summer crop, irrigation is required once a week. With the onset of the monsoon, the rains meet the water requirements of the crop fully, till Setpember. The crop needs irrigation once or twice a month thereafter.

Weeding

Normally, the seedlings are well established in the field one month after planting. The first weeding is done at this stage and a second 4 weeks later. No further weeding is required, as the plants become bushy thereby suppressing the weeds. In large plantations, the expenditure on weeding can be minimised by the use of a cultivator drawn by a tractor.

Inter-culture

Hoeing is done two months after planting. The crop may also be earthed-up at this stage.

Pests and diseases

Pests

The larvae of the leaf eating caterpillar (*Pycharmon caberalis*) cause serious damage to the plants by sticking to the undersurface of the leaves, folding them from midrib length-wise, and webbing them. Finally, the infected leaves fall off. A spray of Thiodon or Malathion (2 ml/lit) with water reduces the menace from this insect. The other pests reported on this crop are aphids (*Aphis gossypii*), white fly (*Dialeurodes* sp.) and tingid bug (*Monanthia globulifera*).

Diseases

The leaf-spot disease on this crop appears as small, water-soaked spots which later turn brown. The disease is caused by *Corynespora cassicola* (Berk and Curt) Wie. Scab disease caused by *Elsinoe arxii* show little defoliation with puckering, dropping of the leaves and distortion of the tender twigs.

The blight of basil caused by *Alternaria sp.* is an important disease which starts with a chlorotic appearance of the leaves which turn purple and finally black. Ultimately, the leaves are shed. *Colletotrichum capsici* also causes leaf blight. The disease appears as small chlorotic spots on the leaves which enlarge rapidly. The older leaves appear to be more susceptible to the infection. The blight disease could be controlled by spraying the crop with 0.2% Dithane Z-78 or Dithane M-45 once or twice.

Basil wilt is caused by *Fusarium oxysporum* at all stages of growth. The disease is more pronounced in the rainy season. Initially, the leaves wilt but soon the disease spreads to the whole plant. To control this disease, the seedlings should be dipped in a solution of Tafason or Agallol at the time of planting.

Harvesting and yield

The crop is harvested 8-12 weeks after planting, at the stage when the plant is in full bloom and the lower leaves start turning yellowish in colour. Harvesting is done with the help of sickles. While harvesting, the whole herb or flowering tops are harvested. Corresponding to the part harvested, two grades of oil are obtained, i.e., herb oil or flower oil. The flower oil has a superior note. To get a high quality of oil only the flowering tops are harvested.

Normally, 3-4 floral harvests are obtained in this crop. The first harvest is taken when the plants are in full bloom and the subsequent harvests are done after every 15-20 days. The whole plant is harvested finally, leaving about 15 cm from the ground level for regeneration of the crop. The harvested produce is allowed to wilt in the field for 4-5 hours so as to reduce the moisture and also the bulkiness.

Floral harvest yields about 3-4 tonnes of flowers and the final harvest of the whole herb is about 13-14 tonnes per hectare.

Packaging and storage of dried leaves

The dried leaves are generally packed in tin containers, which are kept closed. The enzymatic browning of basil has been attributed to oxidation of phenolic compounds by polyphenol oxidase, that are naturally present therein.

Distillation of oil and essential oil yield

The oil of Sweet basil is obtained by hydro-distillation or steam distillation of young inflorescences or the whole herb. The extraction of oil through steam distillation is better than extracting oil by hydro-distillation, as it takes less time and effects better recovery of the oil. While the extraction of oil by hydro-distillation, which is carried on in a direct fire is cheaper and more handy for small plantations, steam distillation is preferred for large plantations. Steam distillation takes about 1-1½ hours to exhaust a charge completely.

The inflorescence contains 0.4% oil and the whole herb contains 0.10 to 0.25% oil. The sweet basil oil yield is about 30-35 kg from the flowers (flower oil) and 18-22 kg from the whole herb. If only the whole herb is harvested and distilled, the oil yield will be about 30 kg per hectare.

Adulterants of sweet basil oil

This oil is frequently adulterated with the much lower-priced Reunion basil oil.

Garlic

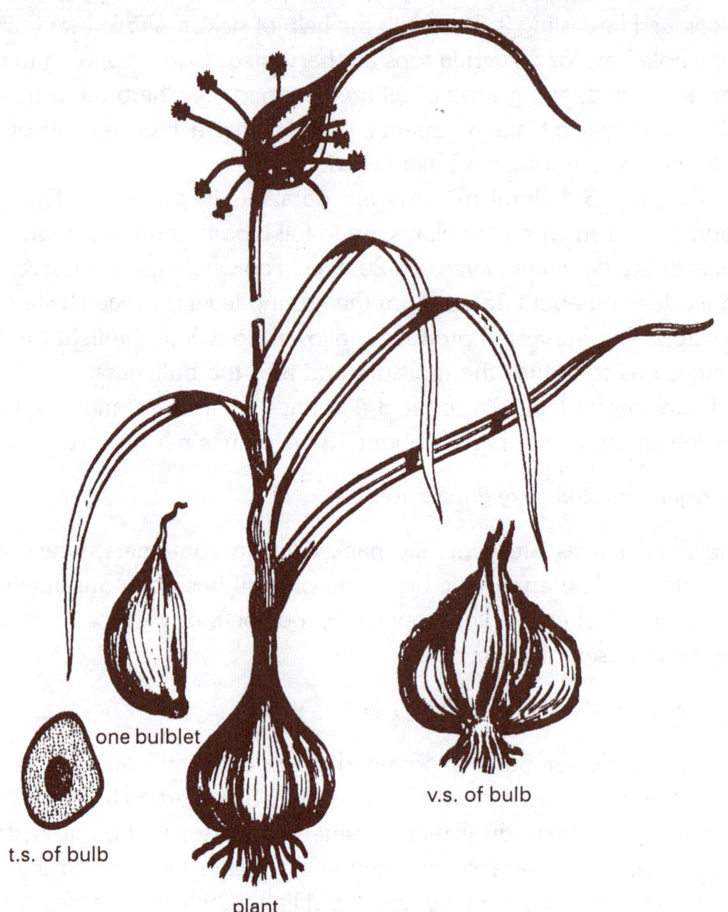

Plant Profile

Family	:	*Liliaceae/Alleaceae*
English name	:	Garlic
Indian name	:	*Lasum, Lassan, Lahsun* (Hindi), *Bellulli* (Kannada), *Vellulli (*Malayalam), *Lashuna* (Sanskrit), *Ullipundu, Vellaippundu* (Tamil), *Velluri* (Telugu).
Species	:	*Alium sativum* Linn., A.Cepa
Distribution	:	Central Asia and Mediterranean region
Uses	:	Food flavouring and Seasoning.

arlic (*Allium sativum*) belonging to the family *Alleaceae,* is an important bulbous vegetable, spice or condiment with medicinal value, commonly used throughout the world. Garlic is a multiple or compound bulb consisting of 10-20 bulblets or segments called cloves, which are covered by a parchment membrane. Garlic is a rich source of carbohydrates, protein, and phosphorus (Table-1). The ascorbic acid content is reported to be very high in green garlic (13 mg/100g) and garlic also contains an amino acid called alliin. On crushing the bulb, the enzyme allinase converts alliin into allicin. Diallyl thiosulfinate (allicin) is the major flavour component in fresh garlic. This was found to undergo non-enzymatic degradation leading to formation of mono-, di-, and trisulfides. It is reported that garlic contained about 0.1% volatile oil. The volatile oil consists of 29 volatile components and the chief constituents of the oil are diallyl disulfide (60%), diallyl trisulfide (20%), ally-propyl disulfide (16%), a small quantity of disulfide, and probably diallyl polysulfide. The main flavour constituents of a water extract of garlic bulbs were sulfur-containing compounds, such as alliin-S-methyl-L-cysteine sulfoxide and γ-L-glutamyl-S-ally-L-cysteine. The bulbs contained a higher percentage of reducing and non-reducing sugars and proteins than did the other two plant parts.

Garlic is chiefly used for flavouring and seasoning vegetable and meat dishes. It is used to flavour the side dishes like *sambār*, *rasam* etc., in South Indian food and also used to prepare soups, pickle and chutneys. It has many medicinal properties and has long been used as a remedy for various ailments like digestive disorders. Garlic preparations are given in whooping cough and other lung diseases, and specifically for sore eyes and ear-ache. Garlic is a good carminative and is recommended to be inhaled for the treatment of pulmonary tuberculosis. It is also used in the treatment of sterility. Spray-dried garlic products are also available. Dehydrated garlic in the powdered or the granulated form is sometimes used in place of fresh bulbs.

Garlic is a gastric stimulant and helps in the digestion and absorption of food. Alliin, which has a hypocholesterolemic property, is present in the aqueous extract of garlic and reduces the cholesterol concentration in human blood. A significant reduction in serum cholesterol levels was observed when 50 g of raw garlic was fed daily to 10 volunteers for one month. Several reports are available on lowering of cholesterol by feeding garlic to animals in experimental situations. A blood sugar-lowering effect has also been reported. This is ascribed to allicin and related disulfide-containing compounds. However, excess consumption may also lead to anemia. Therefore, the consumption of only 7-10 g of garlic is recommended a day.

Garlic leaf extract has fungicidal effects on young fungi. The crude extract of garlic is found to be quite effective against gram-negative as well as gram-positive bacteria. It inhibits the growth of some bacterial cultures that are resistant to commonly used antibiotics. About 10% dilution of the aqueous extract of garlic completely inhibited the growth of these bacteria. In addition, reports showed that garlic extract exhibited promising antibacterial activity against several clinical strains. Garlic possesses insecticidal and repellent properties as well. A formulation containing 1% garlic extract gave protection to a volunteer against mosquitoes and black-fly for 8 hours. It has been reported that the extracts possess nematicidal and larvicidal properties.

Table 1 *Composition of Garlic and its Products*

Nutrients	**Fresh peeled garlic**	**Dry garlic cloves**	**Dehydrated garlic powder**
Moisture (%)	62.80	62.00	5.20
Protein (%)	6.30	6.30	17.50
Fat (%)	0.10	0.10	0.60
Mineral matter (%)	1.00	1.00	3.20
Fibre (%)	0.80	0.80	1.90
Carbohydrates (%)	29.00	29.00	71.40
Calcium (%)	0.03	0.03	0.10
Phosphorus (%)	0.31	0.31	0.42
Potassium (%)	-	-	1.10
Iron (%)	0.001	0.001	0.004
Niacin (%)	-	-	0.70
Sodium (%)	-	-	0.01
Vitamin A (IU)	-	0.40	175.00
Nicotinic acid (mg/100 g)	0.40	0.40	-
Vitamin C (mg/100 g)	13.00	13.00	12.00
Vitamin B_2 (mg/100 g)	-	0.23	0.08

Origin and distribution

Garlic is supposed to have originated from Central Asia and the Mediterranean regions. It is grown the world over, under the same conditions as onion. China is the highest producer of garlic, followed by South Korea, Spain and India. The world production of garlic in 1990 was estimated to be 3,012 thousand tonnes.

Description of the plant

A. sativum L. consists of four botanical varieties namely, *A. sativum* L. var. *sativum*, *A. sativum* L. var. *ophioscorodon* (Link) Doll, *A. sativum* L.var *pekinense* (Prokh) Maekawa, and *A. sativum* L.var. *nipponicum* Kitamura.

Distinguishing between garlic cultivars using the following criteria is relatively simple and may be of practical use.

1. **Morphological characters:** Bolting type, number and size of (the primary) cloves, number of leaf axils forming (primary) cloves, number of secondary cloves formed in a lateral bud (or primary clove), bulb weight, colour of the outer protective leaf of the cloves, number of protective leaves, width and length of foliage leaves, plant height and tenderness of the green leaves.
2. **Physiological and ecological characteristics:** Time of building and maturity, low temperature and long-day requirements for bulb formation, winter hardiness and bulb dormancy.

In addition to the above characteristics, the taste of the cloves may be used in practical classification. Also, adaptation to agro-climatic zones may be of help. This is a complex characteristic determined by the above-mentioned environmental requirements, and is the most important factor for cultural practice, since garlic cultivars adapted to one area are sometimes worthless as a crop in other regions.

Garlic is a bulbous, erect herb, 30-100 cm in height, with narrow, flat leaves and small white flowers and bulbils. It has a superficial adventitious root system. The bulb is tunicate and is composed of disc-like stem, thin dry scales, which are the bases of the foliage leaves, and smaller bulbs (bulblets) or cloves. These are formed from the axillary buds of the younger foliage leaves. The cloves are enclosed by the dry outer scales. Each clove consists of a protective cylindrical sheath, a single thickened storage leaf sheath, and small central bud. The leaf blade is linear, flat, solid, 2-5 cm wide, and folded lengthwise. The flowers are variable in number and sometimes absent, seldom open and may wither in the bud stage itself.

Soil and climate

A wide range of soils with good drainage can be used for garlic cultivation. The soil depth should be at least 45-60 cm. Garlic requires medium black, well-drained loamy soils, rich in humus with a fairly good potash content. Crops raised on sandy or loose soils have poor keeping quality and the bulbs produced are lighter in weight. Bulbs produced in heavy soils are deformed, and during harvest many bulbs are broken and bruised. The bulbs become badly discoloured in ill-drained soil. Acidic soils are not good for clove development; however, a pH range between 5 and 7 has little effect on growth and yield. The crop can tolerate salinity between 5.60 and 7.80 dS m^{-1} EC, depending on the cultivar. Extremely acidic soils as well as heavy clay are not suited for this crop.

Garlic grows under a wide range of climatic conditions. However, it cannot tolerate weather that is too hot or too cold. It prefers moderate temperature in summer as well as in winter. Extremely hot or long dry periods are not favourable for bulb formation. It is a frost-hardy plant requiring a cool (12-18°C)and moist period during growth and a relatively dry period during bulb maturity. Bulbing takes place at longer days and at high temperatures. An average temperature of 25-30°C is most conducive for bulb initiation. Bulbing in garlic occurs with lengthening photoperiods in the spring, with the process being hastened by increasing temperatures up to 25°C. Low temperature during growth may induce the sprouting of the already formed cloves.

Varieties

Two distinct cultivars–Fawari and Rajalle Gaddi–are reported from India. In addition, some local cultivars like Madrasi, Tabiti, Creole, Eknalia and Jamnagar are grown in different parts of India. The research work done by the Associated Agricultural Development Foundation, Nasik, has resulted in the development of superior selections like Agrifound white (G 41) and G50. Besides these, some of the selections like HG1, HG6 from the Haryana Agricultural University are also found to be high yielding. The Ooty 1 variety was developed at the Horticultural Research Station, Ooty, of the Tamil Nadu Agricultural University. The bulbs of this variety are bigger in size with a dull white colour. Each bulb, on an average, possesses 20-25 cloves. It has a potential yield of 17 tonnes per hectare in a crop duration of 120-130 days. It is resistant to leaf thrips and tip burn. The other improved varieties of garlic are Godwari, Sweta, TSS-4, Pusa Sel.10, LCG-1, ARU-52, Yamuna Safed, Yamuna, Safed-2, G-282 and Agrifound Parvathi. All the varieties, except G-282 and Agrifound Parvati, are mostly small cloved and have more than 20-30 cloves. G-282 and Agrifound Parvati have bigger-sized bulbs with bigger cloves that are fewer in number. G-282 develops good bulbs in the Southern hills and Northern plains. Agrifound Parvati is long-day type and can be grown only in the Northern hills.

Odeskii-13 and Imeruli-23 are grown in the CIS. The principal varieties grown in the United States for dehydration are California Late (Pink or Italian) and California Early (White or Mexican and Creole).

Cultivation

Propagation

Garlic is propagated by planting cloves or bulblets that have been stored at 5°C for several months. The size of the seed clove is important; larger cloves give higher yield. The seed should be virus-free and the cloves used for planting should weigh between 3.6 and 5.8 g. Bigger bulbs may be produced for consumption and small bulbs for planting purposes. The size of the bulbs can be regulated by plant spacing; close spacing produces smaller bulbs. Aerial bulbils or topsets are also used as planting material. A yield of 96.10 q/ha has been recorded at Chaubatia (Dist. Nainital, India) using aerial bulbils with closer spacing.

Land preparation

The field is prepared by ploughing or digging and levelling. A quantity of 20-30 t/ha of FYM is applied at the time of field preparation and incorporated into the soil. Beds of convenient size or ridges and furrows are then formed at a spacing of 30 cm.

Planting

Garlic is extremely hardy and survives long periods at temperatures below 0°C. Consequently, in low temperature regions, it may be planted in autumn or in spring. Studies on successional sowing dates in temperate regions have consistently showed that autumn planting gives higher yields than mid-winter or spring planting. In Kalyani, India, mid-November planting gave higher yields than late December planting. In the hills, sowing is done in April and May. In Bangladesh, late October planting gave a higher yield than November or mid-December planting.

The machinery for mechanically separating and grading cloves prior to planting has been developed. Machine planting, using a pneumatic planter, gives more even planting depths and inter-plant spacing than the more widely-used tulip planting machines. The pneumatic planter also gives higher emergence percentages, yields and quality. If the cloves are damaged, as it can happen with machine grading and planting, their subsequent multiplication is reduced.

Garlic is planted using different methods which include dibbling, furrow planting, broadcasting and transplanting. About 500-600 kg cloves are required for planting one hectare. It has been shown that soaking garlic cloves in 0.01, 0.05 or 0.1% solutions of $ZnSO_4$ or $MnSO_4$ induced more vigorous growth of the aerial parts than soaking in distilled water. $MnSO_4$ at 0.01% and $ZnSO_4$ at 0.1% produces the highest yield (95.3 and 93.3 q/ha, respectively). The cloves may be soaked in 1000 ppm CCC solution for 24 hours before planting, to improve the yield. The highest yield would be produced from 5.0 g of cloves planted at 30 × 6 cm, and the best quality crop would be obtained from 5.0 g of cloves planted at 30 × 12 cm. It is reported that the highest yields (16.2 t/ha) of autumn-planted garlic were obtained by planting aerial bulbils of >5 mm diameter with the scapes removed; a yield of 12.8 t/ha was obtained from bulbils planted with the scapes attached. The bulb yield was highest (131.8 and 131.5 q/ha) from planting on September 20th and declined with delayed planting to 59.4 and 59.5 q/ha when planted on October 30th. Numerous studies show that both the yield and the size of the harvested bulbs increases with the weight of the planted clove. A clove weight of 4-5 g has been recommended for planting in Chile and Poland.

Manures and fertilisers

Garlic responds well to organic manure. A good yield was achieved with an application of 40-50 t/ha of FYM. Nitrogen plays an important role in the early vegetative growth. An application of N ranging from 50 to 150 kg/ha has produced positive effects on the garlic yield. Phosphorus application with the nitrogen helps to increase the yield. A combined application of nitrogen, phosphorus and potassium has been found beneficial for the garlic yield as well as the clove flavour. A significantly higher yield has been observed by applying 150 kg N, 250 kg of P, and 75 kg of K per hectare. The application of micronutrients such as boron (0.5-1.0 ppm) was found to increase the garlic yield. The yield also increased significantly as the N,P and K supply increased; N had the greatest effect. The largest bulb diameter (3.67 cm) and the highest yield (7.91 t/ha) were observed with 100 kg of N, 50 kg P_2O_5 and 50 kg K_2O per hectare. Another recommendation is the application of a mixture of 75 kg NPK in the ratio of 18:2:3 with 5 kg B, Zn and Mo. Both organic manure and sulphur were very beneficial for the plant growth, total yield, and quality as well as N, P and K contents in the plant tissues. High levels of nitrogen fertiliser have been reported to increase the percentage of plants showing secondary growth; such plants give rise to undesirable bulbs, with rough surfaces. Of the recommended dose of fertilisers, half the N and the full doses of P and K should be applied basally and the remaining N should be applied one month after planting.

Irrigation

Garlic needs irrigation at 8-day intervals during its vegetative growth and 10 to 15-day intervals during maturation. The last irrigation should be given 2-3 days before harvesting to facilitate harvesting and minimise damage to any bulbs. Irrigation at 60 mm CPE (cumulative pan evaporation) is optimum for a good garlic crop. Irrigation frequencies are decreased towards crop maturity as more moisture causes regeneration and emergence of sprouts, resulting in non-uniform maturity. Irrigation is then discontinued to allow the field to dry for harvest.

In Egypt, the yields of crops receiving between 50 and 150 kg/ha of nitrogen were compared. These were watered when the available soil water declined by 20, 40, 60, 80 or 100%. The maximum yield was attained at the highest rate of nitrogen along with irrigation at the least water deficit. Greenhouse studies in Argentina indicated that irrigation with solutions of electrical conductivity greater than 3.0 dSm^{-1} significantly reduced the yields.

Inter-culture and weed control

Garlic is a closely planted crop; hence, manual weeding is tedious, expensive, and often damages the plant. However, the first weeding and hoeing should be done one month after sowing followed by a second weeding one month after the first inter-culture. Chemical weed control with smazine-Diuron (0.75-1.0 kg/ha), post-emergence, and Simazine-nitrogen (0.75-3.0 kg/ha) or Pendimenthalin (1.5 kg/ha) pre-emergence was 20% more profitable than weeding twice. The effective control of weeds using Linuron (2.25 kg/ha), Pendimethalin (1.5 kg/ ha) and Tribunil (2.1 kg/ha) has been reported. Hoeing the crop just before the formation of bulbs loosens the soil and helps in the setting of bigger and more compact bulbs.

To increase the storage life of the bulbs over a period of 6 months, the plants have to be sprayed with Cycocel @ 1000 ppm during the 3rd, 12th and 18th week of planting.

Pests and diseases

Pests

The occurrence of mites (*Rhizoglyphus echinopus*) and aphids in garlic has been reported. To check mites, a pre-planting dip of the cloves in 0.15% Nematos is recommended. To control aphids, systemic insecticides like Methyl dematon have to be sprayed @ 1 ml/lit. Since garlic is propagated vegetatively, nematode infestation has been reported. These can be controlled by applying nematicides like Temik and Nemafos (30 kg/ha) to the soil before planting. Garlic is also attacked by *Thrips tabacii,* which causes withering of the leaves.

Diseases

White-rot caused by *Sclerotinum cepivorum* and leaf blight caused by *Cladosporium delicatulum* and *Stemphylium botryosum* are the major diseases of garlic. A typical disorder, sprouting of bulbs in the field, has been noticed in Saurashtra, India, mostly during the early stages of the maturity of the bulbs, particularly when there have been winter rains or excessive soil moisture and supply of nitrogen. However, this disorder is not of a permanent nature. The loss due to sprouting in the field was not more than 0.5%. Sprouting is more commonly seen in white cultivars than in pink or purple cultivars. Varietal differences, spacing, and early planting are some of the possible causes of sprouting.

Rubbering in garlic, in which the developing bulbs become spongy and rubbery, is a serious physiological problem relating to the excess application of nitrogen. This can be avoided by judicial application of nitrogen. Foliar spraying of 0.1% $ZnSO_4$ + 0.3% Ammonium sulphate plus 0.1% $MnSO_4$ on the 30th, 45th and 60th day of planting or the application of MH at 2500 ppm on the 30th, 60th and 90th day can help to minimise the rubbering phenomenon.

The important storage and market diseases of garlic include blue mould-rot, bulb decay, *Aspergillus* rot, *Fusarium* rot, dry-rot, and gray mould-rot. The most prevalent disease is blue mould-rot, wherein lesions are produced. In advanced stages of decay the cloves become soft, spongy and covered with powdery spore masses of *Penicullum* sp. Dry rot is caused by *Fusarium solani.* The cloves infected with dry rot do not sprout. Spraying the plant with Bordeaux mixture, Ferbam, Zineb, and Nabam in the field controls dry rot effectively. Fumigation with Methyl bromide at 32 mg/m^3 for 2 hours at 21°C has been recommended for the control of the wheat curl mite (*Aceria tulipae*) in stored garlic. A disorder known as waxy breakdown, in which the clove shrinks and becomes amber, translucent, waxy or sticky to the touch, occurs in stored garlic and occasionally in the field. It may be associated with sun-scald in the field, but it is usually associated with poor ventilation and the lack of oxygen during storage and transport.

Harvesting and yield

Generally, the garlic crop is ready for harvest 130-150 days after planting, depending on the cultivar, soil and season. At this stage, the tops become partly dry and bend to the ground. An early harvest results in poor-quality bulbs, which cannot be stored for long periods. A delayed harvest results in the splitting and resprouting of bulbs in some cultivars. A special cutter, with a horizontal cutting bar passing under the rows, is used to loosen the garlic bulbs from the soil. The whole plant can then be easily pulled out. While piling, the tops are laid over the adjacent bulbs to prevent sun-scald.

Garlic is harvested when the foliage collapses and starts to senesce. In a warm, dry climate, the bulbs may be left to dry and cure in wider rows in the field for about 3 weeks, using the foliage to protect the bulbs from the direct sun. In places where rain or dew are likely occur at harvest time, the tops and roots are removed and the bulbs are cured indoors. Potato root-harvesting machines can be modified and used for garlic. Before lifting the plants, the tops can be cut off to about 13 cm above the bulb using a rotary knife. The trimmed bulbs can be stored in 2 m deep bins or stacks provided with forced ventilation, as well as in shallow containers or sacks. Cold air (below 0°C) with a lower relative humidity (RH) is best for ventilating the stacks.

The curing of bulbs is carried out for about a week in the field, to dry the bulbs thoroughly. After drying, the plants are tied in small bundles and then stored by hanging them on bamboo sticks or ropes. In some areas, after drying the tops, the roots are removed, leaving about 1 cm of root and 2 cm of tops. Normally, 4-12 t/ha of bulbs are obtained. The recovery of cloves from the bulbs would be 86-96%.

Storage

Garlic cloves store best at 1°C and sprout most rapidly at 4-18°C. Hence, prolonged storage at this temperature range should be avoided. The relative humidity should be lower for garlic than for most other vegetables, because high humidity causes rot and mould growth. It is essential that garlic destined for storage should be well cured in the field and be provided with adequate air circulation through storage containers to remove transpired moisture. Incomplete curing results in excessive decay, particularly when the bulbs are stored above 0°C. Garlic can be held at 27-32°C satisfactorily for one month or less if the humidity is kept low as recommended. The storage life of garlic can be extended by treatment with Maleic hydrazide before harvest or gamma irradiation after harvest. These treatments are effective in controlling sprout growth and weight loss, as well as in decreasing external discoloration and diseases.

Dormancy and storage of bulbs

Immediately after harvesting, the garlic cloves are deeply dormant. The number of days for sprout emergence of the cloves planted under optimal conditions for growth varies with the degree of maturity of the cloves and with storage temperatures. The cloves lifted when the foliage is still green require more time to sprout than mature cloves lifted when the foliage has senesced. It appears that the more immature the cloves, the deeper is their dormancy.

Cloves stored in well-ventilated sheds after total leaf senescence resumed sprout growth even later than those immediately transplanted into the field. The rooting of these cloves was much delayed. When such stored cloves were planted at approximately 2-week intervals at 20°C, and the cloves planted immediately after lifting and stored for 2 weeks, they did not differ in the number of days required for the emergence of the sprouts. Cloves stored for 4 weeks sprouted considerably faster. This indicates that the cloves were in total true dormancy for about 2 weeks, after lifting, but that dormancy was in part imposed by the storage conditions of those stored for 4 weeks. Dormancy, as measured by the time for visible sprouting, declined steadily and reached a minimum of about 8 weeks after lifting. The rate at which dormancy declines greatly depends on the storage temperature and on the cultivar.

Low temperature storage

Garlic stores well under a wide range of temperatures, but sprouts most quickly when stored at about 4.4°C. Low storage humidities are generally recommended to discourage mould development and the formation of new rot. To store garlic for atleast 6-7 months, temperatures of –0.6 to 0°C and 80% or less relative humidity are ideal. Higher temperatures of 26.4-32°C are satisfactory to store garlic for one month or less. Intermediate temperatures between 4.4 and 18.2°C are undesirable because they favour rapid sprouting, and high relative humidity results in rot and mould growth. Garlic may be kept in ventilated storage for 3-4 months. The garlic bulbs must be thoroughly cured before they are stored since incomplete curing results in excessive decay, especially when the storage temperatures are greater than 0°C. Dry garlic bulbs can be stored for 28-36 weeks at a temperature of 0°C with 65% relative humidity.

Irradiation

A significant weight loss occurs in stored garlic bulbs due to sprouting. Several investigations have proved the efficacy of ionising radiation in reducing these losses (97-100). It is reported that a 2 krad dose of gamma radiation applied within 8 weeks of harvest inhibited sprouting effectively, reduced the weight loss, and prolonged the storage life of garlic for about one year. When irradiation treatment was given at a later date, sprouting was not inhibited. The respiration rate of the irradiated garlic was higher immediately after the treatment, but within a few days it dropped to values similar to those of non-irradiated garlic.

If garlic bulbs are treated with 50 Gy of ^{60}Cobalt 30 days post-harvest, it can inhibit sprouting and reduce the weight loss during storage. The Diallyl disulfide (DADS) content in garlic bulbs has been evaluated and it was found that doses higher than 100 Gy caused an apparent reduction in the amount of DADS. The irradiation of garlic bulbs cv. Red did not affect the flavour. DNA was the cellular constituent most sensitive to radiation in garlic cloves treated with 10 Gy of gamma radiation.

Processing

A number of garlic products including capsules, extracts, and tablets are now marketed, and odour-free garlic products have become popular.

Dehydrated products

The garlic harvested for dehydration is brought to a dehydration plant in large bulb bins or open mesh bags. The bulb is broken into individual cloves by passing it between rubber-covered rollers, which exert just enough pressure to crack the bulb without crushing the cloves. The loose paper shell is removed by screening and aspiration. The cloves are then washed in a flood washer, at which time the root stubs are floated off.

The garlic is then sliced and dehydrated in a manner similar to that used for onions. After drying, the pink skin adhering tightly to the fresh cloves can be removed by screening and air aspiration. Garlic is commercially dried to about 6.5% moisture. Dehydrated garlic is sold commercially as powder, granules, or in sliced, chopped, or minced forms. The packaging is similar to that used for onions. Dry garlic is widely used in the formulation of spice mixtures for luncheon meats, salad dressings, sauce or soup mixes, and pet foods.

Garlic cloves are dehydrated at 60-75°C in a series of stages, the temperature being reduced as the moisture content falls. A final moisture content of 4% is achieved via warm air circulation. The average yield is about 1 kg of dried product per 5 kg of fresh cloves.

Oil

Garlic oil is obtained by the distillation of the minced cloves. The oil comprises 0.1-0.25% of the fresh weight of the clove, and 1 g of oil is equivalent to 200 g of dried garlic. The pungency of garlic oil makes it difficult to use it directly, and it is usually diluted in vegetable oil, or is encapsulated.

Garlic salt

The specifications for the composition of garlic salt are shown in Table 2. Garlic salt has a much wider culinary potential than garlic powder. One tablespoon is considered to be equivalent to a clove of fresh garlic.

Table 2 *Approximate composition (100 g of Garlic salt)*

Nutrient	Content
Water (g)	1.4
Energy (kcal)	63
Protein (g)	3.2
Fat (g)	0.1
Carbohydrates (g)	13.8
Fibre (g)	0.4
Ash (g)	81.5
Ca (mg)	220
Fe (mg)	1
Mg (mg)	11
P (mg)	79
K (mg)	212
Na (mg)	31.4
Zn (mg)	1

Ginger

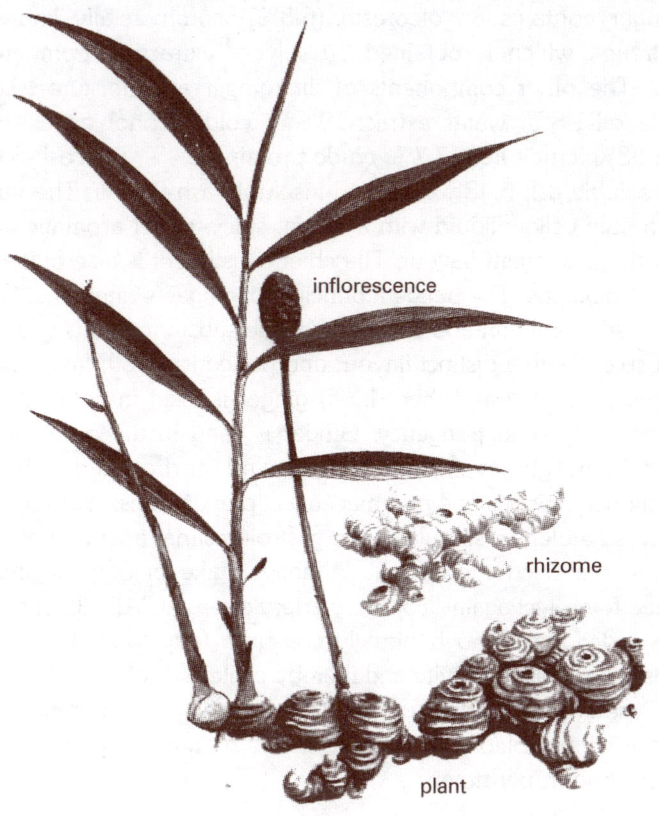

Plant Profile

Family	:	*Zingiberaceae*
English name	:	Ginger
Indian name	:	*Adrak* (Hindi), *Shunti* (Kannada), *Inji* (Malayalam), *Adraka* (Sanskrit), *Inji* (Tamil), *Allam* (Telugu)
Species	:	*Zingiber officinale* Rosc.
Distribution	:	India, China, Taiwan, Nigeria, Jamaica, Thailand, Australia, Brazil, Japan, Indonesia
Uses	:	Culinary spice

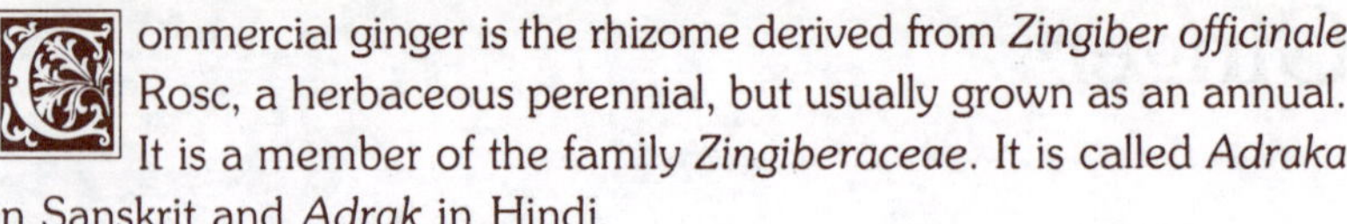

Commercial ginger is the rhizome derived from *Zingiber officinale* Rosc, a herbaceous perennial, but usually grown as an annual. It is a member of the family *Zingiberaceae*. It is called *Adraka* in Sanskrit and *Adrak* in Hindi.

Ginger contains an oleoresin (6.5%), commercially known as 'Gingirene', which is obtained by solvent extraction from ground ginger. The other components of dry ginger are moisture-10.85%, volatile oil-1.8%, water extract-19.6%, cold alcohol extract-6.0%, starch-53%, crude fibre-7.7%, crude protein-12.4%, total ash-6.64%, water-soluble ash-5.48% and acid-insoluble ash-0.14%. The volatile oil is a pale yellow liquid with a warm, spicy, sweet aromatic odour, and a sharp pungent flavour. The chief constituent is a sesquiterpene called zingiberene. The pungent principle of ginger is zingerone. Ginger is reported to contain 200 different volatile substances. It is one of the oldest spices with a distinct flavour and pungency. Both fresh and dry (scraped or peeled and then dried) ginger is used in cooking for its aroma, flavour and pungency. Ginger is used in the production of ginger beer, ginger wine, cordials and carbonated drinks in confectionery, pickles and pharmaceutical preparations. As a medicine, it relieves flatulence, stimulates the gastrointestinal tract and acts as a rubifacient and counter-irritant. A daily intake of 5 g of ginger is reported to protect against coronary artery disease (CAD) that normally afflicts individuals who habitually consume fatty food. Ginger also increases fibrinolytic activity and thereby protects against CAD. Ginger is also used for flavouring soft drinks and medicine preparations. The essential oil and oleoresin are used in the manufacture of flavouring essences and in perfumery.

Origin and distribution

Ginger is indigenous to the Indo-China region but is not known to occur in the wild state. India still enjoys the unique position of being the largest producer of ginger in the world. The other important ginger producing countries in the world are China, Taiwan, Nigeria, Jamaica, Thailand, Australia, Brazil, Japan and Indonesia.

Area and production

In India, ginger in cultivated in an area of 54,000 hectares, out of which Kerala accounts for 60%, amounting to 25% of the country's total production. As far as Karnataka is concerned, the ginger area is spread out in Shimoga, South Kanara, Bidar, Bangalore, Chikkamagalur and Chitradurga districts. The annual national production of ginger rhizomes is estimated to be 31,520 tonnes, which is about 50% of the world's total production. Karnataka produces about 8,460 tonnes per annum out of 1,800 hectares area under ginger. Ginger is also grown in the states of Himachal Pradesh, Orissa, West Bengal, Andhra Pradesh and the North-eastern states.

Description of the plant

Ginger, is a monocotyledon, with a slender, perennial herb-like habit, but is usually grown as an annual. It is 30-100 cm tall, with a robust branched rhizome borne horizontally near the surface of the soil, bearing leafy shoots close together.

The fleshy, sympodial rhizome is hard, thick, somewhat laterally compressed, often palmately branched and usually pale yellow within. It is covered with small distichous roots in the top layers of the soil. The leafy shoots are annual, erect, about 50 cm tall and 5 mm in diameter. They are formed of long leaf sheaths and are glabrous, except for short hairs near the base of each leaf blade. The aerial pseudostems usually bear 8-12 distichous leaves. The lamina is thin, sub-sessile, linear, lanceolate, darkish green above, paler beneath, narrowed evenly to a slender tip, with an obtuse or rounded base.

The inflorescence arises directly from the root-stock, is spicate and 15-25 cm long. The scape is slender, with upper sheaths with or without short leafy tips. The spike is cylindrical and cone-like. The bracts are appressed, ovate or elliptic. They are green with pale submarginal bands and incurved, translucent margins. One flower is produced in the axil of each bract and is fragile and short-lived. The calyx is thin, tubular, and three-toothed. The corolla tube has three yellowish lobes. The labellum or lip, which probably corresponds to three stamens, is nearly circular, with the side lobes free almost to the base, and coloured as the mid-lobe. The filament of the stamen in short and broad and the anther is cream-coloured. The stigma, protruding just below the apex of the appendage, has a circular optical aperture surrounded by stiff hairs. There are two slender free styloids. The inferior ovary is trilocular with several ovules per locule. The fruit, which is very seldom produced, is a thin-walled, three-valved capsule, with small, black, arillate seeds.

Climate

Though ginger is a warm, humid, tropical crop, it can withstand more diverse conditions than any other spice. It is usually grown from sea-level to 1,500 m elevation. However, the optimum altitude is between 300 and 900 m. It is grown both as a rain-fed and an irrigated crop. An annual rainfall of 1250-1500 mm is sufficient for the crop. A moderate rainfall from sowing to sprouting, fairly heavy and well-distributed rains during growing periods and a dry period for a month prior to harvesting are ideal. In areas, where the average annual rainfall is 1500-3000 mm distributed over 8-10 months and the day temperature ranges from 28 to 30°C, ginger is grown purely as a rain-fed crop. But the crop is also seen growing in areas receiving low rainfall, like North India where it is grown as an irrigated crop. Also, the crop is found to be growing in the areas of higher attitudes like Himachal Pradesh, Assam and West Bengal, indicating its wide adaptability to various agroclimatic conditions.

Soil

Ginger thrives best on rich medium loams with a good supply of humus. It can also be grown on a wide range of soils like sandy loams, red loams, clayey loams, lateritic soils and black, rich clay soils. It is very sensitive to water-logging and therefore such situations should be avoided. The ideal soil pH range for the crop is 5.5 to 6.5. Ginger should not be grown in the same site year after year.

Varieties

Ginger is always propagated vegetatively and the number of clones are limited. The recognised clones differ in their size, the fibre and moisture content of the rhizomes and in yield. The cultivars are usually named after the areas where they are cultivated.

The ginger cultivars grown in different states/union territories are as given in the table below.

State	**Cultivars**
Andhra Pradesh	Local types, viz., Medak and Tuni
Arunachal Pradesh	Shillong
Bihar	Desi and Dorabhanya
Gujarat	Local types named after the localities
Haryana and Punjab	Local types named after the localities
Himachal Pradesh	Himachal No.1, SG 666 (Dhariga local), SG 645 and Narag
Jammu and Kashmir	Himachal No. 1
Karnataka	Wynad, Manathodi, Narasapatam, Thaiguppa and Karkala
Kerala	Kuruppampadi, Wynad Local, Valluvanad, Maran, Nadia, Maran/Lodi, Ernad, Thodupuzha, Rio de Janeiro, Jamaica

(continues...)

(...*contd.*)

Madhya Pradesh	Local ginger types found in Tikkamgarh, Chindwara and Bastar districts
Maharashtra	Local types
Manipur	Shing type, Thingpuri and Shingtam
Meghalaya	Nadia, Poona, Rio de Janeiro, Wynad, Thingpuri and Maran
Mizoram	Thingpuri, Maran and Rio-de-Janeiro
Nagaland	Rio de Janeiro
Orissa	Kuruppampadi, Wynad types, Local types, viz., Kuduli, Laxipur, Turia Junagarh, Raikia, Suprapha and Surchi
Sikkim	Gurubathane, Bhaisey, Nadia, Rio de Janeiro, Thingpuri
Tamil Nadu	Rio de Janeiro, Maran, Nadia
Tripura	Himachal No.1, Local types
Uttar Pradesh	Local types named after the localities
West Bengal	Gurubathan, Sambuk-A, Turuksadan, Malli, Rio de Janeiro, Thingpuri, Maran, Tura and Bombay Desi

Some of the cultivars suited for dry ginger are Nadia, Karkala, Maran, Wynad, Mananthody, Valluvanad, Narasapatam and Taiguppa. Two exotic varieties which are exclusively used for green ginger as well as oleoresin extraction are China and Rio de Janeiro. Besides, the local cultivars Ernad and Kuruppumpadi are also suitable for oleoresin extraction. The cultivars Maran, Himachal No.1, Wynad local are also suitable for use as green ginger. The rhizomes of the varieties Jamaica, China and Bangkok contain less fibre while the rhizomes of Himachal No.1, Narasaptam and Ernad contain more essential oil.

Wynad local and Wynad Manathody are the cultivars widely grown in Kerala. Maran, a cultivar of Assam which was introduced into cultivation in Kerala and Karnataka, is becoming popular owing to its high yields and high percentage of dry ginger. The cultivar Rio-de-Janeiro, introduced from Brazil, is found to give higher yields than the local varieties. But it is not well suited for the dry ginger preparation as it gets shrivelled on drying. The other exotic variety 'China' is fairly high yielding, but has a very moisture content. In Taiwan 'Ta-Kuang' and 'Chu-chiang' are cultivated.

The germplasm at IISR, Calicut, includes 437 indigenous, 10 exotic and 12 wild related accessions (total 459).

The descriptions of some improved varieties released from the high Altitude Research Station, Pottangi, Orissa, are given below.

Suprabha

This variety was released in 1988. It is a clonal selection (PGS-35) from the Kunduli local. The average plant height is 51.8 cm, the aerial shoot is green in colour, the leaf length/breadth (cm) is 17.0/1.9, and it produces 11.1 tillers per clump and each tiller has 14.4 leaves. The variety matures in 229 days and gives an average yield of 16.6 tonnes of fresh rhizomes per hectare (potential yield 28.8 t/ha). The rhizome core is whitish-yellow in colour, the average weight of the fresh rhizomes per clump is 200 g, the dry recovery (%) is 20.5 and the crude fibre (%) content is 4.4. The essential oil and the oleoresin content of the rhizomes are 1.9 and 8.9%, respectively. The variety is suitable for use as green ginger.

Suruchi

This variety was released as a clonal selection (PGS-19) from the Kunduli local. The plants attain a height of 51.3 cm, the aerial shoot is green in colour and the leaf length/breadth (cm) is 17.7/1.9. It produces 11.8 tillers per clump and each tiller has 13.5 leaves on an average. The variety matures in 218 days and gives an yield of 11.6 tonnes of fresh rhizomes per hectare (potential yield 21.8 t/ha). The rhizome core is greenish-yellow in colour, the average weight of the fresh rhizomes per clump is 205 g. The dry recovery (%) is 23.5, the crude fibre content (%) is 3.8 and the essential oil and oleoresin contents are 2.0 and 10.0%, respectively. This variety can be used as green ginger.

Surabhi

The variety is an induced mutant (V1K1-3) from the Rudrapur local, released during 1991. The average values of various growth, yield and quality parameters of this variety are as follows: plant height–62 cm, colour of aerial shoot–deep green, length and breadth of leaf (cm)–15./1.6, number of tillers produced per clump–14.5, number of leaves per tiller–14.3, maturity–225 days, average yield of fresh rhizomes (t/ha) 17.5, potential yield–21.6, colour of rhizome core-deep yellow, weight of rhizomes per clump–220 g; dry recovery (%)–23; crude fibre (%)–4, essential oil–2.1% and oleoresin 10.2%.

The other improved varieties of ginger are Himgiri, IISR Varada, ACC No.117 and ACC No.35 whose important characteristics are given the following table.

Variety	Pedigree	Maturity (days)	Average yield t/ha (fresh)	Dry recovery (%)	Crude fibre (%)	Oleoresin (%)	Essential oil (%)	Features
Himgiri	Clonal selection from Himachal collection	-	13.5	-	-	-	-	Best for green ginger, less susceptible to rhizome-rot disease, suitable for rain-fed condition
IISR Varada	Selection from germplasm	200	22.2	20.7	3.29	6.7	1.75	High yielder with tolerance to rhizome rot. Adaptable to various agroclimatic conditions
ACC.No.117	Selection from germplasm	-	22.8	18.5	3.26	4.48	1.72	Bold rhizome, low fibre
ACC.No.35	Selection from germplasm	-	20.5	20.8	4	6.34	2.36	Performs well in Kerala and Karnataka

The recently released varieties are IISR Mahima and IISR Rajatha, which are reported to yield 23.2 and 22.4 t/ha of fresh rhizomes, respectively.

Cultivation

Propagation

It is a general practice to propagate ginger by rhizomes. The synseed technology protocols for rapid multiplication and also somatic embryogensis and regeneration methods have been developed at the IISR, Calicut. For propagation by rhizomes, carefully preserved, healthy and disease-free seed rhizomes or sets, cut into small pieces of 2.5 to 5 cm length, weighing 25 to 50 g, with a minimum of 1 or 2 good buds or growing points should be used. At the time of planting, the rhizome pieces are treated with 0.25% organo-mercurial compound or 0.3% Dithane M-45 for 10-30 minutes as a control measure against *Furasium* root-rot. If required, they may be treated with 0.05% Malathion or 0.1% Quinolphos and 200 ppm Streptocycline. About 1,500-1,800 kg sets are required to plant one hectare area. Early sowing, with the onset of pre-monsoon showers, ensures good growth and yield.

Land preparation

The land should be ploughed 4-5 times to bring the soil to a fine tilth. Beds of 1 m width, 15 cm height and 3 m length, or of any convenient length, are prepared at a 40 cm spacing. About 2,000 beds of 3 m × 1 m size are prepared in one hectare of land. Being an irrigated crop, the ridges are formed 40 cm apart. The width of the channels between the beds is about 30 cm.

Planting

The planting is done in shallow pits, 4-5 cm deep, on 15 cm-high beds. For an irrigated crop, the planting is done on flat beds of convenient size for irrigation, or on ridges. The recommended spacing for this crop is 30 cm × 30 cm under the bed system of planting. A bed of 3 m × 1 m can accommodate about 40 plants.

Manures and fertilisers

As ginger is a heavy feeder, it should be well manured. Usually, it is the practice to apply 25-30 tonnes of well rotten FYM or compost per hectare at the time of planting. The manure may be applied to each hole over the seed, which is then covered with soil, or it may be broadcasted over the entire field and mixed with the soil during the last ploughing. In addition, a fertiliser dose of 75:50:50 or

100:50:50 kg. NPK per hectare is recommended. The whole of phosphatic and half of potassic fertilisers may be applied at the time of planting. Half the Nitrogen is applied 40 days after planting and the remaining nitrogen and potassium a month after that. The application of neem cake @ 2 t/ha as a basal dressing helps to reduce the incidence of soft-rot disease and increases the yield.

Irrigation

In areas of low rainfall, irrigation is a must. In most areas in India, where it is grown as an irrigated crop, watering is done at fortnightly intervals, usually during the period from the middle of September to the middle of November.

Weeding

Usually, two to three weedings are done in the whole of the growing period. The first weeding is done just before mulching and then repeated at monthly intervals, depending on the intensity of weed growth. The application of 2,4-D has been found to be effective as a pre-emergent herbicide. Manual weeding consists of either pulling out the weeds, chopping with a hoe or cutting the roots with a knife.

Mulching

A mulch of green leaves is provided twice or thrice as and when required, to prevent the soil from drying and to check weed growth. It also breaks the force of the rains and provides organic matter and also enhances germination. Mulching thrice in pure crop has been found to be better, but usually the mulching is done twice. The first mulching is done at the time of planting with 12.5 tonnes and the second after 40 days with 5 tonnes of green leaves per hectare.

Rotation and inter-cropping

The crops most commonly rotated with ginger in Kerala are tapioca, chilli, rice, gingelly, ragi, groundnut and maize. Ginger is also grown with maize as a mixed crop and as an inter-crop in arecanut and coconut gardens.

Pests and diseases

Pests

Shoot borer (*Conogethes punctiferalis*) attacks the ginger crop. The larvae bore into the shoots, and the plants wilt and eventually die. One or two prophylactic sprayings with Monochrotophos (0.1%) or Rogor (0.01%) or Dipel (0.3%) should be given as a precautionary measure and also the seed rhizomes may be dipped in 0.1% Quinalphos for 5 minutes before planting. The usual practice is a bio-monthly spray of either of these chemicals. *Hexamermis* sp. and hymenopteran parasitoids could be used for bio-control of this pest.

Rhizome scale

Another important pest of ginger is the rhizome scale. (*Aspidiella hartii*) which can be controlled by the application of Quinolphos or Fenthion (0.025%) before storage of rhizomes.

Leaf roller (Udaspes folus)

The larvae cut and fold the leaves, remain within and feed on them. The adult larvae are medium-sized and have brownish-black wings with large white spots. Spraying with Carbaryl (0.1%), Dimethoate or Phosphomidon (0.05% each) checks the pest infestation.

Nematode damage has also been reported. The control measure recommended is application of Thimet @ 10-15 kg/ha or Phorate 10 kg/ha at the time of land preparation. Denematisation of ginger rhizomes at 45°C for 3 hours plus summer ploughing and covering of soil with polythene in May (15-30) is reported to reduce the population of root-knot nematode (*Meloidogyne incognita*).

In addition, stored ginger is attacked by the common storage pests like dung store beetle, the tobacco borer beetle, etc. Fumigation of the store houses should be carried out regularly to control these pests. The rhizome maggot, *Mimegralla coerulifrous,* is reported to cause rotting of the rhizomes.

Diseases

Soft-rot or rhiozme-rot (*Phythium apanidermatum, Phythium myriotylum* and *Fusarium solani*) is the most serious disease of ginger. The bases of the aerial shoots become soft, watery and they rot. The affected plants become pale. Their leaf tips turn yellow, followed by complete yellowing and drying. The shoots fall and cease to produce rhizomes. The infection also spreads to the rhizomes and the inner tissues are reduced to a soft, black, putrefying mass. The disease is favoured by high soil moisture content of the soil with insufficient drainage.

In order to control this disease, the first and foremost measure is the eradication of the infected plants and avoiding the monocropping of ginger. For fresh planting, only sound, disease-free and healthy seed rhizomes should be selected. They are treated with 0.25% organo-mercuric compounds for 10-30 min. before planting or with Indofil M-45 (0.25%) plus Bavistin (0.1%) for 60 min. and drying in shade for 24 hours before sowing or storage, or the rhizomes may also be dipped in Blitox 50 (0.3%) before storage. *Trichoderma harzianum, T. hamatum* or *T. viride* @ 50 g (2×10^7 cfu)may be used for biological control of the disease pathogen. The cultivars resistant to soft-rot are 76.1, OCP-855 and OCP-822.

The leaf-spot disease is caused by *Taphrina* sp., *Colletotrichum* sp. and *Phyllosticta zingiberi*. This disease is controlled by spraying Dithane M-45 (0.2%) or Bordeaux mixture (1%) at 15-day intervals. The bacterial wilt disease is caused by *Ralstonia solanacearum,* for which the control measure is spray of streptocycline (200 ppm) during the month of June. The fungus *Khuskia oryzae* is also reported to cause leaf-spot disease.

'Yellows' disease is caused by *Fusarium oxysporum* f.sp.*zingiberi*. The disease is characterised by foliar yellowing, starting from the base. The rotting of the pseudostem, stunting of the plant, poor rhizome development and rotting of the rhizomes are the other symptoms. The control measures besides phytosanitation are application of compost/FYM @ 2 t/ha and seed treatment with Mancozeb (0.25%) plus Bavistin (0.1%).

The nematode (*Aphelenchus* sp.) and the fungus (*Fusarium* sp.) together are reported to cause basal sheath rot.

Harvesting and Yield

Ginger is a 7-8 months' duration crop. It is ready for harvest when the leaves turn yellow and start withering, after which the rhizomes become more fibrous and pungent and are better suited to the production of dried ginger. A light irrigation is provided 4-5 days before harvesting and clumps are lifted carefully with the help of a pick-axe or a digging fork or hoe. If the crop is meant for green ginger, it is harvested in 5-6 months.

An average crop of ginger may yield about 10-30 tonnes of fresh rhizomes per hectare.

Processing

The five main products of the ginger rhizome are fresh (green) ginger, dried ginger spice, bleached ginger, preserved ginger in syrup or brine and crystallised ginger.

Green ginger

Fresh ginger, is consumed as a vegetable spice, both when immature and mature. It is of lesser importance in world trade, but in the local market it is a major commodity. For this purpose, green ginger rhizomes, soon after harvesting are thoroughly washed in water 2-3 times to remove any soil and dirt. Then, the damaged and diseased rhizomes are separated and graded depending upon the requirement in the market.

Dried ginger

In order to obtain a more pungent and aromatic spice, ginger should be harvested at the time of proper maturity and dried properly. For this purpose, the harvested rhizomes are soaked in water overnight and rubbed well between to remove the adhering soil. After thorough cleaning, they are removed from the water. The outer skin is removed with a split bamboo with a pointed end, taking care to see that only the outer skin is peeled as otherwise the essential oil glands beneath the skin will also get damaged and thus the oleoresin will be lost. The peeled rhizome is washed and uniformly sun-dried for a week. The dry ginger is rubbed together to remove dry foreign matter. This type of dried produce is known as unbleached ginger. Depending upon the variety, on an average, 16-25 kg dried ginger is obtained from 100 kg green ginger, i.e., the yield of dry ginger is 16-25% of the green ginger.

Dried ginger forms the major bulk of international trade. It is exported as whole or in split forms and is ground at the consuming centres. It is also used for preparation of its extractives, ginger oleoresin and ginger oil.

Bleached ginger

To prepare bleached ginger (white), the fresh rhizomes are peeled off and soaked in 2 per cent lime water for about 6 hours. The produce is then removed and dried for 5-6 days and rubbed with a piece of gunny cloth to give the desired dried appearance.

Preserved ginger

This is prepared from immature rhizomes, mostly for export purposes. Discounting green ginger preserved in brine, the two forms of processed, preserved ginger entering the market are:

1. Preserved ginger in sugar syrup
2. Dry or crystallised ginger, wherein ginger is impregnated with sugar syrup, dried and coated with crystalline sugar and must be harvested while they are still immature, tender, succulent, and mild in pungency, usually under 7 months of age. For processing, the salted ginger is removed from the acidified brine, washed and soaked in cold water for 2 days, changing the water several times. It is then placed in cold water, which is heated in syrup for 10 minutes, after which the ginger is removed and pricked with a fork. The ginger is then boiled in syrup for 45 minutes. It is left in the syrup to soak for 2 days or more, and then reboiled for 45 minutes after which it is packed in fresh syrup. The quantity of syrup used depends on the market for which the ginger is intended.

Ginger oleoresin and essential oil

Ginger oleoresin (3.5–9.5%) is obtained by the solvent extraction of ginger powder using organic solvents like acetone, ethylene dichloride, etc., and it possesses the full organoleptic properties of the spice such as aroma, flavour and pungency. It is a blend of oil and resinoids and finds similar applications as the ground spice in the flavouring of processed foods. The oleoresin is also used in certain beverages and to a limited extent in pharmaceutical preparations.

On the other hand, ginger oil (0.5–3.0%) is distilled from the dried spice. The product is characterised by the aroma and flavour of the spice but lacks the pungency. It finds its main application in flavouring of beverages and in the confectionery and perfumery industries.

Grading

Dried ginger is marketed on the basis of geographical origin and the form of preparation. The chemical and physical characteristics of the spice differ from one producing region to another.

Indian dried ginger is classified as Malabar ginger, Assamese ginger and Himachal ginger. But the two types of the spice which are in great demand in the world market are Cochin and Calicut ginger, which come under the Malabar ginger type. The bulk of our exports are of rough-scraped, whole rhizomes. Sometimes, coated ginger is also exported. Bleached or limed Calicut ginger is mainly exported to the Middle Eastern countries.

Cochin dried ginger is about 20 mm long and has a light brown to yellowish-gray colour, whereas, Calicut dried ginger is orange to reddish-brown in colour and is generally considered to be inferior in quality to the Cochin spice. Both types are graded prior to export into the following categories, according to the number of fingers on the rhizomes–B (three fingers), C (two fingers) and pieces (individual finger). The grades of whole ginger and the specification are given in the table below.

Grade	Specification	Size of rhizome
Calicut (NGK)	Garbled (distorted), non-bleached	Not less than 20 mm in size
Calicut (NUGK)	Ungarbled, non-bleached	Small, cut pieces, < 20 mm in length
Cochin (NGC)	Garbled, non-bleached	Not less than 20 mm in length
NUGC	Ungarbled, non-bleached	Small, cut pieces, not less than 20 mm in length
BGC	Garbled, non-bleached	Not less than 20 mm in length
Cochin (BUGC)	Ungarbled, bleached	Small, cut pieces of less than 20 mm in length

The technical and physical specifications of ginger as prescribed by ASTA-(1) and FADA mould (% by weight) are: (1) insect defiled (% by weight) (2) volatile oil (%) (min.) 2; moisture (% max.) 12; ash (% max.) 5, acid-insoluble ash (% max.) 1.0; average bulk index (mg/100 g) N/A.

Value-added products

Ginger oil, oleoresin, candy, preserves, vitaminised effervescent ginger powder, plain effervescent powder, starch from spent ginger, ginger brandy, wine, beer, medicinal beverages, encapsulated ginger oil, and delydrated ginger are some of the value-added products of ginger.

Storage of seed ginger

For good germination, the seed rhizomes should be stored properly in pits under the shade. The rhizomes usually harvested during December-January, have to be preserved for about 4 months before planting (April-May). Good, disease-free, big, plump rhizomes should be selected for seed purposes and treated with 0.25% organo-mercurial compound for 10 minutes, against soft-rot disease, and then dried in the shade. If the seed rhizome is infested with rhizome scale, it is advisable to treat it with 0.05% Malathion or Dimethoate. Then, the seed rhizomes are loosely placed in pits of convenient size to a height of 10-15 cm from the top and covered with a wooden plank. The remaining surface is plastered with mud. In certain areas, the rhizomes are loosely heaped over a layer of sand or paddy husk and covered with dry leaves.

Greater Galangal

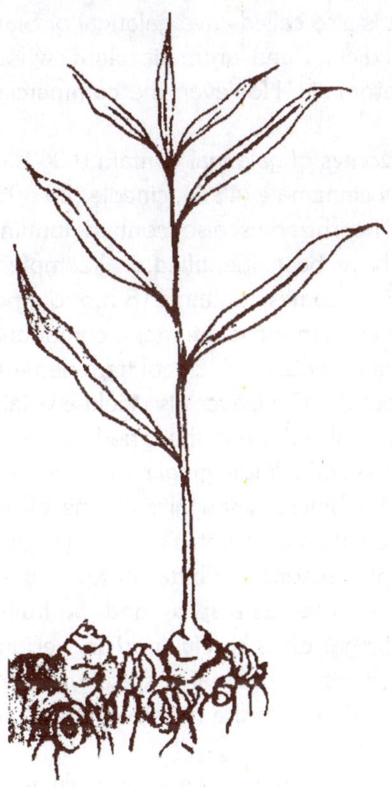

Plant Profile

Family	: *Zingiberaceae*
English name	: Greater galangal, Java galangal, Siamese ginger
Indian name	: *Kulanjan* (Hindi), *Rasmi, Sugandha vachi* (Kannada), *Aratta, Perasatta* (Malayalam), *Kulanja, Kulanjana* (Sanskrit), *Sungandam Tittiram* (Tamil), *Pedadumparashtram, Kachoramu* (Telugu)
Species	: *Alpinia galanga* (Linn.) Willd.
Distribution	: China, India
Uses	: Spice, Culinary purposes

alangal, popularly known as *Kulanjan* (Hindi) and *Kulanja, rasna* and *sugandhamula* (Sanskrit), is the dried rhizome or root of *Alphinia galanga* (Linn) Willd. of the family *Zingiberaceae*. It is also called Java galangal or Siamese ginger. It is a highly prized medicinal and aromatic plant, whose roots, rhizomes and leaves are aromatic. However, the commercially important part is the rhizome.

The green rhizomes of galangal contain 0.04% essential oil, which consists of methyl-cinnamate (48%), cineole (20-30%), sesquiterperoid and d-pinene. The rhizomes also contain tannins and flavonoids, some of which have been identified as kaempferide, galangin and alpinin. The fresh rhizomes contain 118 monoterpenoids of which a-pinene, b-pinene and limonene are major compounds and 17 oxygen-containing monoterpenoids with cineol, terpinen-4-01 and a-terpineol are minor compounds. The leaves also yield a volatile oil. The volatile oil that is steam distilled from the dried comminuted rhizomes is known as 'galangal oil', 'false ginger oil', 'lesser galangal oil' and sometimes called 'Chinese galangal oil'. The oil is a pale yellow to olive-brown liquid with a eucalyptus-cardamom-ginger-like odour and a warming, camphoraceous-like bitter taste.

The rhizomes are used as a spice, and the fruits as substitutes for cardamoms. Galangal oil is rarely used in perfumery but it could, however, be of value in an oriental or spice-type perfume formulation. The oleoresin obtained from the rhizomes is used in medicine and in flavouring foods, etc.

The rhizomes are used in indigenous medicines to treat rheumatism and catarrhal problems, especially bronchial catarrh. The drug is a depressant of the cardiovascular system. The respiration was stimulated by small doses but depressed by larger doses in experimental animals. The rhizome and its essential oil are useful in respiratory troubles, especially in children. The rhizomes are also a carminative and stomachic. They are also useful in treating rheumatoid arthritis, inflammations, stomatopathy, pharyngopathy, cough, asthma, hiccoughs, dyspepsia, stomachalgia, obesity, diabetes, cephalagia, tubercular glands and intermittent fevers.

Origin and distribution

Galangal has its botanical origin in South China. It is distributed in China and India and grows mainly in the Eastern Himalayas and South-east India.

Description of the plant

The plant is 1.8-2.1 m high and bears perennial rhizomes which are deep orange-brown in colour, aromatic, pungent and bitter. It produces around 24 suckers per clump per year. The fruits are about 13 mm long, constricted in the middle and contain 3-6 seeds. The seeds are slightly pungent, with an aroma similar to that of the rhizome. The cut pieces of the rhizome of this species are known as 'greater galangal'. The rhizome of 'lesser galangal' is smaller and reddish-brown in colour and has a stronger odour and taste. Galangal root is also known as 'small galangal' and 'radix galangal minoris'.

Climate

It is a tropical plant. It can be grown as a rain-fed crop in assured rainfall areas, with well distributed rainfall in hilly areas and places with an altitude of 1,400 m. A temperature range from 26°C to 34.5°C and rainfall of about 750-3000 mm are suitable. In areas receiving low rainfall the crop needs supplementary irrigation during the dry period.

Soil

It prefers fertile red loams to forest soils. It also comes up well in laterite soil. However, very heavy soil with a tendency for water-logging should be avoided while cultivating this crop.

Varieties

There are no named varieties in this crop. At present, only local types are available for cultivation. Five geographical races namely, Koothattukulam, Palghat, Thodupuza, Ponnukkara and Vallanikkara have been identified. Thodupuza is reported to give the highest fresh rhizome yield. However, the dry recovery is maximum (32.5%) in the Koothattukulam variety.

Cultivation

Propagation

The crop is propagated through rhizomes. Fresh, healthy, disease-free rhizome bits of 5 cm length with atleast one shoot are preferred for planting. The older roots have to be removed before the rhizome pieces are planted in the main field. The seed rate is about 1000-1500 kg/ha.

Land propagation

The field should be ploughed to a good titlth. All the stones and pebbles should be removed. Then beds should be prepared at 1 m breadth, 15 cm height and of convenient lengths.

Planting

The rain-fed crop is planted with the onset of monsoon in May-June; the irrigated crop can be planted any time. While planting, small pits are made in the bed to plant the rhizomes. The rhizome pieces are covered with FYM and the beds are immediately mulched with leaves or straw. The optimum spacing is 30 × 20 cm under good fertility and 40 × 30 cm under poor fertility conditions.

Manures and fertilisers

At the time of bed formation FYM is incorporated into the soil @ 10-15 t/ha. Besides, fertilisers should also be applied at 100:50:50 kg NPK per hectare per year in 2-3 split doses. The application of bio-fertilisers, *Azospirillium* at 10 kg/ha and cowpea green manuring *in situ* are also beneficial for the crop.

The mean nutrient uptake by the crop is reported to be 22.8 kg N, 28 kg P_2O_5 and 36 kg K_2O per hectare.

Irrigation

The crop should be irrigated immediately after planting. Gap filling, if any, has to be carried out within one month. Weeding should be done two months after planting followed by top-dressing, earthing-up and mulching. Thereafter no weeding is required as the crop smothers the weeds.

Inter-cropping

Galangal can be cultivated as an inter-crop in coconut or rubber plantations.

Pests and diseases

Usually pests and diseases are not serious enough to any control measures. Occasionally, shoot borers and leaf eating caterpillars and blight disease are observed. *Pseudomonas solanacearum* has been reported for the first time to cause bacterial wilt to galangal in Kerala.

Harvesting, processing and yield

The crop comes to the harvesting stage after 18 months. Though the crop can be harvested at this stage, the optimum stage to harvest for obtaining the the maximum rhizome and oil yield is 36-42 months after planting. While harvesting, the shoot portions are cut and the rhizomes and roots are dug out carefully. Harvesting is very arduous due to the strong and extensive root ramification. After the rhizomes are removed from the soil, the roots are separated, the rhizomes cleaned and cut into 5 cm-long pieces, and dried in the sun for 3-5 days to 10% moisture for marketing.

The average yield of fresh rhizomes is 23 t/ha, which on drying gives 25% recovery. The fresh rhizomes on steam distillation for 3-5 hours give 0.22% essential oil. The oil recovery on a dry weight basis is 0.93%. The roots also contribute significantly to the essential oil yield.

Value-added products

The value-added products of Kulanjan are galangal oil and galangal oleoresin (4.10-4.91%).

Indian cassia

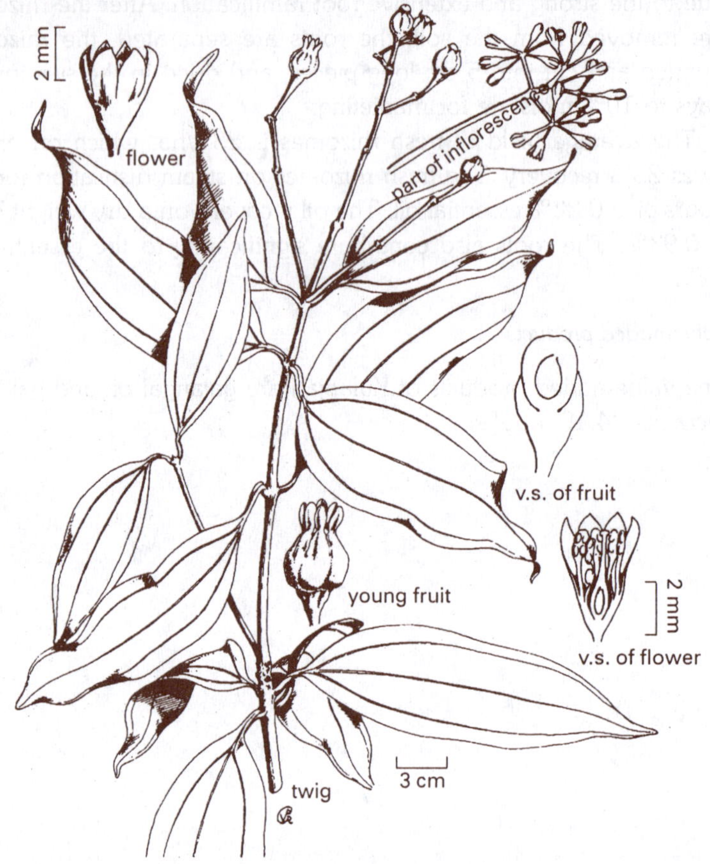

Plant Profile

Family	: *Lauranceae*
English name	: Indian cassia
Indian name	: *Tejpat* (Hindi, Bengali, Punjabi and Urdu), *Tamalapatra (*Gujarati), *Darchini (*Marathi), *Tamalaka* (Tejpatra) *(*Sanskrit), *Talishappattiri* (Tamil), *Talisapatri* (Telugu)
Species	: *Cinnamomum tamala* (Buch-Ham)
Distribution	: India
Uses	: Flavouring curries, medicine

ndian cassia or *Cinnamomum tamala* (Buch-Ham) T. Nees and Eberm of the family *Lauraceae* is commonly known as *Tejpat, Tamalaka* or *Tejpatra*.

The leaves of this plant, called *Tejpat,* are used mainly as a flavourant or spice especially in Northern India. The leaves yield about 2% essential oil, which contains about 80-85% eugenol. The oil resembles cinnamon-leaf oil and also contains d-phellandrene. The bark, also known in trade as Indian cassia bark or Indian cassia Lignea, yields a pale yellow essential oil containing 70-85% cinnamic aldehyde. The bark is collected from trees growing at the foot of the Sikkim Himalayas.

The leaves are used as a substitute for *Pan* or betel leaves in Kashmir. In Indian cookery it takes the place of Bay leaves in Europe. The leaves can be used with myrobalans in dyeing and in the manufacture of vinegar. In dyeing, they are used as clarifiers.

The leaves are carminative and are used in colic and diarrhoea. The essential oil of the leaves is used in flavouring curries and other dishes. The bark is aromatic, coarser than the bark of cinnamon and is one of the common adulterants of true cinnamon.

Origin and distribution

The tree is native to India. It is distributed in the tropical and subtropical Himalayas (915-2440 m altitude), Khasi and Jainta hills (915-1220 m altitude), and in Eastern Bengal, Meghalaya, Assam, Manipur, Arunchal Pradesh, Tamil Nadu and Karnataka. Indian cassia is grown on plantation scale in the Khasi and Jaintia Hills, Garo Hills, Mikit Hills, Manipur and Arunachal Pradesh. Tejpat is grown mainly in the Jaintia paraganas of the Sylhet district. Many plantations in this tract are self-sown, while a few are planted. The exact statistics on the area and production are not available. The total area covered is approximately 240 hectares, mostly in the North-eastern Indian region, distributed largely in the Himalayas at 250 to 700 m above sea level.

Description of the plant

It is a moderate-sized, evergreen tree attaining a height of up to 8 m and a girth of 1.4 m. The leaves are glabrous, three-nerved from the base, long and pointed. The flowers are unisexual and white in colour, blooming in panicles. The flowering and fruiting occurs between July and August.

Climate and soil

Tejpat trees grow best in a warm humid climate at an elevation of 1500-4000 m from MSL, with an average temperature of 30°C and 1,500-2,000 mm rainfall per annum.

It can be grown in poor soils, however, sandy loam soils with a mixture of humus or vegetative mould is the best for the production of fragrant leaves.

Varieties

There are no improved varieties available for this crop, only local types are being cultivated.

Cultivation

The method of cultivation is the same as that for cinnamon. *Tejpat* is propagated by seeds. The seedlings are raised before the onset of the monsoon. The seeds are sown in beds by March-April and they sprout within one or one-and-a-half months. The seedlings should be transferred to polybags at the 2-4 leaf stage. Pits of 60 cm^3 in size are dug at the recommended spacing and should be filled with top soil and well-decomposed cattle manure or compost in the 1:1 ratio.

The seedlings are planted in the centre of the pits in the main field, only when they are 4-5 years of age. The spacing varies from 3 m–4.5 m. The plants require another 5-6 years to grow and they start bearing only from the 10th year onwards. Once the plantation reaches maturity, the fallen seeds from the tree will germinate and natural regeneration occurs. These self-sown seedlings are also transplanted when they are 30 cm tall. The transplanted seedlings need special care and are planted in the shade to protect them from the hot sun.

Manuring

No manuring is required in established plantations. However, to boost the growth of the young plants, the application of 10-12 kg of organic manure to individual plants every year has been found to be a good practice.

After care

Weeding is done twice a year in the plantations for first 8-9 years, but afterwards only one weeding per year is sufficient. The seedlings are irrigated till they are established, if there is a long drought period.

Pests and diseases

Pests

No pests of a serious nature have been reported on this crop. However, scale insects and mealy bugs have been found to infest the plants sometimes. These insects can be controlled by spraying any contact insecticide.

Diseases

Leaf-spot and die-back diseases caused by *Collitotrichum gloeosporoides*, grey blight caused by *Pestalotiopsis palmarum* and sooty mould caused by *Phragmocapinus beetle* are the common diseases of *Tejpat*. The spraying of 1% Bordeaux mixture at frequent intervals will control these diseases.

Harvesting and yield

The plants start bearing harvestable leaves in 4-5 years' time and the trees continue to bear for a century. The season of harvest is the dry days of October to December and in some places the collection of leaves extends up to March. *Tejpat* leaves are collected once a year from young vigorous trees and every alternate year from old and weak ones. While harvesting, small branches are cut along with the leaves, dried in the sun for the 3-4 days and then tied into bundles for marketing. Sometimes, the dry leaves are packed in *bora* or *gungra* (cylindrical bamboo nets) before they are taken to the market.

There are at present no quality standards for *Tejpat*. But recently, quality standards for 3 grades of *Tejpat,* based on systematic physico-chemical analysis of 41 samples collected from the local Nagpur and North Indian markets, have been fixed.

The annual production of dry leaves fall within the range of 8-20 kg per tree, but on an average a yield of 13 kg per tree is available. The plants remain productive up to 50 to 100 years.

Essential oil from leaf

An essential oil (0.7%) is obtained by the hydro-distillation of the fresh leaves. Fourteen oil components constituting 93.41% of the oil have been identified. Linalool has been found to be the main constituent (60.73%) of the oil, whereas eugenol and cinnamic aldehyde have been detected in trace amounts (<1%). The other components of significant occurance (3%) in the oil are α-pinene (10.54%), β-pinene (10.42%), limonene (3.21%) and camphene (3.06%).

Adulterants

The *Tejpat* bark oil is sometimes adulterated with terpenes.

Kokum

Kokum Fruits
(*Garcinia indica*)

Plant Profile

Family	:	*Guttiferae*
English name	:	Kokum, Kokum butter tree, Brindonia tallow tree, Mangosteen oil tree
Indian name	:	*Kokam* (Hindi), *Murgala* (Kannada), *Punampulli* (Malayalam), *Murgal* (Tamil)
Species	:	*Garcinia indica* Choisy, *G. gummigatta*, *G. mangosteena*, *G. morella.*
Distribution	:	India, Malaysia.
Uses	:	Acidulant, cookery.

okum or *Kokam* (*Garcinia indica* Choisy), of the family *Guttiferae* is also known as the 'kokum butter tree', 'Brindonia tallow tree' or 'mangosteen oil tree'. Its fruits, seeds, dry rind and bark are the economic parts. Out of more than 200 species, 36 species of *Garcinia* are found in India. Among these, *G.gummigatta* (combogea) and *G.indica* have medicinal value and are also used as a spice. *G.mangosteena* and *G.morella* are also cultivated for their fruit and gambodge, respectively.

The fruits, seeds and dry rind mainly contain tartaric acid, garcinol, reducing sugars and fat. *Kokam* bears small to medium-sized fruits whose percentage composition of the fresh rind is: moisture-80.0, protein-(N × 6.25) 1.9, crude fibre-14.3, total ash-2.6, tannins-2.8, pectin-5.7, starch-1.0, crude fat hexane extract-10.0, acid (as hydroxy citric acid)-22.8, pigment-2.4, ascorbic acid-0.06, carbohydrates by difference-35.0. All the values are expressed on a moisture-free basis. The seeds yield 31% edible fat.

Kokam has been traditionally used as an acidulant in certain foods and is locally called *Punarpuli* and *amsul*, in some regions of Karnataka and Maharashtra, respectively. It is specially used in fish fries/curries, as it imparts a delicate flavour with its minimal use. It is also used to make a red, pleasantly-flavoured extract for use as a beverage. A very popular product, 'Amrut kokam', which is a syrup, is usually pressed from the rind of *kokam*. Besides *kokum* nectar, squash, agal are also prepared from the rind. The fruit has an agreeable flavour and a sweetish acid taste and is used in the *Konkan* area chiefly in the form of *kokam* prepared by drying the outer rind, soaking it repeatedly in the juice of the pulp and sun drying it. *Kokam* contains about 10% malic acid and a little tartaric and citric acids. It is used as a garnish to give an acid flavour to curries and also to prepare cooling, attractive, red-coloured syrups during the hot months.

Kokam rind contains 2-3% anthocyanin pigments and is a promising source of a natural colourant for acid foods. The processing conditions have been established at CFTRI, Mysore, for the extraction and purification of the pigment concentrate on a commercial scale. As a colourant, *kokam* finds use in the area of processed fruit products, alcoholic and non-alcoholic beverages, preserves and instant foods.

A process has also been standardised for making a concentrate which can be used as an acidulant in various food preparations. The acid present in the *kokam* rind has been identified as hydroxy citric acid (HCA), present to the extent of 15%. The isolation of the acid in pure form has been carried out on a small scale. The acid finds a potential application in medicine, and is widely used against obesity and weight management. HCA is said to block the enzyme responsible for storing fat in the body. It is also reported that the liver produces more glycogen when HCA is present and suppresses the appetite. HCA also burns fat. Further, HCA slows down the oxidation of protein and, hence, the loss of clean tissue reducing while reducing obesity.

A new fat-soluble, yellow pigment, called garcinol has also been isolated from the fruit rind. The chemical identity of this pigment has been established by chemical and spectral studies, and an integrated process to extract it from *kokam* has also been developed at the CFTRI. Thus, the *kokam* fruit appears to be a promising industrial raw material for commercial exploitation in view of its interesting chemical constituents.

The fruit is an antihelmintic and cardiotonic, and is useful in treating piles, dysentery, tumours, pains and heart complaints. A syrup made from the fruit juice is given in bilious problems. The root is an astringent. The seeds (8-10%) of the fruit yield 23-26% on the weight of seeds and about 33-44% on the weight of kernels, a valuable edible fat known commercially as '*Kokam* butter'. It is extracted mostly on a cottage industry basis by crushing the kernels, boiling the pulp in water and skimming off the fat from the top; or churning the crushed pulp with water. It is white in colour with a yellowish tinge. It remains solid at room temperature. Most importantly, it has good export potential.

Kokam butter is considered as a nutritive, demulcent, astringent and emollient. It is suitable for ointments, suppositories and other pharmaceutical purposes. It is used as a local application to ulcerations and fissures of the lips, hand, etc. The cake left over after the extraction of the oil is used as manure.

The physico-chemical characteristics of the fat are as follows–melting point 39°-43°C, Ref. index at 40°C-14565-14575, sap value-187-191.7, iodine value-25-34, Unsap mat-14 to 23%, R.M. value-0.1-1.0, and free fatty acids (as oleic acid) 7.2%.

Kokam butter as sold in bazars, consists of egg-shaped lumps or cakes of light grey or yellowish colour with a greasy feel and a bland, oily taste. It is used mainly as an edible fat and is also used as an adulterant of ghee. As ordinarily seen, it contains seed particles as impurities. The refined and deodorised fat is white in colour and compares favourably with high class hydrogenated fats. It is also used in the preparation of chocolate and confectionery.

The butter, like other *Garcinia* fats, is rich in combined stearic and oleic acids. It contains about 75% of mono-oleodisaturated glycerides. It is suitable for use as confectionery butter. However, as it solidifies forming a rough surface, the addition of another fat is necessary to correct the defect. It is also suitable for candle and soap manufacture.

A method has been developed for the production of stearic acid from the fat with a yield of 4.57%. It possesses properties similar to those of piney tallow (from *Vateria indica*) and may be used in the sizing of cotton yarn.

The fat is also used in the manufacture of soap, candle and ointments. An ointment made out of *kokam* fat, with dammar resin (a resin exuded by the *Varteria indica* tree) and wax is said to be effective in treating carbuncles. It is reported that Italy and some other foreign countries are importing *kokam* fat from India for use in confectionery.

Origin and distribution

The *kokum* tree originated from the tropical rain forests of the Western Ghats of Kerala and Malaysia. Its cultivation is confined to the coastal hilly regions of Maharashtra and Goa and is popularly known as '*Ratamba*'. Some *kokam* trees are also observed in Tamil Nadu, the Western Ghat forests of Karnataka, Kerala, as well as in parts of West Bengal, Assam and Gujarat. It is reported to flourish well on the lower slopes of the Nilgiri hills.

Kokam is reported to be exported to Japan, Taiwan, USA, Zanzibar, etc., from India.

Description of the plant

Kokam is a slender, pyramid-shaped evergreen tree with drooping branches. The leaves are ovate or oblong, lanceolate, 6.25 to 8.75 cm long and 2.5 to 3.75 cm broad, dark green above and pale green beneath. It is an androdioceious tree, producing male and bisexual flowers in separate plants. It flowers during December-January on the past season's shoot in the axil of the leaves. The flowers are pale yellow in colour and borne either singly/solitary or in clusters. The female flowers are 2-3 in number and the male flowers are 3-4 in number. The male buds are short and roundish, whereas the female buds are oval in shape and show variation in the size. Their weight varies from 50 g to 180 g. The fruits are globose or spherical, 2.5 to 3.75 cm in diameter, dark purple when ripe, enclosing 5-8 large seeds, which have 7-10 ridges and take 4-5 months to ripen. It takes about 8-10 years for commercial bearing when grown from seeds. The fruits are edible with an agreeable pleasant flavour and sour–sweetish taste inducing a cool feeling.

Climate and soil

In its natural habitat, it grows well in a hot humid climate under partial shade. The soil should be well-drained, deep and have a good water-holding capacity. However, the crop can also be grown on sandy, clayey, loamy and lateritic soils which can hold moisture. It tolerates water- logged conditions and also drought conditions, to some extent. It thrives well all along the river banks and seashores. When grown in transition belts and drier tracts, the plants should be irrigated, preferably through drip irrigation. Clear-cut details are not available for its cultivation in traditional areas. The tree favours dry and warm conditions, and also comes up well under the shade to some extent.

Varieties

After studying and evaluating the local types collected from all over the Konkan area for several years, a variety '*Konkan Amrita*' (S-8) was released in 1998 from the Konkan Krishi Vidyapeeth, Dapoli, Maharashtra. A full-bearing tree of this improved variety attains a height of 10.3 m with a volume of 48.58 m^2. This is an early bearer and the fruits could be harvested well before the onset of the monsoon. It has the shortest harvesting period (78 days) and the lowest number of harvests (3). It bears attractive apple-shaped fruits with a maximum shelf-life of 15 days. The average fruit yield is 138.28 kg/tree/year.

For the development of superior genotypes, the germplasm of *kokam* are being evaluated for: (1) earliness (2) big size fruits with high juice content (3) better processing characteristics and (4) attractive colour and shape of fruit at the Regional Fruit Research Station, Vengurla, Dist. Sindhudurg, Maharashtra. Some of the accessions collected at this Research Station recently are described below.

Accession No.	**Characters**
KV-18	Big size fruits with an attractive red colour
KV-19	Early type with big size and attractive, coloured fruits
KV-20	Red attractive colour
KV-21	Attractive blood-red colour
KV-22	Red with white striped fruits
KV-23	Early type
KV-27	Early type, heavy yielder
KV-29	Early bearing type
KV -32	Early flowering male plant
KV -35	Very early type, male
KV-38	Very early type, male

Trade varieties

There are several trade varieties of *kokam* available in the Indian markets, viz., (i) Plain *kokam* (ii) Satted *kokam* (iii) Lonawala *kokam* (iv) *Pakali kokam* (v) *Khane* or edible *kokam* and (vi) *Khola kokam*.

Cultivation

Propagation

Traditionally, *kokam* is propagated by seeds. In this method, the main problems are a long gestation period, bigger growth (plant) size, and the inability to differentiate between male and bisexual plants. Further, plant selection for quality and yield is not possible. The seeds take a longer time to germinate (more than 7 months) when they are in a dormant condition and this can be overcome by treating the seeds in 500 ppm GA_3 for about 12 hours.

However, recently high yielding clones are used for soft-wood grafting. Grafting is done using one-year-old gamboge root-stock and nine-month-old orthortrophic scion shoots of 6-8 cm length for soft-wood grafting. June is supposed to be the best month for grafting. The other root-stocks used are *Garcinia indica* and *Garcinia tinctoria*. The planting material is available with the Indian Institute of Spice Research, Calicut, Indian spice associates, Koppa (Chikkamagalore) and the Konkan Krishi Vidyapeeth, Dapoli (*Garcinia indica*).

Land preparation

The field should be prepared thoroughly by ploughing and breaking up the clods of earth. About 25 tons of FYM can be incorporated into the soil at the time of the final ploughing.

Planting

Though it is suggested to plant seedlings at a spacing of 8 × 8 m or 10 × 10 m and grafts at a spacing of 5 × 5 m, it is not advisable to opt for seedlings. The grafts are of a compact size and are suitable as an intercrop and for high density planting (200 plants/ha) both with coconut and arecanut. The pit size should be 1 m^3 and the pits should have sufficient organic manure and the plants should be mulched. One-year-old plants are the best for planting. Another possibility is *in-situ* grafting. The seedlings can be planted prior to the monsoon or even in summer under drip irrigation. These can be grafted using elite clones from bisexual plants.

After care

Kokam has to be irrigated once in 7-10 days during the summer and the planted area should be kept weed free. It does not require any fertilisers. However, about 40 kg of FYM needs to be mixed with the soil to get higher yields.

Inter-cropping

Under protective drip irrigation it is possible to grow *kokam* as an inter-crop in a coconut field. Generally, a 5 × 5 m spacing is suggested so that one row of *kokam* can be planted in between two rows of coconut. It is also possible to grow them in arecanut fields, but arecanut should be widely spaced (10 × 10 m). However, under closer spacing at a later stage, the yield of arecanut may be affected. The crop can also be tried in cardamom plantations.

Pests and diseases

Kokam is not seriously affected by any pests or diseases.

Harvesting and yield

Kokam plants start bearing about 7-8 years after planting and are harvested when the fruits are mature and ripe, which takes about 5 months after the time of flowering. *Kokam* flowers in November-February and the fruits ripen in April-May. Usually, the fruits fall after they are fully ripe. Since the male and female flowers are borne on separate trees (diecious), the grown-up seedlings are separated as productive female plants and unproductive male plants.

Immediately after the harvest, the fruits are cut open and the seeds along with the pulp are separated out, then the rind can be dried under the sun or smoked for about 1-2 weeks. The rind now turns black in colour and imparts a characteristic flavour. The dried rinds used for culinary purposes can be preserved by rubbing them with salt and coconut oil, which further improves appearance and quality. The dried rind fetches about Rs.200/kg in the local market. A well-maintained bearing tree (5-7 years) gives an yield of 300 to 400 kg of fresh fruits. Farmers can get very high returns from this plant with minimum inputs and little care.

Packaging and storage

Kokam is packed in gunny bags. The life of the fresh fruits is about 5 days at room temperature. Hence, sun-drying is practiced for its longer preservation.

Value-added products

Dried green kokam rind

This product is prepared from the native green *kokam* fruits. The fruits are cut into four longitudinal pieces and kept in a solution of potassium metabisulphide for two hours and then dried at 50 to 55°C in a dryer or under the sun.

Kokam syrup (Amrit kokum)

This product is prepared from the fresh ripe *kokam* rind by adding cane sugar in a 1:2 proportion. The mixture is kept as it is preferably in the sunlight, after packing it in a glass container for 8 days, stirring occasionally. The mixture is strained through a muster cloth and the syrup is stored in food-grade plastic containers or in glass bottles. If required, the preservative *viz.*, sodium benzoate, is used at the rate of 160 mg/kg of syrup. This product is required to be diluted 5 to 6 times with water before consumption as *sharbat*. A pinch of common salt and cumin powder are added for taste.

Dried ripe kokam rind

The ripe *kokam* rind is dried under the sun or in a cabinet drier at 60°C. It is treated with brined *kokam* juice and dried again for about five minutes. The finally dried product is used in curries in the place of tamarind.

Kokam agal (brined kokam juice)

This product is prepared either from ripe *kokam* fruit rind, unripe *kokam* fruit rind and pulp or only from the pulp. *Kokam* agal is prepared by mixing 15 to 16% common salt with the fruit part. The mixture is kept for 8 days with occasional stirring and the extracted *kokam* juice is strained and stored in food-grade plastic containers or glass bottles. This product is used for making '*skolkadhi*', a refreshing drink taken after lunch.

Kokam butter, *kokum* oil, stearic acid from *kokam* fat, *kokam* concentrate, *kokam* rind acid, *kokam* colour ointment, garcinol, and gambodges are the other value-added products of *kokum*. Stearic acid is obtained from *kokam* fat. The *kokam* fat is rich in combined stearic and oleic acids.

Constraints for area expansion

Kokam trees are found scattered on roadsides, in jungles, in the backyard of houses, waste lands and in coconut and arecanut plantations. *Kokam* is not cultivated systematically. This is mostly because of certain specific problems with this crop. These are:

- The dioecious nature of the plant, where the sex of the plant is confirmed only after completion of the juvenile period; i.e, after 7 to 10 years.

- Fifty per cent of the plants being male are unproductive
- The large variation in quality and quantity of fruits due to seedling progenies
- Late bearing and harvesting periods, which extend into the rainy season leading to the loss of fruits.
- The grafted plant cannot attain a normal shape and becomes bushy, irregular and viny. The internodal length is reduced.
- Lack of scientific studies on sex variability and productivity of the different types.

Large Cardamom

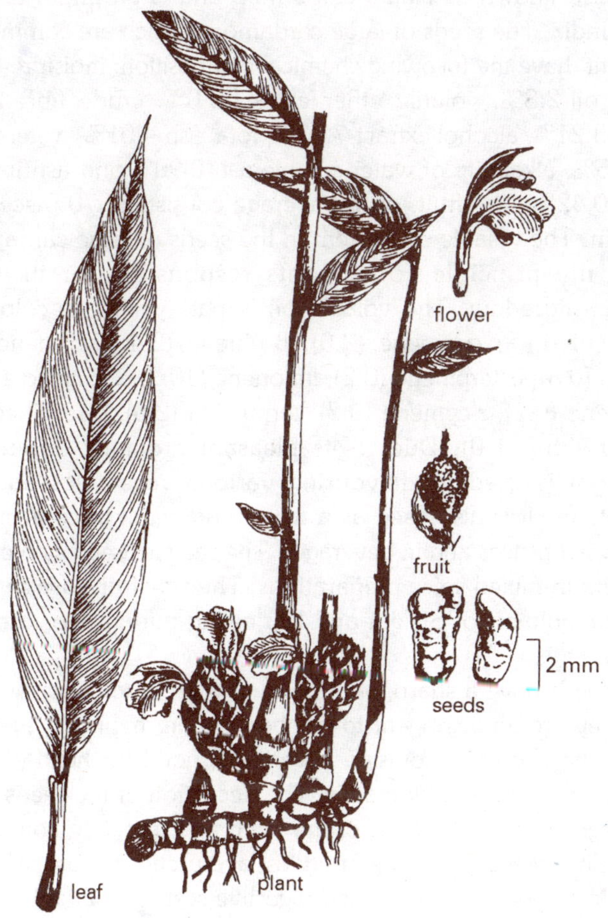

Plant Profile

Family	:	*Zingiberaceae*
English name	:	Large cardamom, Greater cardamom, Nepal cardamom
Indian name	:	*Badi Elaichi* (Hindi, Bengali, Punjabi and Urdu), *Perelam* (Malayalam), *Brihadaela* (Sanskrit), *Periya yalam* (Tamil), *Peddayelaki* (Telugu).
Species	:	*Amomum subulatum* Roxd.
Distribution	:	India, Nepal, Bhutan
Uses	:	Antidote to snake venom or scorpion venom, spice, medicine

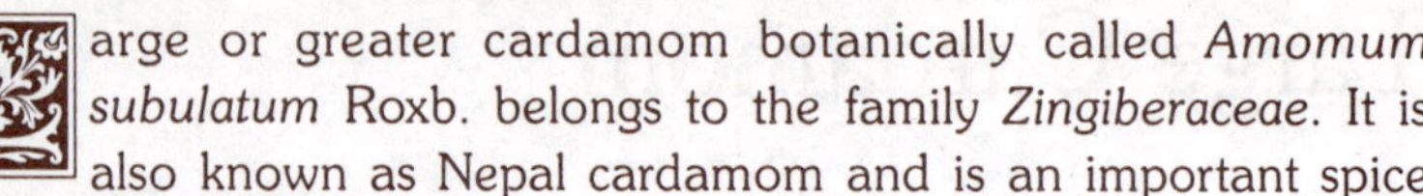

Large or greater cardamom botanically called *Amomum subulatum* Roxb. belongs to the family *Zingiberaceae*. It is also known as Nepal cardamom and is an important spice crop of India. The seeds of large cardamom, which are commercially important, have the following chemical composition: moisture-8.49%, volatile oil-2.8%, volatile ether extract-3.0%, crude fibre-22.0%, starch-43.21%, alcohol extract-7.02%, total ash-4.01%, water-soluble ash-2.15%, alkalinity of water-soluble ash-0.90% and ash insoluble in acid-0.42%. The fruit, on an average consists of 70% seeds and 30% skin. The volatile oil present in the seeds of large cardamom is one of the principle constituents responsible for its typical characteristic odour. The volatile oil is pale yellow in colour and consists of (in%) α-pinene (2.0), β-pinene (2.4), sabinene (0.2), myrcene (0.3), α-terpinene (0.2), limonene (10.3), 1,8 cineole (74.0), γ-terpinene (0.2), r-cymene (0.2), terpin-4-d (2.0), d-terpineol (5.6) and nerolidol (1.0). Due to its pleasant aromatic odour, large cardamom is used for flavouring various vegetables and meat preparations. It is also used as a flavouring agent in confectionery, hot or sweet pickles and in beverages. The seed is used as an essential ingredient in mixed spice preparations. The ripe fruits are eaten raw by the inhabitants of Sikkim and Darjeeling during September and October.

The seeds have a sharp, good taste and are a tonic for heart and liver. They are an astringent to the bowels, are hypnotic, appetizing and aid digestion. The outside covering is good for headaches and for the teeth and heals stomatitis. The decoction of the seeds is used as a gargle to cure problems of the teeth and gums. In combination with melon seeds, they are used as a diuretic in case of kidney stones. In certain disorders of the digestive system, marked by scanty and viscid secretion from intestines, they promote elimination of bile and are useful in congestion of the liver. The seeds are used in treating gonorrhoea and as an aphrodisiac. They have also been found useful for the cure of neuralgia, in large doses in conjunction with quinine.

The seeds are used as antidotes to snake or scorpion venom. The seeds are also widely used in India as a spice or condiment and in the preparation of sweets. In medicine, they are fragrant adjuncts to other stimulants, bitters and purgatives. The oil extracted from the seeds is applied to the eyelids to allay inflammation. It is an aromatic cardiac stimulant, stomachic, diuretic, carminative and is a remedy for throat and respiratory ailments.

Origin and distribution

Large cardamom is native to the moist, deciduous and evergreen forests of the sub-Himalayan region. It is an important crop of Sikkim from where it has spread to the North-eastern states and parts of West Bengal, Assam and neighbouring countries like Nepal and Bhutan.

Area and production

Sikkim is the leading state in area (26,151 ha) and production (4,200 t) followed by Darjeeling in West Bengal. Arunachal Pradesh, parts of Himachal Pradesh and the Garhwal hills of Uttaranchal have just initiated its cultivation with the present average yield being only 160 kg/ha. The bulk of the large cardamom produced is consumed in the country and a limited quantity (about 200 to 400 Mt) is exported to Singapore, Pakistan and the United Kingdom. Among the other countries, Bhutan produces about 3000 to 3400 Mt and Nepal around 1000 Mt of large cardamom annually.

Description of the plant

Large cardamom is a herbaceous perennial with subterranean rhizomes, which gives rise to leafy shoots and spikes. The mature plant height ranges from 1.5 to 3.0 m. Leafy shoots are formed by long sheath-like stalks encircling one another. The leaves are green or dark green, glabrous on both the surfaces, with an acuminate apex. The inflorescence is a dense spike on a short peduncle bearing 40 to 50 flower buds in an acropetal sequence. The fruit is a trilocular, many-seeded capsule. The capsule wall is echinated and is reddish-brown to dark pink in colour. Wild species of *Amomum* also exist in Sikkim. Their distribution and uses are given in the table on pg 238.

Soil and Climate

Large cardamom is a shade-loving plant (seophyte) and grows luxuriously from 600 to 2300 m above sea level in a humid, subtropical natural habitat on slopes under the canopy of lofty, evergreen forest trees. Moderate shade, high humidity and cool surroundings are essential for its satisfactory growth and production. The temperature varies from 10-30°C in the growing areas. The average annual rainfall ranges from 200 to 350 cm spread around 200 days. It is highly sensitive to wind and drought.

The crop grows best on well-drained, rich forest soils with plenty of humus and leaf mould. Since the plant has a lot of surface feeding roots, the soil should essentially be well-drained. Cardamom plantations established on rich and well-drained soils, besides giving

Distribution and uses of different Amomum *species*

Species	**Common name**	**Country**	**Uses**
A. aromaticum Roxb.	Bengal cardamom or Nepal cardamom	Eastern India, Pakistan	Rhizomes are used as a spice and flowering shoots are used in curries
A. compactum Soland	Round cardamom	Malaysia, Java	Fruits are used as spice
A. glabosum Cour	Round Chinese cardamom	China	Seeds are used as spice
A. krervanw Pierre	-	Cambodia, Indo-China	Fruits are used as spice and to flavour curries, sausages and cordials
A. maximum Roxb.	Java cardamom	Malaysia	Used as spice
A. xanthioidies Wall.	Wild bastard Siamese cardamom	Myanmar, India	Used as spice

higher yields, maintain their productivity for a considerably long period, sometimes for more than 20 years. Water-logged areas are not suitable and hence adequate drainage is essential for better growth along with a pH of 5.8 to 6.5.

Varieties

There are mainly 6 varieties of large cardamom. In addition, there are several sub-varieties or strains which are named in the local dialects of Lepchas, Bhutias and Nepalese. These varieties are *Bebo, Bharlangy, Golsey, Ramla, Ramsey* and *Swaney*. Among these, three popular cultivars grown in Sikkim are *Ramsey, Sawney* and *Golsey*. Other cultivars are derivatives of these varieties. These cultivars are white *Ramla, Chivey Ramsey, Garday Seto Ramnag, Madhusay (Madhusai)*, *Seto Golsey*, *Slant Golsey*, *Red Sawney* and *Mingney*. The same cultivar is called by different names in different localities, as variations in growing conditions cause changes in plant colour, whereas, harvesting at different stages of fruit development changes the fruit-skin colour.

The characters of the three major cultivars are

Ramsey or Ramsmey

It is a Bhutia word which means yellow colour (*ram*–colour, *shai* or *sey*-yellow). Plants of this cultivar are tall (approx. 2.5 m) having many shoots (10-15) with thin and long leaves. The fruits are the smallest in size and are of inferior quality. The cultivar is grown mostly on higher altitudes (around 1,500 m and above). In lower altitudes, the plants are severely affected by the Forkey virus disease.

Sawaney or Sawney

This is a Nepali word which means that the cultivar is harvested in *sawan*, i.e., August. The plants are tall (approx, 2.5m) but the leaves are wider and shorter compared to those of *Ramshai*. The fruits are bold in size and brown in colour. This cultivar is very popular in the lower altitudes (below 1,500 m) as it is comparatively less susceptible to the Forkey virus disease.

Golshai or Golsey

This is a derivative of the word *golraund* in Hindi and *shai* or sey-yellow (Bhutia), indicating the round shape and yellow colour of the fruits. The plants are short with fewer shoots (5-6). The leaves are comparatively short and broad. The fruits are big and bold and are sold at premium prices because of their good appearance.

The salient features of different cultivars, studied at the ICAR Complex for NEH Region and Sikkim are given in the table below.

Salient characters of large cardomom cultivars

Sl.No.	Particulars	Golsey	Sawney	White Ramna
1.	Number of bearing shoots per clump	11.4	25.2	21.0
2.	Number of flower panicles per clump	5.8	12.2	26.0
3.	Number of capsules per panicle	8.9	9.9	7.1
4.	Total number of capsules/clump	51.6	120.8	185
5.	Weight of fresh capsules/clump	243.6 g	263.8	425.5 g
6.	Weight of dry capsules/clump	50.0 g	64.1 g	98.0
7.	Single fresh capsule weight	4.7 g	2.1 g	2.3 g
8.	Single dry capsule weight	0.97 g	0.53 g	0.53 g
9.	Number of plants/ha	12.345 (90x90 cm)	6.410 (1.25 × 1.25 m)	6.410 (1.25 × 1.25 m)
10.	Total weight of dry capsules/ha	617.8 kg	410.8 kg	628.6 kg
11.	Volatile oil content	2.8%	1.8%	1.5%

Dried capsule characters of large cardamom

Cultivars	Av. Weight of 100 capsules (g)	Av. Number of seeds/ capsule	Length of capsule (mm)	Diameter of capsule (mm)	Remarks
Red Sawney	64.00	40	20.6	13.5	Large-sized, greyish-brown capsules, with seeds of dull brown colour and moderate pungency
White Green Golsey	77.0	46	20.3	12.4	Medium to large-sized dark black capsules, with black seeds and high pungency
Sawney	54.3	38	19.9	12.1	Medium-sized, light brown capsules with brown seeds of lesser pungency. The volatile oil content of the seeds is 1.8-2.5%
White Golsey	80.0	59	22.0	14.0	Large-sized, pinkish-brown capsules with large black seeds of high pungency
Golsey	104.0	58	23.3	14.8	Large-sized, roundish, brown capsules, with light brown seeds that are highly pungent. The seeds contain 2-3% volatile oil
White Ramna	53.0	24	17.0	10.0	Small–sized, yellowish capsules with light grey seeds of less pungency

Five promising types of large cardamom, *viz.*, ACC. No-3 (Ramea from Singhik), ACC. No.-4 (*Ramea* from Singhik), ACC. No.-9 and (Ramasey green type), ACC. No 12 (Burlangey) ACC, No 14 (*Bebo*) and five relatives, viz., *Churumpa* (*A. dealbatum*), *Baklok* (*A. kingii*), *Belak, Tali* and *Jaker* (all *Amomum sp*) have been identified from 34 germplasm types maintained at the experimental farm of the ICAR complex at Tadung, Sikkim.

Cultivation

Propagation

Large cardamom can be prepared by seed as well as by vegetative methods. The seedling plantation starts bearing fruits after five years, whereas plants raised by separating the rhizomes start bearing after three years of planting. Therefore, the vegetative method is preferred for the commercial plantation of large cardamom.

Vegetative propagation

Large cardamom is generally propagated by planting a section of the rhizome, called a bulb or slip, from an established clump. Care should be taken to select rhizome slips only from disease-free, healthy and high-yielding mother plants. The plants propagated by this method come to the yielding stage two years earlier than those grown by seed propagation. However, repeated vegetative propagation reduces plant vigour and hence should be avoided.

Seed propagation

Seeds and sowing

The seeds for sowing in the nurseries should be collected from well-matured capsules selected from high-yielding mother plants. The capsules are dehusked just after harvesting and the seeds are thoroughly washed in water to remove the mucilage coating. The seeds are then mixed with wood ash and dried in the shade for 2 to 3 days. Now the seeds are ready for sowing. Generally, about 30 to 50% germination is obtained in large cardamom. To increase the percentage of germination, treating the seeds with concentrated sulphuric acid for 2 minutes or concentrated nitric acid for 10 minutes is advised.

Primary nursery

Beds of 5 m × 1 m and about 15 cm to 25 cm in height are prepared. A thin layer of humus-rich jungle soil is spread on the beds. The seeds are sown in lines about 10 to 15 cm apart. They are covered with a thin layer of fine soil and the beds are mulched with paddy straw or any other dried grass to a thickness of 2 to 3 cm. The beds are irrigated daily to keep them moist. The mulch should be removed after germination and the beds should be kept free of weeds. There are two seasons for sowing the seeds: pre-winter (October-November) and post-winter (February-March). The seeds sown in October-November remain viable till the end of winter and start germinating in March. However, it is observed that the seedlings obtained from the pre-winter sowing are healthier and more vigorous compared to the post-winter sowing, hence pre-winter sowing of seeds is preferred. *Pandals* (thatched structures) may be provided to protect the seedlings from the sun. Regular weeding and watering should be done. The seedlings are ready for transplanting in June-July.

Secondary nursery

Here, the beds are prepared as in the primary nursery. The seedlings are transplanted to these beds in rows at a spacing of 10 cm between plants and 15 cm between rows. After about 12 months, the seedlings will be ready for planting in the main field.

The following steps are involved

Primary nursery (sowing during November-December)
↓ Germination (March-April)
Transplantation (August-September)
Secondary nursery
↓ Transplantation (the following, June-July)
Main field

Land preparation

Virgin forest areas, if selected for plantation, should be cleared of all undergrowth and trees not suitable for the relatively heavy shade needed by cardamom. Where the tree growth is sparse and shade insufficient or unsatisfactory, quick-growing trees are planted for temporary shade as well as other trees with a tall and spreading habit for permanent shade. The trees selected for providing permanent shade in cardamom plantations should have the following characteristics:

1. A tall and fast-growing habit, so that within 3 years of planting the tree attains sufficient growth to provide shade for the already-flowering cardamom plants.
2. The tendency to develop ample canopy so that sufficient shade is provided
3. The tree should provide maximum shade during the dry period.
4. The leaves should decompose quickly when they fall on the ground.

Utis (*Alnus nepalensis*) has been recommended as an ideal shade tree. Other important shade trees are Chillowne, *Schima wallichi*, *panisaj*, *Bucklandea populnea*, Malato, *Macaranga denticulate* and *Edgeworthia gardneri*. The shade trees are planted with a spacing of 7-10 m.

Planting

Bulbs or slips or seedlings along with 1-2 shoots are planted in the prepared pits (30 cm × 30 cm) at 150 cm × 150 cm distance for *Ramshai* and *Sawaney* and at a 90 cm × 90 cm distance for *Golshai*, by digging a hole in the soil. The planting is done at 8-10 cm depth. After a few showers, the pits may be filled with surface soil. Well-rotten cattle manure, compost or leaf-mould should be mixed thoroughly with the top soil before planting. It is also advisable to add 100 g rock phosphate per pit and mix it with the top soil before filling the pits. Planting is done in June-July when there is sufficient soil moisture, atmospheric humidity and optimum temperature for growth.

Mulching

Soon after the planting, the base of the plant should be mulched during November-April with dried leaves. Mulching will preserve the soil moisture and provide a source of nutrients after decomposition. Organic manures like FYM, compost, leaf-mould and humus, rich forest soil may be applied. As the soil is rich, generally no fertiliser is recommended.

Manures and fertilisers

As the inherent capacity of the soil to supply nutrients is high due to the presence of high organic matter, fertilisation is not a common practice in cardamom. But it has been observed that the application of 30 kg N along with 30 tonnes of FYM or compost per hectare every year increases the yield of large cardamom. Wood ash is also effective. About 10 kg cowdung manure or compost along with 10 g N and 25 g superphosphate per clump are added each year for higher production.

Irrigation

Irrigation may be provided once in every 10 days during December-April through ground channels or by drip or sprinkler sets.

Weed control

Large plantations require regular weeding during the first (3-4 times) and second (2-3 times) year of planting and, thereafter, the problem lessens as the foliage covers the entire field. The intensity and frequency of weeding depends upon the weed species, population and diversity. Weeding by scraping the soil surface is a common practice and it should be discouraged because it adversely affects the soil structure. Instead, weed threshing should be carried out twice a year once during May-June and the second time before harvesting.

Shade regulation

Tall growing trees are trimmed regularly to a height of 4-5 m to encourage the spreading habit with renewed vegetative vigour to provide uniform shade and higher quantity of leaf-litter.

Replanting

Large cardamom starts bearing from the 3rd year after the final transplanting into the main field and continues to yield up to 25 to 30 years; but the economical production period is from the 5th to 10th year after which the yield decreases. Thus, replanting has to be done after the 9th year, gradually, as it is difficult and uneconomical to destroy the entire plantation at one time. However, the removal of the entire plantation is resorted to when virus diseases appear.

Pests and diseases

Pests

Although 21 insect species are associated with the crop, only a few of them cause severe damage.

Leaf-eating caterpillars (Clelea plumbioda)

The caterpillars feed gregariously on the chlorophyll of the leaves, leaving only the veins and midribs. *Eupterote fabia* is a sporadic polyphagous pest, causing severe damage to the crop. The larvae feed voraciously on the leaves causing complete defoliation and reduction in yield. Spraying of Fenitrothion 50 EC (0.05%), Dichlorvos 40 EC (0.03%) and Carbaryl (Sevin 50 WP 0.02%) is effective in controlling the leaf-eating caterpillars.

Borers (Glyphepteuix sp.)

This pest is specific to large cardamom. The young larvae bore into the shoot and feed on the central leaf core causing 'dead heart' symptoms, finally resulting in the death of the shoot.

Scolytid beetle

Scolytid beetles are serious pests of immature capsules. They feed on the soft seed mass and make the capsule hollow. Spraying of Thiodan

(0.1%), Nuvacron (0.1%), Ekalux (0.1%) and Dimecron (0.1%) controls the pests.

Leaf-eating beetles and weevils

Chrysamela chlorine and *Basiolepta femoratums* scrape away the green matter of the leaves resulting in the formation of numerous perforations on the leaf surfaces. The application of Thiodan (0.1%) and Nuvacron (0.1%) controls the pest.

Aphids

Most of the aphids are associated with the transmission of virus diseases, hence the damage caused by them is mostly indirect. However, a few of them cause different adverse effects by colonising at the base of clumps. Among the aphids, the banana aphid (*Pentalonia nigronervosa*), the cardamom aphid (*P. caladi*), pink aphid (*Micromyzus kalimpongensis*) and corn aphids (*Ropholosiphum maidis* and *R.padi*) are important. They can be controlled by spraying Rogor or Dimecron (0.1%).

Root grubs

Polyphagous white grubs, which are c-shaped, infest the roots and rhizomes. Dusting Heptachlor (5%) and Aldrin dust (5%) controls the grubs. The leaf hoppers, *Kolla mimica* and *K.opponens*; the thrips, *Rhipiphorothirps cruentatus;* and mealy bugs are considered as minor pests. Cardamom moth and psocids damage the cardamom in storage.

Nematodes

Root-knot nematodes, such as, *Meloidogyne* sp. infest large cardamom. The affected seedlings are stunted with numerous thin grassy shoots. Drenching the nursery beds with Formaldehyde (2%) and covering them with polythene sheets for 72 hours and treating the beds with granular nematicides like Carbofuran or Aldicarb was found effective in controlling nematode infestations.

Other pests

Rodents, monkeys and wild cats are also serious pests attacking plantations during the harvest period and it is suggested that poison baits and traps are the best methods to control these pests.

Diseases

Viral disease

Chirkey

This is a major disease of large cardamom. The corn aphid *Rhopalosiphum maidis, R. padi* and *Brachycaudus helichrist* are the vectors for the disease. The symptoms of the disease are very similar to that of '*Katte*' disease in small cardamom. In the young leaves, discrete pale green or yellow longitudinal stripes running parallel to each other can be seen, but in the mature leaves these symptoms are masked. In the advanced stages, flecks measuring about 2.5 mm and the characteristic mosaic symptoms are also seen. The affected plants produce flowers and spikes but without capsules.

Forkey

The Forkey disease causes severe damage in the Darjeeling district of West Bengal. It is transmitted through the vector *Micromyzus kalimpongensis*. This disease is characterised by production of bushy growth of stunted shoots which are sterile. Diseased plants survive for a few years but remain sterile and unproductive.

Control

(1) The infected plants should be uprooted and the surrounding area drenched with systemic insecticides (Rogor or Dimecron or Metasystotox at 0.1%) to check the aphid population as well as the movements on the main host and collateral hosts such as corn, banana, peach and squash, (2) Collateral host plants in the vicinity should be destroyed. (3) Planting rhizomes/slips collected from diseased plantation should be avoided. (4) Healthy seedlings raised from good seeds should be used.

Fungal diseases

Seedling-rot and collar-rot

Collar-rot, caused by *Fusarium oxysporum,* is very destructive to large cardamom plantations. At the nursery stage, it is called seedling-rot and at maturity stage, it is called collar-rot, clump-rot or spike-rot. The primary symptoms include paleness of the leaves followed by yellowing and decaying of the collar, which gradually extends towards the base. Normally, the flowers also remain unopened, and even if a few do open, they do not bear fruits and ultimately the spikes rot or wilt. This disease is locally called 'agulata'. Continuous rainfall during the flowering period, water stagnation and old plantations with degenerated clumps favour the spread of this disease.

Leaf-spot

This is also widespread in occurrence and is caused by two pathogens, viz., *Pestalotiopsis vericolor* and *P. royenae.* The former produces angular, irregular spots with reddish-brown margins surrounded by chlorotic holes and is characterised by numerous, elongated, transluscent streaks appearing on the young leaves along the veins. The streaks turn reddish-brown within 3 to 4 days with a central straw coloured necrotic area surrounded by prominent dark brown margins.

Leaf rust

The rust disease, caused by *Phakospora eleuaria*, is observed on the ventral surface of the lower leaves in the form of numerous minute, brown uredosori mainly in May and June. In the early stages of development, the uredosori are surrounded by chlorotic holes. In the advanced stage, the whole leaf is covered by uredosori causing premature drying of the leaves.

Capsule-rot

This is caused by *Pestalotiopsis* sp. The affected capsules become slimy, water-soaked and have a discoloured appearance. The seeds from diseased capsule emit a foul smell.

Control of fungal diseases

Cultural practices, such as rouging of plants, field sanitation by providing drainage in nurseries and plantations, removal of diseased debris and weeds are effective in controlling the diseases. Drenching/spraying once or twice before the onset of the monsoon and just after flowering with Bavistin (0.3%), Rovral (0.5%), Captafol (0.2%), Bordeaux mixture (1.0%) or Dithane M-45 (0.25%); and drenching the soil with Thiram (0.2%) or Dithane M-45 (0.25%) during February-March in and around the clumps are a few general control measures.

Harvesting and yield

The crops raised by seeds start fruiting after 5 years, while vegetatively propagated crops fruit after 3 years of planting. Flowering starts from April-May and continues up to June-July. After fertilisation, the fruits ripen within 5-6 months. Harvesting is done by cutting the mature panicles with the help of a long narrow knife. The harvesting span is between August and November. The economic yield period varies between 12-15 years; however, some well-managed plantations can yield profitably up to 20 years. During the third year, when the first harvest is done, the yield is negligible (<25 kg/ha). The yield gradually increases and reaches the maximum in the sixth or seventh year. The yield at this stage varies greatly (3 to 10 q/ha of dry cardamom), depending on the management.

Curing

The fruits are separated from the harvested panicles for drying and curing. The freshly-shelled fruits are dried on a kiln. Over the kiln, a thin bamboo structure, just like a mesh, is placed, on which the freshly-harvested fruits are spread out for curing. Wet and freshly-cut wood which does not burn well and gives out a lot of smoke is used for fuel in the kiln, so that temperature does not rise too high. For good curing, a low temperature is required. Once the kiln is started, it takes about three days of continuous smoking for complete curing or drying. A kiln of 2.4 × 2.4 × 2.4 m has the capacity to dry 12 bags (40 kg each) of fresh fruits. Approximately 25% of the fresh weight of the fruits is recovered after the drying is completed. The dried capsules are dark pinkish-brown in colour and have a smoky odour and the dried material cannot be stored for a long time. The flue-pipe system of curing may be used as it is better than the kiln system, since the colour of the capsules remains a shiny pink in the former method. In this system, a flue from the furnace passes through the pipes. The harvested wet cardamom is spread over wire meshes fixed above the flue pipes. The product is dried by the movement of hot air. It takes about 24-30 hours for the produce to dry. The cardamom cured by this system retains its colour and flavour, which fetches a good price in the market.

Storage

The cured produce needs to be packed in insect-proof bags. Coal-tar coated and polythene-lined gunny bags are effective against insect attack during storage.

Value-added products

The value-added products of large cardamom are the essential oil of the seeds and the oleoresin.

Long Pepper

Plant Profile

Family	:	*Piperaceae*
English name	:	Long pepper
Indian name	:	*Pimpli, Pipli, Piplamul* (Hindi), *Hippali, Hippaliballi, Kuna,*(Kannada), *Tippali, Pippali* (Malayalam), *Pippali* (Sanskrit), *Tippili, Sirimulam* (Tamil), *Tippili, Pippuloo* (Telugu).
Species	:	*Piper longum* Linn. *P. pepuloides,* P. officinarum, P. betle, P. nigrum.
Distribution	:	India, Malaysia, Nepal, Sri Lanka, Rhio, Timor, Philippines.
Uses	:	Spice, pickles and preserves.

Long pepper is the dried unripe spikes of *Piper longum* Linn., a slender, aromatic, perennial climber belonging to the family *Piperaceae*. The spikes of this plant contain the alkaloids piperine (4-5%) and piplatin, and two new alkaloids one of which is designated as alkaloid A and is closely related to pellitorine and three more new alkaloids piperolactam A,—piperolactam B and piporadione. The roots contain the alkaloids piper longuminine (0.2-0.25%), piper longumine (0.02%) and piperine. Further purification of piper longumine has yielded six known alkaloids—cepharadione B, cepharadione A, Cepharanone B, aristolactam A 11, norcepharadione B and 2 hydroxy 1 methoxy 4 H dibenzoquinoline-4,5 (6 H) dione; lignins, *viz.*, pluriatilol, fargosin, sesamin, asarinine, guineensine and pipercide. Besides, the dried spikes on steam distillation yield 0.7% essential oil with a spicy odour similar to pepper and ginger oil.

The fruits are used as a spice and also in pickles and preserves. They have a pungent peppery taste and produce salivation and numbness of the mouth when eaten or mulched. The root is used to ferment rice-beer. The leaves are reported to be chewed like betel leaves in the Andaman Islands.

The roots constitute the important drug known as *Piplamul,* used both in *Ayurvedic* and *Unani* systems of medicine. The root is pungent, heating, stomachic, laxative, antihelmintic, carminative, improves the appetite, useful in bronchitis, abdominal pains, diseases of the spleen, tumours, ascites and causes biliousness. The unripe fruit is sweetish; cooling and useful in biliousness. The ripe fruit is sweetish, pungent, stomachic, aphrodisiac, alternative, laxative, anti-diarrhoeic, anti-dysenteric; useful in '*vata*' and '*kapha*', asthma, bronchitis, abdominal complaints, fevers, leucoderma, urinary discharges, tumours, piles, diseases of the spleen, pains, inflammations, leprosy, insomnia, jaundice, hiccoughs, tuberculous glands and increases biliousness (*Ayurveda*).

The root and fruit are used in palsy, gout and lumbago. The root has a bitter, hot sharp taste; it is a carminative, tonic to the liver, stomachic, emmenagogue, abortifacient, aphrodisiac, haematinic, diuretic and digestive. It is a general tonic, useful in inflammation of the liver, pains in the joints, lumbago, snake-bite, scorpion sting, night blindness (*Yunani*), etc.

In Travancore, an infusion of the root is prescribed after parturition, to facilitate the expulsion of the placenta. It appears to partake, in a minor degree, of the stimulant properties the fruit.

As an alternative tonic, long pepper is recommended for use in a peculiar manner. An infusion of three long peppers should be taken with honey on the first day, then for ten successive days the dose should be increased by 3 peppers every day, so that on the tenth day the patient will take thirty at one dose. Then the dose should be gradually reduced by 3 daily, and finally the medicine should be stopped. Thus administered, it is said to act as a valuable tonic in paraplegia, chronic cough, enlargement of the spleen and other abdominal viscera. Long pepper enters into the composition of several irritating snuffs; boiled with ginger, mustard oil, buttermilk and curds it forms a liniment used in paralysis. In the Konkan areas, the roasted fruits are beaten up with honey and given in cases of rheumatism; they are also given powdered with black pepper and rock salt (2 parts of long pepper and 1 of salt) in half *tola* doses.

The dried immature fruit and the root in the form of a decoction are extensively used in acute and chronic bronchitis with cough, and are found to give gradual relief in all such cases.

The alcoholic extracts of the dry fruits and aqueous extracts of the leaves are reported to show activity against *Micrococcus pyrogens* var. *aureus* and *Escherschia coli*. The ether extract of the fruits shows larvicidal properties.

Origin and distribution

Long pepper is a native of the Indo-Malayan region. It is found growing wild in the tropical rain forests of India, Nepal, Indonesia, Malaysia, Sri Lanka, Rio, Timor and the Philippines. Indian long pepper is mostly derived from the wild plants, but is also grown in small areas in the Khasi hills, the lower hills of West Bengal, Eastern Uttar Pradesh, Madhya Pradesh, Maharashtra, Kerala, Karnataka, Assam, Meghalaya, West Bengal, Orissa, Gujarat and Tamil Nadu. It occurs wild in the forests of Andhra Pradesh and the Andaman and Nicobar Islands.

Description of the plant

Piper longum Linn (2n=24, 48, 52, 96) is a slender, climbing undershrub, creeping and rooting below. The young shoots are downy; with leaves 5-9 cm long, 5 cm wide, ovate, cordate with broad rounded lobes at the base, sub-acute, entire, glabrous; and bearing unisexual flowers in solitary erect spikes during or just after the rainy season. The male spikes are larger, slender and are 2.5 to 7.5 cm long while the female spikes are sessile and blunt at the tips and erect, 1.25 to about 2 cm long when in the flowering stage and growing to about 3 cm when in the fruiting stage. The fruits (drupe) are ovoid, yellowish -orange, minute, sunk in the thick rachis, about 0.25 cm in diameter.

Wild species and varieties

The Indian long pepper appears to be derived from two or three species, including the one which is imported from Malaysia and Indonesia. It is a product of either *P. longum or P. pepuloides,* while the Java long pepper is from *P. officinarum* DC. (Syn. *Chavica officinarum* Miquel). The long pepper of English commerce is chiefly the produce of Java.

In the different agro-climatic regions of the country, a number of geographical races exist. The Assam, West Bengal and Nepal races are the most popular among them. The West Bengal type is the *Gol Thippali* (ball-round type). The Maharashtra type is *Pipal Nonsori*. The races found in Assam are *Asali* (true) and *Suvali* (female in Assamese).

Under the All India Coordinated Research project on Medicinal and Aromatic Plants, a selection from 'Cheemathippalli', which bears female type spikes possessing the highest spike weight and total yield, has been released under the name 'Viswam' from the Kerala Agriculture University for cultivation as an inter-crop in irrigated coconut plantations.

Soil

It flourishes well in a rich, nicely-drained loamy soil. It is also cultivated on a large scale in limestone soil below the Cherrapunji region. Laterite soils rich in organic matter content, with good moisture-holding capacity, are also suitable.

Climate

The plant requires a hot moist climate and an elevation between 100 and 1000 m and a rainfall of 160-200 cm for its successful cultivation. At still higher elevations, the yield will be less. It can be successfully grown even in areas which receive heavy rainfall with high relative humidity. In its natural habitat, the plant is found growing as an undershrub. Hence, it should be grown under partial shade with 20-25% shade intensity.

Cultivation

Propagation

Long pepper is propagated through seeds, suckers or cuttings, or by layering of mature branches at the beginning of the rainy season. The establishment percentage of 3-5 noded rooted cuttings is reported to be very high. The suckers establish very well in heavy rainfall areas. Vine-cuttings can be rooted in polythene bags filled with the common pot mixture. The nursery can be raised during March and April. In the nursery, mealy bugs attack the roots of the rooted cuttings. This can be avoided by mixing 10% DP with the potting mixture. Similarly, excessive irrigation may lead to *Phytophothora* wilt of the rooted cuttings. Irrigation should be done on alternate days to avoid dampness and also to control the wilt disease. The cuttings planted in March-April will be ready for planting in the main field by the end of May. Recently, a method of large-scale multiplication in long pepper has been developed in Kerala. This has been achieved by rapid plant regeneration through indirect adventitious shoot development.

Land preparation

The land should be ploughed 2-3 times and levelled properly. Then, raised beds of 3 m × 2.5 m are prepared in which pits are dug at a spacing of 40 cm × 40 cm or 60 cm × 60 cm. The pits are filled with soil mixed with 100 g per pit of well-decomposed FYM or compost. In order to avoid water stagnation in the pits and beds, channels are made to drain away the excess rain water.

Planting

The rooted cuttings or suckers are planted in the pits at the rate of 2 per pit, with the onset of the monsoon. The pits are gap-filled one month after planting. One year after planting, each plant is reported to bear 7 primary and 6 secondary roots with lengths of 12 and 7 cm, respectively. Long pepper is planted as an inter-crop with subabul, Eucalyptus and coconut in different parts of the country.

Manuring

This crop needs heavy manuring. Nutritional studies at the University of Agricultural Sciences, Bangalore, have shown that an application of 20 t of FYM along with 50-80 kg of N, 20-60 kg of P_2O_5 and 30-70 kg of K_2O per hectare is required for the optimum growth and yield of the crop. Of this, 50% N and the entire dose of P_2O_5 and K_2O is given as a basal dose and the remaining N is given as a top-dressing.

Irrigation

The crop should be irrigated once a week if it is grown as a pure crop. If the crop is grown as an inter-crop, the irrigation provided to the main crop is sufficient. The sprinkler system of irrigation may be adopted for economising on water. In Kerala, it is the unirrigated crop which has shown more vigour and hardiness than the irrigated crop. It is observed that summer irrigation induces continuous spike formation during the off-season.

Inter-culture

During the first year of planting, weeding is done whenever weed growth is noticed in the beds. After the application of FYM to the beds, earthing-up is done from the channels. Once the crop grows and covers the inter-spaces of the bed it is difficult to do any inter-cultural operations.

Mulching

When the crop is not irrigated during summer, the beds may be mulched with dry leaves or straw.

Pests and diseases

Pests

Mealy bugs infest the roots of plants and suck its sap. The infested plants look stunted in growth. The severity of the damage caused by this insect is more in the summer. The insect can be controlled by the application of systemic insecticides like Rogar. *Helopeltis theivora* is another insect whose adults and nymphs feed on the tender foliage causing necrotic lesions around the feeding punctures, leaving shot-holes on the lamina. The application of neem kernel extract at 0.25% concentration controls this insect.

Diseases

During the monsoon season, the fungus *Collitotrichum* sp. causes leaf and vine rotting. This fungus and *Cercospora* sp. cause necrotic spots and blight on the leaves during the warm months. Sometimes, these two organisms may cause partial or total crop loss. A spray of Bordeaux mixture at 1% concentration during May, and 2 or 3 sprays subsequently during the rainy season is the suggested control measure. Besides this, a virus-like disease resulting in yellowing and crinkling of the leaves with stunted growth of the plants is also reported. The affected plants produce spikes of reduced size and inferior quality.

Harvesting, drying and yield

The spikes are picked from the vines six months after planting; they are ready for harvest 2 months after their formation on the plants. The spikes are picked when they are blackish-green and most pungent. If the spikes are not picked at this stage, their pungency is lost to a great extent. In Bengal, picking is done annually in January. while in Kerala the spikes are picked 3 to 4 times. The harvested spikes are dried in the sun for 4 to 5 days until they are perfectly dry. The green-to-dry spike ratio is around 10:1.5. The dried spikes are then stored in moisture-proof containers. During the first year, the dry spike yield is around 400 kg/ha. The yield increases thereafter upto 3 years and it is about 1000 kg/ha during the third year. After three years, the productivity of the vines decreases and they should be replanted.

Besides the spike, the thick parts of the stems and roots, which have medicinal value, may also be harvested 18 months after planting. While harvesting, the stems are cut close to the ground, the roots dug up, cleaned and heaped in shade for a day, after which they are cut into 2.5 to 5 cm long pieces. The average yield of roots is 500 kg/ha.

Grading

The dried thicker parts of the stems and roots are called *piplamool*. There are three grades of *piplamool*. Grade-I with thick roots and underground stems fetches a higher price than Grade-II and III, which consist of either roots, stems or fragments.

Malabar Tamarind

Plant Profile

Family	: *Guttiferae*
English name	: Malabar tamarind
Indian name	: *Malabar Imli* (Hindi), *Kodampuli* (Kannada)
Species	: *Garcinia cambogia* (Cambojea), Syn. *G. gummigatta*
Distribution	: India, Thailand, Sri Lanka
Uses	: Condiment, marinade for fish, herbal medicines

alabar tamarind [*Garcinia cambogia* (Cambojea)] Syn. *G.gummigatta* of the family *Guttiferae*, popularly known as *Kodampuli* or cambojea, is an under-exploited fruit tree.

The fruits are valued for their dried rind. The fruit rind is rich in carbohydrate, (15% reducing sugars), fat, protein, minerals and acids. It contains 30% hydroxy citric acid (HCA) on a dry weight basis. The acid is in the isomeric form. The rind also contains tartaric acid (10%) and phosphoric acid (1.52%).

The dried rind is used as a condiment for flavouring curries, as a substitute for tamarind or lime. In Kerala, South Tamil Nadu and in parts of Karnataka, the fruit rind is used for making curry and to marinate fish. The extracts from the rind form a major ingredient of herbal medicines against various diseases. It lowers blood lipids, such as cholesterol and triglycerides by triggering fatty acid oxidation in the liver *via* thermogenesis (raising body temperature to speed up the body's metabolism which increases the burning of fats). It burns the fats softly and gently without the stimulation of the central nervous system. It also blocks the enzymes responsible for storing fat in our body, from glucose. It mobilises the body's fat stores and dissolves fat in the liver and throughout the body. It paves the way for gradual weight loss (obesity reduction) and supports the body's natural appetite suppression mechanism. In addition, it promotes the growth of lean muscles and is also safe for diabetics. A decoction of the fruit rind can be given against rheumatism and bowel complaints. The rind is also used for polishing gold, silver and as a substitute for acetic and formic acid in the coagulation of rubber latex. The translucent yellow resin of this tree has purgative properties and is also used in making varnish. The decoction of the bark is used against rheumatism and also has antiseptic properties. The leaves are astringent and sour and are used in treating diarrhoea, dysentery and dyspepsia.

Origin and distribution

This plant is indigenous to South Asia, particularly Thailand and India, and is fairly common in the forests of western Sri Lanka. In India, it thrives best in the evergreen forests of the Konkan stretching southwards to the Kerala coast and the Western Ghats of Karnataka and Goa and also up to 180 m in the Nilgiris in Tamil Nadu. It is commonly found in the backyards of homesteads.

Description of the plant

It is a small- to medium-sized tree with a rounded crown, having orthographic and plagiotropic branches. It is an andromonoecious, species producing male and bisexual flowers on separate trees. It flowers during December-February on the past season's shoot in the axils of the leaves. The flowers are solitary or in groups of 2-3 in bisexual trees and in clusters of 3-5 in male trees. Generally, the male flower buds are longer and slender, while bisexual ones are shorter and plump. The fruits are ovoid, 2-3" in diameter, yellow or reddish-yellow, 60 to 200 g in weight when ripe with 6-8 grooves. It has 4-8 seeds surrounded by a succulent aril. It takes 4-5 months for ripening; usually the ripening coincides with the onset of the monsoon.

Climate

Kodampuli is a hardy crop and can be grown in tropical and subtropical climates. The crop thrives in partial shade with good rainfall.

Soil

Soil is not a limiting factor for the cultivation of this crop. However, well-drained sandy loam soils, with good water-holding capacity, are ideally situated for its cultivation.

Varieties

At present, no improved varieties are available in this crop. A lot of variability is seen in *kodampuli* with respect to the size of the tree, bearing age, flowering season, fruit size, fruit colour, fruit shape, rind recovery and the yield. The identification and selection of early-bearing and high-yielding trees can benefit the growers so that the monsoons do not coincide with the ripening season. Such trees can be multiplied vegetatively.

Cultivation

Propagation

Generally, it is propagated through seeds. However, in this method the main problems are the long gestation period, bigger plant size and the fact that it is not possible to differentiate the male and bisexual plants. Further, plant selection for quality and yield is not possible. For seed propagation, plump and heavy seeds give about 80% germination. The seeds have a dormancy period of 8-9 months and then they start sprouting. The seeds with the seed-coat intact remain dormant in the soil for 8-9 months and then begin sprouting. On removing the seed-coat, germination occurs earlier (2-3 months of sowing). About 10-12% polyembryony is also noticed. The seedlings segregate into productive bisexual and unproductive male trees. The sex is revealed only after flowering, for which the seedlings take 7-8 years. The unproductive male trees should be removed after they show their sexuality. To overcome this problem, the vegetative propagation method can be followed.

The vegetative method of soft-wood grafting has been standardised. One-year-old *Garcinia cambogia* root-stocks are used for grafting 9-months-old scion sticks of 6-8 cm length. Soft-wood grafting done during June gives maximum success. Besides, vegetatively propagated plants come to flowering earlier (about 3-4 years after planting). It also results in dwarf, bushy trees, favouring high-density planting. To get normal growth habits, orthrotropic shoots should be used as scion material. If plagiotropic shoots are used, the resulting grafts are dwarf, bushy and drooping, with a reduced fruiting area. But orthotropic shoots are limited in number on the mother plants. The other root-stocks used are *G. indica* and *G. tinctoria*.

Land preparation

Pits of 1 m^3 size should be dug at the recommended spacing and filled with sufficient organic manure well in advance of the planting season.

Planting

Planting can be done during the onset of the first monsoon showers. One- year-old seedlings or soft-wood grafts can be used for planting. The seedlings are planted at a spacing of 10 m × 10 m and grafts 5 m × 5 m (200 plants/ha) apart.

Manures and fertilisers

Generally, inorganic fertilisers are not applied. However, the application of 50 kg FYM per tree per year at the beginning of the monsoon gives very high yields and keeps the trees healthy.

Inter-culture and irrigation

The bases of individual plants are mulched and the soil is turned over occasionally. Weeds, if any, should be removed regularly. It is suggested to provide water through drip irrigation. The dead and diseased wood should be removed from time to time.

Inter-cropping

Kodampuli can be grown as an inter-crop in coconut plantations under protective drip irrigation. Generally, 5 m × 5 m spacing is suggested, so that one row of cambogia can be planted in between two rows of coconut. The crop can also be tried in cardamom estates.

Pests and diseases

There is no serious pest or disease attacking this crop.

Harvesting and yield

For getting good quality, non-astringent rind, the fruits should be harvested in about 5 months at the ripening state. In order to escape the prevailing monsoon rainfall during ripening, the farmers collect the fruits prior to the monsoon at an under ripe-stage. But in this case, the rind is left with slight astringency. A well-managed, fully-grown tree yields 400-500 fresh fruits, from which about 58 kg dried rind can be recovered. However, in the compact grafts, the yield will usually be low.

Processing and preservation

Immediately after harvesting the fruits are cut longitudinally. The seeds along with the pulpy mass are removed in the case of ripe fruits. In the unripe fruits, only the seeds are removed leaving behind the mucilage along with rind, which further adds weight to the rind. Then they are dried under sun or smoked or both. The dried rind can be preserved by rubbing it with salt and coconut oil, which further improves the appearance and quality.

Mango Ginger

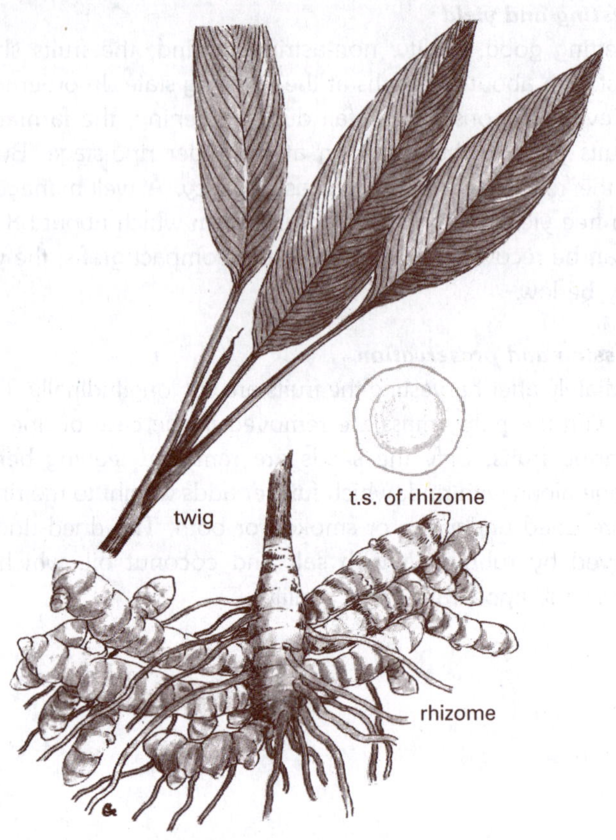

Plant Profile

Family	: *Zingiberaceae*
English name	: Mango ginger
Indian name	: *Amba haldi* (Hindi and Urdu), *Amaragandha haridra, Karpoora haridra, Suranayika* (Sanskrit), *Mavinasunthi* (Kannada), *Maingi, Mangai inji* (Tamil), *Mamidiallam* (Telugu), *Ambe haldi* (Marathi)
Species	: *Curcuma amada*
Distribution	: Asia, Asiatic Islands
Uses	: Vegetable, spice, condiment, medicine

ango ginger, botanically called *Curcuma amada,* of the family *Zingiberaceae* is a less-known spice. It is commonly called *AmrArdraka* and *karpooraharidru* in Sanskrit and *Ama-haldi* in Hindi. The underground rhizome is the economic part of this plant. The dry rhizomes on steam distillation yield a volatile oil, a pale-yellow fluid with a camphoraceous ginger aroma (0.8-1.1%). The rhizomes have less fibre (2-3.5%) compared to ginger (5-7%). The essential oil contains α-pinene, α and β-curcumene, camphor, curminyl alcohol, mystic acid and turmerione, car-3-ene and cis-ocimene, which contribute to the characteristic mango odour of the rhizome. The colouring matter is curcumin. Numerous sesquiterpenoids of germacrome and guaiane skeletons also are present.

The fresh rhizomes find use in many culinary preparations like pickles, salads, *chutneys*, preserve, candy, sauces and in meat and other products. Besides these, mango ginger finds extensive use in the indigenous systems of medicine. The rhizome is bitter, sweet-sour, and aromatic. It is useful in biliousness, itching, skin diseases, bronchitis, asthma and hiccoughs. The crushed pulp is applied over contusions, sprains and bruises for rapid healing. The rhizomes are cooling and can be applied over inflammations caused by injuries, for rapid cure. It is a good appetizer and has antipyretic properties and helps to clear the bowels (laxative), to increase the secretion of gastric juices (stomachic) and prevents formation and helps in the expulsion of gases from the stomach and intestine (carminative). The rhizomes are alexiteric and aphrodisiac. The roots also have many medicinal uses. They are reported to increase the flow of urine (diuretic), produce smoothening and softening influence on the applied part (emollient), increase liquidity and expel bronchial secretions (expectorant), reduce fever and increase appetite. The roots are a maturant and are useful in inflammations, troubles in the mouth, ear, gout, scabies, lumbago and stomatitis. They are astringent, useful in diarrhoea and prurigo and can be topically applied over contusions and sprains. They possess stomachic and carminative properties. The alcohol extract of mango ginger was reported to decrease the blood cholesterol level of experimental hyper-cholesterol rabbits. In another study, the rhizomes of mango ginger were found to have inhibitory activity against trypsin. The oil is found to have antifungal properties as well.

The wide variety of uses of mango ginger as vegetable, spice, condiment and medicine make the crop a promising one for commercial exploitation. Over the centuries, man has had a tendency to concentrate on fewer and fewer plant species and he is more or less dependent on only 20-25 plant species for his daily needs. This has resulted in the abandoning of crops like mango ginger. Since there is a need for producing more food for the rapidly increasing population, the role of cultivating under-exploited crops assumes more significance. The shade tolerance capacity of mango ginger is an added advantage which has to be exploited. Therefore, it can be safely included as one of the crop components for inter-cropping situations or in homesteads where many species of crops are grown.

Origin and distribution

Mango ginger is a native of China and various other parts of Asia and the Asiatic Islands. The crop is cultivated to a limited extent in certain parts of West Bengal, Assam, Western Ghats, Konkan coast, Andhra Pradesh, Tamil Nadu, Orissa and Kerala.

Description of the plant

Mango ginger is not botanically related to mango or ginger, but to turmeric (*Curcuma longa*). Though a perennial herb, it is cultivated as an annual for its underground rhizome.

Morphologically, the mango ginger plant is similar to turmeric, but has only a short crop duration of six months. The rhizomes more or less resemble those of true ginger at first sight and have the sweet smell of unripe mango, but lack the pungency of ginger. The rhizomes are pale yellow or green outside with a lighter colour inside.

Soil and climate

The crop prefers fertile soil with good drainage. It also grows well in sandy loam soils. Since, it is a shade-tolerant crop, it can be grown in partially shaded situations or as a crop component in homesteads where multi-species crops are grown. It can also be grown in open conditions. The crop is suited for rotation with vegetable crops. Hot, humid tropical climate with a high rainfall of more than 1500 mm is essential.

Varieties

Mostly, local varieties are used for cultivation. But recently, a variety called 'Amba' has been released for cultivation from the High Altitude Research Station, Orissa University of Agricultural and Technology, Pottangi.

Propagation

This crop is propagated through the underground rhizomes. The whole or split rhizomes or well-developed, healthy and disease-free rhizomes are suitable for propagation. The micro-propagation in mango ginger has been standardised. The rhizome explant is reported to produce both the shoot and root simultaneously when cultured on B5 medium containing NAA (0.5 mg/l) and BA (4 mg/l). The field establishment of the *in vitro* produced plants may range from 60-70%.

Land preparation

The field should be prepared by ploughing and thoroughly breaking up the clods, followed by levelling during February–March. After the first pre-monsoon showers in April, beds of 1.2 m width, 25 cm height and of convenient length are laid out, with a spacing of 40 cm between them. Cattle manure or compost at the rate of 30 to 40 t/ha is spread over the beds and mixed well in the soil.

Planting

The whole or split mother rhizomes or finger rhizomes, weighing 15-20 g, are suitable for planting during April or May-June. For local and 'Amba' varieties, small pits are made in the beds at a spacing of 25 × 30 cm and at a depth of 4-5 cm. About 1500 kg of seed rhizomes are required to cover a one hectare area.

Manures and fertilisers

In order to obtain the maximum yield from the crop, inorganic fertilisers should be supplied at the rate of 30 kg N, 30 kg P and 60 kg of K per hectare. The full dose of P and half dose of K may be applied as a basal dose. Two-third N can be applied 30 days after planting, and one third nitrogen and the remaining potash may be applied 60 days after planting.

Irrigation

Mango ginger is a rain-fed crop, but supplementary irrigation may be necessary during dry weather and also at the time of harvest.

Weeding and mulching

The rhizomes will sprout within 3-4 weeks after planting. After the complete emergence of the sprouts, weeding is done after 45 days of planting and is repeated if necessary. The crop should be mulched with green leaves at the rate of 15 t/ha, immediately after planting. Repeat the mulching after 50 days with 7.5 t/ha of green leaves, if it is necessary. The crop is earthed-up 60 days after planting.

Pests and diseases

Compared to the related crops, ginger and turmeric, this crop is relatively free from pests and diseases. But when large-scale cultivation is taken up, the shoot borer (*Dichocrosis punctiferalis*) causes damage to the crop. The affected plants with the appearance of dead-heart with the larvae inside have to be uprooted and burnt. If the infestation is severe Dimethoate or Quinalphos (0.05%) could be sprayed.

Harvesting and yield

The drying up of the leaves is the indication of harvesting time. From the sixth month onwards, harvesting can be done by lifting the entire plant with a spade and then cutting away the top portion. If a few mango ginger plants are grown in homesteads or home gardens, it may serve as an immediate remedy for some of the common stomach disorders, besides being used in food. Partial harvesting (removing a portion of the rhizome from the plant) is also practised in homesteads in Kerala and Karnataka to suit the family requirements. After the harvest, roots and soil particles are removed from the rhizomes by washing with water and it is advisable to dry the rhizomes in the shade for one or two days before storage or transplanting.

The yield of fresh rhizomes will be around 30-40 t/ha. A plant can yield upto 1-1.5 kg.

Mint

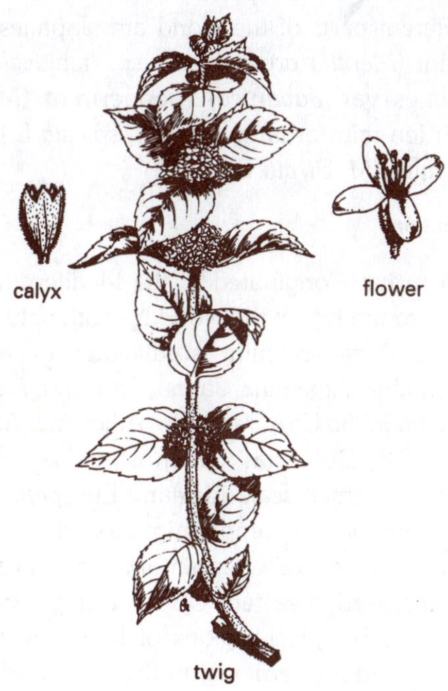

Plant Profile

Family	:	*Laminaceae/Labiatae*
English name	:	Japanese mint, Peppermint, Spearmint, Bergamot mint
Indian name	:	*Pudina* (Hindi, Bengali, Gujarati, Marathi, Punjabi and Urdu), *Pudyanu* (Kashmiri), *Muthina* (Malayalam), *Pudina(*Tamil & Telugu)
Species	:	*Mentha arvensis,* sub sp. *haplocalyx* Briquet, var. *piperscens* Holmes, var. *Javanica;* Peppermint (*M. piperita* L.); Spearmint or Garden mint or Lamb mint (*M. spicata* L.) and Bergamot mint or orange mint (*M. citrata* Ehrh.,)
Distribution	:	India, Brazil, Paraguay, USA
Uses	:	Cosmetics, culinary purposes, flavouring, perfumery

ints are a group of perennial herbaceous plants belonging to the family *Laminaceae*; *Labiatae,* which yield essential oil on distillation. The various species of mint which are commercially cultivated in different parts of the world are: Japanese mint or corn mint or field mint (*Mentha arvensis* sub sp. *haplocalyx* Briquet var. *piperscens* Holmes var. *Javanica*), peppermint (*M. piperita* L.), spearmint or garden mint or lambmint (*M. spicata* L.) and bergamot mint or orange mint (*M. citrata* Ehrh.).

Origin and distribution

Mint is believed to have originated in the Mediterranean basin and spread from there to the rest of the world by both natural and artificial means from there. Japanese mint is cultivated on a large scale in Brazil, Paraguay, China, Argentina, Japan, Thailand, Angola and India. Peppermint is grown in the USA, Morocco, Argentina, Australia, France, the erstwhile USSR, Bulgaria, Czechoslovakia, Hungary, Italy, Switzerland and on a small scale in many European countries. The USA is the major producer of peppermint oil in the world. Spearmint is mostly cultivated in the USA. Bergamot mint is a native of Europe and has been naturalised in eastern USA. It also grows successfully in the temperate and subtropical regions of India. In India, Japanese mint is grown mainly in the *Terai* region (Nainital, Badayun, Bilaspur, Rampur, Moradabad and Bareilly), extending to parts of the Indo-Gangetic Plains (Barabanki and Lucknow) and parts of Punjab (Ludhiana and Jalandhar) in North-western India.

The total area under mint cultivation, which is mostly confined to Uttar Pradesh and Punjab, is around 10,000 hectares.

Description of species of Mint

M. arvensis (Japanese mint) is a downy, perennial herb, spreading by root-stocks, which creep along the ground or just under the surface and root at the nodes. There are three horticultural varieties: one with a reddish-purple stem and broad obtuse leaves and the other two with green stems, broad or narrow leaves and with a purplish-green stem and narrow leaves, respectively. They have rigid branches that are pubescent and 60-90 cm tall. The leaves are lanceolate to oblong, 3.7-10 cm long, sharply toothed, or shortly petiolated and hairy. The flowers arranged in cymes which are usually sessile or rarely pedunculate. The flowers are purplish and minute. The calyx is 2.5-3.0 mm long, deltoid and acuminate. The corolla is white to purple coloured and 4-5 mm long.

M. piperata spreads by a system of branching, underground root-stocks and grows to a height of 45 to 90 cm (1½ to 3 ft). The square, branching stems are a dark-green (var. *officinalis* sole sub var. *pallescens* Camus) to deep purple colour (var. *officinalis* sole sub var. *rubescens* Camus) and bear opposite, broadly lance-shaped, slightly-toothed leaves of a deep-green colour that are 2.5 to 5 cm (1 to 2 in.) long. Small purplish flowers are borne in terminal spikes on the main stem and branches from June to September.

M. citrata grows up to 30-60 cm in height, with decumbent branches and erect ends. The leaves are 1.25-5.0 cm long, thin, bronzy-green, petiolate, smooth, broadly ovate to elliptical with an obtuse apex. The flowers are purple, appearing in the upper axils on short, dense terminal spikes. The calyx is glabrous with subulate teeth, while the corolla is glabrous and about 0.5 cm long. *M.citrata* is a hybrid between *M.aquatica* and *M.viridis*.

M. spicata (2n=41) is propagated by stolons, from which 30-60 cm-long, erect, ascending branches arise. The leaves are sessile, smooth, lanceolate or ovate-lanceolate, sharply serrate, smooth above and glandular below, with an acute apex and grow up to 6.5 cm long. The flowers are sharply pointed, long and narrow and hence the name spearmint is attributed to the plant. The calyx teeth are hirsute or glabrous and the corolla is about 3 mm long.

Another variety of spearmint known as scotch spearmint *M. cordiaca* (S.F. Gray Baker) is cultivated in the USA to a limited extent.

M.arvensis was first introduced into India in 1954 by the Drug Research Laboratory, now called the Regional Research Laboratory, Jammu and Kashmir whereas, peppermint was introduced into India during the British period on a very small scale. Scientific experiments were started in 1959 in the Kashmir valley. The bergamot mint was introduced into the country at the Regional Research Laboratory, Jammu and Kashmir, during 1959 and spearmint during the British period at the Regional Centre of the Central Institute of Medicinal and Aromatic Plants, Pantnagar.

Varieties

Japanese mint

Himalaya (MAS-1)

This is a selection from the Thai bud sprout released by the Central Institute of Medicinal and Aromatic Plants, Lucknow, which contains 0.8 to 1.0% oil (FWB), with 81% menthol and a low congealing point. Thus, 65% of it can be easily removed by chilling. The yield of oil is around 290-293 kg/ha, which is very high. This variety has a high regeneration capacity in the second and subsequent harvests and also a delay in harvesting time does not affect the oil production and quality. This variety is highly resistant to rust, leaf-spot and powdery mildew diseases.

Kalka (Hyb-77)

This is a tall, vigorous variety evolved by the Central Institute of Medicinal and Aromatic Plants, Lucknow, through a cross between MAS-1 and MAS-2. It produces 762 q of fresh herb and 468 kg oil/ha, containing 81.5% menthol. It is highly resistant to leaf-spot, powdery mildew and rust diseases.

Shivalik

This is an introduction from China and was released by the Central Institute of Medicinal and Aromatic Plants (CIMAP), Lucknow. The plant has a compact, bushy growth with thick leathery leaves. It compares very favourably with other cultures in terms of high oil yield and the menthol content in the oil.

Koshi (Sel.3)

This is a recently released variety from the Central Institute of Medicinal and Aromatic Plants, Lucknow, which has features like high leaf density, moderately broad and light green leaves, wide and globular plant canopy, very synchronous growth habit, high biomass, early maturing habit, high oil yield and high menthol content and resistance to diseases like leaf-spot, rust and powdery mildew.

Gomati

This is an early variety released by the Central Institute of Medicinal and Aromatic Plants, Lucknow, for higher yield and menthol content.

EC-41911

This is a progeny selection of an inter-specific cross between *M.arvensis* × *M. piperita*. It is less affected by rain, root-rot and aphids and is reported to yield 236.5 q/ha of herbage and 12.2 kg/ha of oil with a menthol content of 70%.

Pepper mint

Kukrail

This is a high-yielding variety developed and released by the Central Institute of Medicinal and Aromatic Plants (CIMAP), Lucknow.

Bergamot mint

Kiran

This is a mutant selection developed by mutation through gamma irradiation at CIMAP, Lucknow. It has a high oil content and quality. This variety can yield 239 kg of oil/ha with 48% linalool, over 172 kg oil and 39% (linalool) in the parent cultivars.

Spearmint

MSS-1

This is a selection from the spearmint cultivars introduced from the USA. This variety is released from CIMAP, Lucknow. It gives a fresh herbage yield of 535 q and an oil yield of 253 kg in 3 cuttings, containing 60% carvone.

MSS-5

This is a selection from MSS-1 made at CIMAP, Lucknow. This variety has potential yield of 600 q fresh herb and 349 kg/ha of oil from 3 cuttings. The carvone percentage in the oil is 70.

Punjab spearmint-1

This variety is a clonal selection made at CIMAP, Lucknow. This variety has an erect growth habit with a quadrangular purplish-green, hairy stem. It produces 209 q/ha of herb, 0.57% oil, and can yield 119.29 t/ha oil with 68% carvone.

Arka and Neera are the recently released varieties from CIMAP, Lucknow.

Supriya is a newly-released variety of garden mint (*M.viridis*) from CIMAP, Lucknow.

Chemical composition and uses

Japanese mint (M.arvensis)

Japanese mint is a primary source of menthol; the fresh leaves contain 0.4-6.0% oil. The main constituents of the oil are: menthol (65-75%), menthone (7-10%), menthyl acetate (12-15%) and terpenes (pipene, limonene and comphene). The menthol content of the oil varies depending on the climatic conditions. Generally, it is higher in tropical regions.

Menthol is used in the flavouring of a large number of pharmaceutical and oral preparations like toothpastes, dental creams, a number of confectionary products, beverages and other items like tobacco, cigarettes, *pan masala,* etc. Menthol itself finds extensive use in medicine (usually in combination with eucalyptus oil) for the treatment of nasal and bronchial catarrh. It is also occasionally employed as a carminative, and as an analgesic in the treatment of neuralgia and rheumatism.

Peppermint (M. piperita)

The fresh herb contains essential oils ranging from 0.4 to 0.6%. The constituents of peppermint oil are almost similar to Japanese mint oil. However, the menthol content is lower in peppermint oil and varies between 35 and 50%. The other constituents are menthyl acetate (14-15%), menthone (9-25%), menthofuran, and terpenes like pinene and limonene.

Peppermint oil is one of the most popular and widely used essential oils. It is usually considered as an *arvensis* type of oil and is commonly used in the flavouring of pharmaceuticals and oral preparations like toothpastes, dental creams, mouthwashes, cough drops, chewing gums, confectionery products and alcoholic liquors. Medicinally, it is an excellent carminative and gastric stimulant. When applied externally, it acts as a mild analgesic.

Bergamot mint (M.citrata)

Linalool and linalyl acetate are the main constituents of bergamot mint oil. The oil is directly used in perfumes and cosmetics. It is also used in cosmetic preparations like scents, soaps, aftershave lotions, colognes, etc.

Spearmint (M. spicata)

The principal constituent of spearmint oil is carvone (57.71%) and the other minor constituents are phellandrene, limonene, pinene and cineole. The oil is used mostly as a flavouring agent in toothpastes and as a food flavourant in pickles and spices, chewing gums, confectionery products, soaps and sauces.

Climate

Japanese mint can be grown in all tropical and subtropical areas under irrigation. However, it does not tolerate damp winters, which cause root-rot. A temperature of 20-25°C promotes vegetative growth, but the essential oil and menthol content are reported to increase at a higher temperature of 30°C or above under Indian conditions. Peppermint and spearmint cannot be grown profitably in tropical and subtropical areas especially those with very high summer temperature (41°C), and the ideal yield is obtained only in humid and low temperature conditions like Kashmir and the hills of Uttar Pradesh and Himachal Pradesh. Open, sunny situations, without excessive rains during the growing period, are congenial for good growth and development of the oil.

Bergamot mint can be grown both in temperature as well as subtropical areas. However, the yield is higher in a temperate climate.

Soil

Medium to fertile, deep soil, rich in humus is ideal for the cultivation of mint. The soil should have a good water-holding capacity, but water-logging should be avoided. A pH range of 6-7.5 is optimum.

Season

In the plains, planting is done during the winter months, whereas, in a temperate climate, planting is done in autumn or spring from the last week of December to the first week of March or from the first week of January to the third week of February. Late planting always gives a poor yield.

Cultivation

Propagation

Mints are propagated through the creeping stolons or suckers. In the case of peppermint and bergamot mint, sometimes even runners are planted. The stolons are obtained from the previous year's plantings. A hectare of well-established mint, on an average, provides enough planting material for ten hectares. On an average, about 400 kg of stolons are required for planting one hectare of land. The best time for obtaining stolons is during the months of December-January.

Land preparation

Mints require thoroughly ploughed, harrowed, fine soil. All the stubble of the weeds should be removed before the crop is planted. Manuring may be done at the time of land preparation by adding FYM @ 25 to 30 t/ha. Green manuring may also be done before mint is planted. Sun hemp (*Crotalaria juncea* L.) is an ideal green manure crop. Mints are planted on flat land or ridges. Thus, flat beds of convenient sizes or ridges are made according to the recommended spacing.

Planting

The stolons are cut into small pieces (7-10 cm) and planted in shallow furrows, about 7-10 cm deep, at a distance of 45-60 cm from row to row, manually or mechanically. While planting on ridges, the stolons are planted half-way down on the inner sides of the ridges. The plot is irrigated immediately after planting.

Fertiliser application

Mint responds very well to the heavy application of nitrogenous fertilisers. The increase in herbage by the application of phosphorus is not as remarkable as is the case with nitrogen. Generally, nitrogenous fertilisers @ 80-120 kg, phosphorus and potassium at 50 kg P_2O_5 and 40 kg K_2O per hectare is required for a good crop of mint. However, in *M. arvensis,* an increase up to 160 kg N/ha and in *M. piperata* 125 kg N/ha has given increased fresh herbage and essential oil yield. An amount of 100-120 kg N/ha is recommended for producing the optimum herb and oil yield in *M.citrata* under Pantnagar conditions. The split application of 75 kg N/ha in combination with phosphorus at 60 kg P_2O_5/ha is recommended under Kodaikanal conditions. Potassium application has no significant effect on the herb and oil yield. In *M. spicata,* the maximum herb yield is obtained with the application of 100-120 kg N/ha. Nitrogen may be applied in three split doses at 1½-2 and 3 months after planting and the third dose after the first harvest of the crop.

The deficiency of boron reduces both the yield of green herb and the essential oil in peppermint. Higher yields of peppermint oil are obtained by applying boron fertilisers. Increased yields of herb of peppermint, menthol content and essential oil content have been obtained by using a combination of boron and zinc fertilisers. The visual symptoms have been documented for some cultivars of Japanese mint towards Fe and Zn deficiencies. With respect to Zn, the crop response was maximum at 20 kg/ha of Zn applied during planting. Similarly, experiments conducted at CIMAP, Lucknow, have shown that the application of 20 kg sulphur per hectare will increase the herb and oil yield in *M. spicata*. Among the different sources of S, calcium sulphate was found to be the best followed by ammonium sulphate and elemental sulphur.

Irrigation

The water requirement of mint is very high. Depending upon the soil and climatic conditions, the crop is irrigated 6-9 times before the first monsoon. The crop requires another three irrigations after the monsoon during the months of September, October and November. Sometimes another irrigation is required during winter, if the growth is dormant and there are no winter rains to encourage the proper growth of the underground stems. Experiments conducted at Pantnagar have revealed that 15 irrigations are required to get the maximum herb and oil yield in Japanese mint. When mints are grown under a temperate climate, only 3-4 irrigations during the period from July to October are required.

Inter-culture and weed control

Uninterrupted weed growth causes about 60% reduction in herb and oil yields. Hence, mints require weeding and hoeing at regular intervals in the early stages of crop growth. One hand weeding is required after the first harvest. Sinbar is the only herbicide which controls a large number of weeds effectively, when applied as a post-emergence spray @ 1 kg/ha. However, combining organic mulch with a combination of 0.5 kg Oxyfluorfen herbicide per hectare and weeding or application of Pendimethion herbicide at 1 kg/ha and weeding are found to give excellent weed control throughout the period of crop growth. Dalapon (4 kg/ha), or Gramaxone (2.5 l/ha) as post emergent; Diuron (2 kg a.i/ha) or Terbacil (2 kg a.i/ha) as pre-emergent applications are also recommended for chemical weed control in mints.

In temperate areas, the plants become dormant in November. In order to give a perennial crop (of 3 years), only in the case of peppermint, the recultivation is done either in autumn (November-December) or in spring (March-April). When peppermint is grown as a perennial crop, the first year crop is called 'Row mint', while the second and third year crops are called 'Meadow mint'. This practice is not followed for the other mints which are planted every year.

Crop rotation

Crop rotations help achieve a reasonable control of weed growth, maintain the fertility of the soil and provide higher returns from the land. The following crop rotations are practised in Uttar Pradesh: (a) Mint - Maize and potato (b) Mint - early paddy and potato and (c) Mint - late paddy and sweet pea whereas, in Punjab, farmers practise mint - maize and rape seed/mustard, and mint - maize and potato, or mint and paddy rotation.

The recommendation for the *Terai* region of Uttar Pradesh is a 2 year rotation of mint with a summer fallowing or millet (fodder) followed by mint on poor fertility lands; and mint - wheat or paddy and mint on medium fertile lands.

Pests and Diseases

Pests

A large number of insect pests attack mints. Among them, the important ones are the leaf roller, pyralid, hairy caterpillar and termites.

The caterpillar of the leaf roller (*Syngamia abruptalis* Walk) attacks the mint crop during August-September. This can be controlled by spraying systemic insecticides like Monocrotophos or Dichlorovas (1 ml/lit.). The larvae of the Army worm-pyralid (*Laphygma exigua* Hub.) bore into the plant and eat the leaf buds during February-April.

The spraying of systemic insecticides like Monocrotophos will also control this pest. The attack of the hairy caterpillar (*Diacrisia obliqua* walk) during the months of April-May causes rapid defoliation. This can be controlled by using 5% Dipterex. During the dry months, termite attack is often observed. This can be effectively controlled by the soil application of 3% Heptafan @ 50 kg/hectare before planting. Other pests, like cutworm (*Agrotis flammetra*), semi-loopers and red-pumpkin beetle (*Aulucophora favicollis*), also cause damage to the crop. They can be controlled by spraying Thiodon. *Nisia atrovenosa,* the cabbage semilooper (*Tysanoplusia orichalsia*), gram caterpillar (*Helicoverpa armigera*), and aphid (*Aphis affinis*) are the other pests reported on this crop. The effective control of pests can be achieved by spraying Monocrotophos.

Nematodes like *Pratylynchus scribneri, Belonolaimus longicaudatus* and *Dolichodorus heterocephalus* severely damage the foliage yield of mint. These nematodes can be effectively controlled by the soil application of Fenamiphos @ 11.2 kg/ha.

Diseases

Rust, powdery mildew, wilt, leaf blight and stolon-rot are the five fungal diseases which have been reported to affect mint to a significant extent under Indian conditions.

Rust is caused by *Puccinia menthae* Pers., due to which severe leaf shedding is observed. *Erysiphe cichoracearum* DC causes powdery mildew, which is a major disease. These two diseases can be controlled by spraying Hexasul (Wettable sulphur @ 0.4%) and Karathane. In Japanese mint, stolon-rot is caused by *Macrophomina phaseoli* Hob. Ashy, *Rhizoctonia bataticola* and also by *Thielavia basicola* which adversely affects the production of stolons and mentha oil. Dithane M-45 (0.3%) and Brassicol (0.5%) can be used to control the disease. A new leaf-spot disease caused by *Curvularia lunata* and *Corynopsora cassicola* (Berk and Curt), Sclerotium-rot (*Sclerotium rolfsii*) and leaf blight (*Alternaria* sp) are also reported. Severe wilting of Japanese mint during the months of February and March has been observed in different parts of India; the disease is incited by *Verticillium arboratum*. The use of disease free stolons for planting, deep summer ploughing and well-drained field conditions help minimise its incidence.

Harvesting

Japanese mint is generally harvested after 100-120 days of planting, when the lower leaves start turning yellow. If the harvesting is delayed, the leaves start falling resulting in the loss of oil. Further harvesting should be done in bright sunny weather. Harvesting consists of cutting the green herbs that are 2-3 cm above the ground with a sickle. A second harvest is obtained about 80 days after the first harvest, and the third one after about 80 days from the second harvest. Whereas, in peppermint, spearmint and bergamot mints, which are grown in temperate climates, the first crop is ready by the end of June and the second in September or October.

Yield

A good crop of Japanese mint can give as high a yield as 48 tonnes of fresh herb per hectare. However, the average yield of mints from three cuttings is 20-25 t/ha. The fresh herb contains 0.4% oil.

Distillation of oil

Mint oil is obtained by distilling fresh or dry herbs. The distillation is done both in primitive and modern stills. In the former approach, the principle of water and steam distillation is followed, while in the latter, steam generated in separate boilers is employed. The stems are removed from the dried material prior to distillation, because they constitute 30 to 50% of the material and contain only traces of the oil.

The average yield of oil is 50-70 kg/ha, although bergamot mint as well as Japanese mint give an average yield of 70-100 kg/ha. The yield of peppermint oil is lower, with an average of 50 kg/ha.

Storage of oil

Mint oil is a light and golden coloured, motile liquid and should be completely free from moisture before storage. It should be stored in large, galvanized steel or aluminium containers, filled up to the brim to protect against any air remaining inside and placed in cool storage godowns away from light and humidity.

Mustard

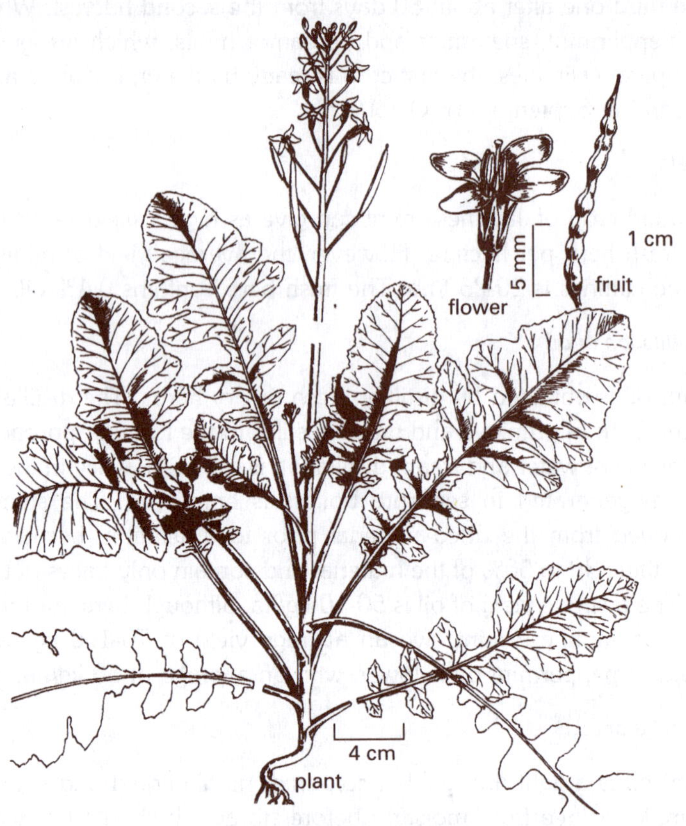

Plant Profile

Family	:	*Cruciferae*
English name	:	Mustard
Indian name	:	*Rai, Banarasi rai, Sufed rai, Kalee sarson* (Hindi), *Sasave* (Kannada), *Asuri, Bimbata* (Sanskrit), *Kadugo* (Tamil), *Avalu* (Telugu)
Species	:	*Brasica nigra* *B. alba / B. hirta / Sinapis alba* *B. juncea* (Linn.) Zernjajev
Distribution	:	Europe, Central and Southern Asia and China
Uses	:	Spice, condiment

he genus *Brassica,* belonging to the family *Cruciferae,* consists of over 150 species of annual or biennial herbs, several of which are cultivated as oilseed crops (as *rape, sarson, toria,* etc.) or as vegetable or fodder crops. Only the seeds of *B. nigra* Koch (*Banarsi Rai*), 'True Mustard' or 'Black Mustard'; *B. alba* or *B. hirta* or *Sinapis alba* L. (*Safaid Rai*), 'White Mustard' or 'Yellow Mustard' and *B. juncea* (Linn.) Zernjajaev, 'Indian Mustard' or 'Black Mustard' are valued as a condiment.

Origin and distribution

The plants of the genus *Brassica* are natives of the North Temperate Zone, but have passed into the sub-tropics and tropics as cold weather crops. There appear to be three major places of origin: Europe, Central and Southern Asia and China.

Area and production

Mustard occupies an area of 7.1 million hectares in the country and the production is 6.9 Mt. This crop is commonly cultivated in areas of marginal and sub-marginal productivity, either as pure, mixed or as an inter-crop. Assam, Haryana, Jammu and Kashmir, Madhya Pradesh, Rajasthan, Uttar Pradesh and West Bengal are the major states that have the maximum area and production in the country. Recently, the crop is also being cultivated in the northern Telangana zone of Andhra Pradesh. The crop is also grown in China, Canada and Pakistan.

Although in recent years, there has been an increase in the area and production of mustard, the average productivity (977 kg/ha) of the country is quite low in comparison to the developed countries.

Description, composition and uses

A brief description, composition and uses of the various mustard species is given below.

1) White mustard (Sinapis alba or B. hirta)

White mustard is a self-sterile species, easily recognised by its hairy stem, devoid of any bloom. It has irregularly pinnate leaves, large yellow flowers that are spreading, and few-seeded and hairy pods. These have long, empty, knife-like beaks. The seeds are large, white and lightly pitted and give a considerable amount of mucilage with cold water. *S.alba* is a native of southern Europe and Western Asia. It is grown only as a garden crop, in temperate northern India, during the winter. In India, it does not contribute to the supplies of mustard or mustard oil.

Composition

The seeds contain—moisture: 7.2%; protein: 27.6%; crude fat: 29.7%; N-free extract: 20.8%; fibre:10.3%; and ash: 4.5%. The seeds also contain the glucoside, sinablin, and the enzyme myrosin. In the presence of water, myrosin hyrdrolyses sinablin, producing acrinyl isothiocyanate, (which is only slightly volatile), sinapine acid sulphate and dextrose. The yield of volatile oil is low (0.16%).

Uses

The seeds of *B. alba* are rarely used alone, they are usually mixed with black mustard in the preparation of mustard. They are not much used for the extraction of oil. In Europe, the oil-cake is used for fattening sheep. The young leaves and tender shoots are used as a pot-herb. The species is sometimes cultivated for green manure.

***2) Black mustard or True mustard (B. nigra* Koch)**

Black mustard is highly self-sterile and is quite distinct from other brassicas. The fruits at maturity are closely appressed to the inflorescence axis. The seed-coat shows fine reticulations under a lens, and is mucilaginous. The outer epidermis is covered with a thin cuticle. *B. nigra* has been cultivated in Europe since the 13th century and is now reported to be growing wild. It appears to have been introduced only comparatively recently into India and is a cold season crop, grown to a limited extent in Uttar Pradesh, Punjab and Tamil Nadu. It does not contribute to the supplies of mustard oil.

Composition

The seeds contain—moisture: 7.6%; N-substances: 29.1%; N-free extract: 19.2%; ether extract: 28.2%; crude fibre: 11% and ash: 5%. The seeds yield 23-33% of the fixed oil. The volatile oil of mustard (Sp.gr.: 1.015-1.025; Ref. Index 1.5267-1.5291) is obtained in yield of 0.7-1.2% after the hydrolysis of the glucoside, sinigrin, (potassium myronate), by the enzyme myrosin (Indian seeds are reported to yield 0.68% volatile oil). The oil is optically inactive and consists almost entirely of allyl-isothiocyanate (93.99%). The specifications for the pharmaceutical oil are (BPC), Sp.gr.: 1.014-1.025; n/20°: 1.525-1.530; and allyl-isothiocyanate content: not less than 92%. For the preparation of volatile oil, the fixed oil is first expressed from the seeds, which are subsequently macerated with tepid water for several hours, and steam-distilled. The oil obtained is an extremely powerful irritant owing to its volatility and penetrating power, and is responsible for the painful nature of alcohol. It is also used in cases of pleurisy and pneumonia.

Uses

Black mustard is ground with white mustard for preparing table mustard and also various medicinal mustard preparations, such as a mustard bath, mustard bran, and mustard flour. The expressed oil has mild rubefacient properties and is used as a liniment. The technical oil obtained during the preparation of mustard also contains the oil from white mustard seeds. In Europe, it is used for making soap, for burning and as a lubricant. In India, seeds of black mustard are used in pickles and curries.

3) Indian mustard (B. juncea)

Indian mustard is also a self-fertile species, and is a very variable annual. Its narrow-based leaves are not stem-clasping like those of a *toria* and *sarson*. *Rai* matures later than either. The seeds are rugose, reddish-brown, and generally smaller. There are two races of *rai*: a tall, late race and a short, early race. The latter is again divided into rough-leaved and smooth-leaved types.

B. juncea is a common field crop of Bihar, Uttar Pradesh and Bengal and is found in almost all the brassica areas of India. Its cultivation extends westwards to Egypt and Europe and eastwards to China. It is also cultivated in Afghanistan.

Composition

The seeds contain—moisture: 6.2%; fat: 35.5%; nitrogenous substances: 24.6%; N-free extract: 20.4%; crude fibre: 8.0%; ash: 5.3%. The oil content of the seeds is usually 30-38%. Certain forms cultivated in Uttar Pradesh and locally known as *lahi, latha*, have a higher oil content (42-43%). Commercial *rai* also contains some *sarson*. Some samples contain the seeds of var. *crispifolia* Bailey, which resemble those of var. *rugosa*. Widely varying figures have been given for the volatile oil content of *B. juncea* seeds (upto 2.9%). The volatile oil has the following characteristics—Sp.gr.:0.995; Refractive Index: 1.5185: Optical Rotation: 0°12'. It is reported to contain allyl-isothiocyanate and related compounds, among which crotonyl isothiocyanate has been identified. A re-examination of the volatile oil content of the seeds and their constituents is necessary.

Uses

Rai is an efficient substitute for black mustard. The USP recognises this species also as mustard. 'Brown mustard' and the 'expressed oil of mustard prepared from the seeds of *B. juncea* are included in the IPC. The former should contain not less than 0.6% of allyl-isothiocyanate. In studies of the preservative properties of spices, condiments and herbs with respect to their effect on yeast fermentation in wines, etc., mustard flour was easily found the most efficacious.

Climate and soil

Mustard is a cool season crop. To get a better yield of mustard, sowing should be done when the maximum temperature is around 30°C with a relative humidity of 40%. If the temperature is high, sowing should be delayed. However, delayed sowing reduces the seed yield drastically. The best time for sowing is mid-October to mid-November; October sowing gives an increased yield. Sowing beyond November not only favours the incidence of aphids and other sucking pests but also causes reduction in the total phytomass production and ultimately results in lower yields.

Mustard is found growing on a wide variety of soils from heavy clay to silt loams. However, it grows well in medium to heavy soils, which have good drainage. Saline–alkaline soils are not suitable for its cultivation. Soils with a neutral pH (6.5-7.5) are ideal for its cultivation, but it can be grown on saline soils as well.

Varieties

The improved varieties recommended or released at the national or state level get 20-25% increase in yield over the traditionally grown varieties. The important varieties/hybrids recommended for cultivation in different parts of India are given in the table below.

Dominating varieties

State	Crop	Varieties
Punjab	Indian mustard	RLM 514, RLM 601, RL 1359, PBR-91, PBR-97, PHR-2*
Haryana	Indian mustard	RH 30, RH 8812 (Laxmi), RH 781, RH 8113, RH 819
Himachal Pradesh	Indian mustard	RCC 4, RL 1359
IARI, Delhi	Indian mustard	Pusa Bold, Pusa Bahar, Pusa Jai Kisan (Bio-902), Agrani (SEJ-2), Jagannath (VSL-5)

(continues...)

(...contd.)	Dominating varieties	
State	**Crop**	**Varieties**
Madhya Pradesh	Indian mustard	Kranti, Narendra rai, Pusa Bold, Rohini, Vardan, Varuna
Uttar Pradesh		
Pantnagar	Indian mustard	Karnti, Krishna, PHR-7*
Faizabad	Indian mustard	NDR 8501
	Yellow sarson	NDYS-2
Kanpur	Indian mustard	Varuna, Vardan, Vaibhav, Rohini, Urvashi, PHR*
Rajasthan	Indian mustard	Arawali (RN 393), Rajat (PCR-7)
Gujarat	Indian mustard	GM-1, GM-2, Pusa Jaikisan, Rajat
	Yellow sarson	Gujarat Sarson-1
Bihar	Yellow sarson	Rajender Sarson-1, Swarna
	Indian mustard	Kranti, Pusa Bold, Varuna
West Bengal	Indian mustard	Seeta (B-85), Bhagirathi, Sarma, Pusa Agrani, Sanjucta asech
	Yellow sarson	Benoy (B-9), Subenoy, Jhumka (YSBNC-1)
Assam	Indian mustard	Pusa Agrani, TM 4, Varuna
Maharashtra	Indian mustard	Pusa Bold, Pusa Jaikisan, Rajal
Andhra Pradesh	Indian mustard	Bhavani, PT 303, GMI, Divya, Kranthi, Vardan, Varuna

*Hybrid

Aravali mustard, RH 30, RH 819 and Vaibhav are drought-tolerant. The variety RH 781 is frost-tolerant, whereas the varieties S 52 and Narendra *rai* are salt-tolerant. The variety Sourabh (RH 8113) is moderately tolerant to *Alternaria* blight and white rust. The varieties Pusa bold and Sarma are suitable for late sowing. The variety RT-11, evolved in Uttar Pradesh, is the most aphid resistant. The variety RH 30 is also suitable for late-sown conditions, while Kranthi is resistant to shattering.

Cultivation

Propagation

Mustard is propagated through seeds. About 8-12 kg ha of seeds are required for sowing by broadcast and 5-10 kg/ha for line sowing. The seeds are treated at the time of sowing with Apron SD 35 (@ 6 g/kg) for the control of white rust and downy mildew at the seedling stage. To protect the crop from other seedling diseases like root-rot, wilt etc., the seeds are treated with Carbendazim or Thiram or Captan (@ 2 g/kg) or Mancozeb (@ 3-4 g/kg of seed).

Land preparation

The land is ploughed by giving 5-6 criss-cross ploughings followed by plankings. In areas where clodding and their breakage is a problem, a three, strip bamboo ladder should be used. About 20-30 tonnes of FYM or decomposed cowdung should be mixed with the soil at the time of the last ploughing. One pre-sowing irrigation should be given at a depth of 5 cm, if there is less moisture in the field.

Sowing

The seeds are broadcasted over the well-prepared land and planked in the broadcast method of sowing. But line sowing is preferred over the broadcast method, as line sowing gives a better yield and helps in better management. A spacing of 30 cm between rows and an inter plant spacing of 10-15 cm has been recommended for this crop.

Under dryland conditions, the establishment of optimum plant stand is a real hurdle in realising a good harvest. To overcome the problem, a ridger seeder should be used for line sowing. This practice, besides maintaining a perfect crop stand, helps in better conservation of moisture.

While sowing, care should be taken so that the seed does not come into contact with the drilled fertiliser, as this affects germination. For this purpose, the seed should be sown 3-4 cm deep, whereas the fertiliser should be drilled at 7-10 cm depth. To ensure good germination and early seedling vigour, the seeds must be soaked in water before sowing. This is done by covering the seeds with a moist gunny bag or directly with damp earth overnight. In the line-sown crop, the seedlings are thinned within the row two weeks after sowing to maintain the optimum plant population.

Fertiliser application

A crop yielding 13.25 q/ha requires 118 kg N, 17.39 kg P and 94.36 kg K. The crop responds to both organic and inorganic fertilisers. The fertilisers should be applied at the rate of 40 + 35 + 15 kg NPK per hectare at the time of sowing. In hilly areas, it is recommended to apply 45 + 35 + 10 kg NPK per hectare. In acidic soils, especially in rice growing areas, lime should be applied atleast 30 days ahead of sowing. In the North Bank plains and the upper Brahmaputra valley zone, Borax should be applied to the soil @ 10 kg/ha.

Inter-culture and weed control

In the early stages, the crop should be kept free of weeds. In orobanche-infested areas, one inter-culture is sufficient at 20-25 days after sowing. The predominant weeds of mustard are *Chenopodium album, Anagalis arvensis, Convolvolus arvensis, Melilotus alba, Cynodon dactylon, Cyprus rotundus,* etc.

Weeding should be done soon after thinning. Three herbicides, viz., Fluchloralin, Isoproturan and Pendimethalin, are found effective in controlling the weeds of mustard. A pre-plant application of Fluchloralin @ 1.25 kg/ha and a pre-emergence application of Pendimethalin @ 1.25 kg/ha would prove effective. If the weeds emerge after planting, Isoproturon @ 0.75 kg/ha may be used 30 days after sowing.

Protective irrigation

Mustard has a low water requirement. In command areas, atleast one irrigation is given at flowering, but two irrigations are beneficial. In Gujarat however, six irrigations have given better monetary returns. These should be given on 1, 5, 33, 50, 63 and 79 days after sowing. The requirement is 400 mm of irrigation water. The first two irrigations should be light and the remaining with 75 mm water each. It is better to delay the first irrigation as much as possible. This helps the plants to branch well which, in turn, results in profuse flowering and fruiting. The best time for the first irrigation is 25-35 days after the sowing. The second irrigation should be given at the fruiting stage; i.e., 55-60 days after sowing. It is recommended to irrigate the crop at 0.8 IW/CPE with mulch to obtain the maximum yield.

Inter-cropping

Mustard can be inter-cropped with wheat, barley, gram, pea, sugarcane, lentil, *toria*, potato, etc., for better economic returns. It can also be grown sporadically as a sprinkle crop along with finger millet, groundnut, etc., or as a mixed crop along with chickpea, coriander, jowar, etc.

In orobanche endemic areas, it is recommended to adopt crop rotation with cereals or legumes.

Pests and diseases

Pests

One of the serious constraints in rapeseed-mustard production is the lack of varieties resistant to diseases and insect pests. To protect the crop from diseases and insect pests, it is essential to undertake timely plant-protection measures.

The mustard sawfly, painted bug and mustard aphid normally attack the mustard crop. The aphid is the most destructive in late-maturing mustard.

Mustard aphid

The mustard aphid (*Lipaphis erysimi* Kalt.) is pale-green, soft bodied and 1-2 mm long. The nymphs and adults suck the cell sap from various parts of the plant and live in colonies. Due to the sucking of the plant sap, plants remain stunted, the leaves show curling and yellowing, the pods shrivel up and the seeds do not develop in the infested pods. The infestation due to mustard aphid is more serious from the end of December to the middle of March. They become innumerable, completely covering the inflorescence and pods.

To control this important pest, the following corrective measures are required: (i) wherever the aphid problem is serious and acute, the mustard varieties may be sown instead of rapeseed which is more susceptible to the pest; (ii) early sowings, i.e., in the first fortnight of October, to escape the mustard-aphid infestation; (iii) the removal of early infested, twig-bearing stems by the end of December is very useful; (iv) the crop may be sprayed with 0.025% Methyl oxydemeton or 0.03% Dimethoate or 0.025% Quinalphos or 625, 850 and 1,000 ml of either of these insecticides or Phosphomidon 0.025% i.e., 150, 210 and 250 ml mixed with 625, 850 and 1,000 litres water, depending upon the crop-growth stage. Spraying should be done thrice at 15-day intervals, starting from the end of December or early January when 9-10 aphids/twig and 30% of the plants are infested. The number of sprays may be reduced, depending upon the severity of the infestation. The above-mentioned quantity of water is for high-volume sprayers such as foot-sprayers.

The precautions to be taken are: (i) If the crop is to be used for *saag* (leafy vegetable)/fodder purposes, it may be sprayed with 0.05% Malathion 50 EC (500 ml Malathion 50 EC mixed in 500 litres water/ha), 5 times at 10-day intervals. Do not pluck the leaves within 7 days after spraying; (ii) insecticides should be sprayed in the afternoon to avoid any adverse effect on pollination; (iii) the quantity of water will be one-tenth of the normal spray of liquid if a motorised sprayer is used (the quantity of insecticides remain the same).

Mustard sawfly

The Adult fly (*Athalia proxima* Klug) is orange yellow with smoky wings and black veins. The young larva is green with a black head, about 4 mm long. The full-grown larva is greenish-grey, having 5 strips of black dots on the lower side of the body and is about 1 cm long. The larva attacks the young crop. All the leaves are damaged under the heavy incidence of the pest. The pest appears in October and causes more damage during October-November.

The application of 10% Carbaryl dust @ 25 kg/ha or spraying the crop with 0.2% Carbaryl 50 WP (2 kg Carbaryl 50 WP mixed in 500 litres water/ha) controls the pest.

Painted bug

The nymphs of *Bagrada cruciferarum* Kirk have a number of pale, brown and red markings on their bodies. The nymphs and adults suck the cell sap from the foliage and pods, making them dry. The sucking of the leaf sap causes white blotchy spots. Its high population becomes a nuisance at the threshing floor also. This pest appears at the seedling stage and again at the pod-formation stage.

The application of 10% Carbaryl dust @ 25 kg/ha at seedling stage or spraying the crop with Malathion, 50 EC (0.05%), i.e., 500 ml Malathion 50 EC mixed in 500 litres water per hectare at the seedling stage, controls the pest. The second round of spraying with Malathion 50 EC may be given in March (after the pod formation), if necessary.

Cabbage butterfly (Pieris brassicae)

The pest appears at the flowering and pod-formation stages. The caterpillars feed on the leaves, buds and developing pods. The insect can be controlled by spraying *Endosulfan* @ 1.2 lit or Dichlorvos @ 300 ml in 600 litres of water, per hectare.

Bihar hairy caterpillar

The abdomen of the moth (*Spilosoma oblique* Walker) is bright red with black spots; the thorax, antennae and eyes are black. The fore-wings have a reddish tinge and the hind-wings have dark brown spots. The newly-hatched larvae are brown and feed gregariously (in groups) on the leaves of a few plants. The fully grown caterpillars are 4-5 cm long and have numerous dark brown, long hairs on the body. The mature caterpillars spread to the whole field and cause huge losses. The pest is more serious on *toria* during periods of droughts.

To control the pest, the affected leaves should be destroyed in the initial stages of infestation. This would reduce the spread of the damage. The destruction of the larvae in the gregarious phase is quite easy and effective. The young caterpillars can be easily killed by dusting the infested crop with Carbaryl 10% (25 kg/ha). Spraying the crop with 0.07% *Endosulfan* (Thiodan 35 EC) 1200 ml or 0.05% Dichlorvos (Nuvan 70 EC) 300 ml mixed in 600 litres water/ha controls the pest. The rhizosphere nematode problem has been noticed in *B. juncea*, which can be managed by different weed control practices. The beet cyst nematode (*Heterodera sachactii*) has also been reported to cause damage to white mustard.

Diseases

The major fungal diseases affecting rapeseed and mustard crops are *Alternaria* blight, white rust, downy mildew, powdery mildew and phyllody.

Alternaria blight

The blight caused by *Alternaria brassicae* is characterised by brown or greyish spots on the leaves, stems and pods, and is the most destructive and widespread. Compared to mustard, *sarson* is more prone to this disease; moist and warm weather favours the disease.

To control this disease, the following measures should be taken: (i) the debris of the previous crop should be burnt (ii) removal of weeds like *Chenopodium album* to reduce the disease. (iii) the crop should be sprayed with Mancozeb or Captafol (1.5 kg/ha) at intervals of 15 days, 40-45 days after sowing. This fungicide is cheap and helps in reducing the intensity of another important disease, white rust, which is characterised by the white pustules of *Albugo candida* on the lower surface of the leaves in the early stage, i.e., when plants are 45-50 days old. Later, the malformation of the floral parts takes place, which is known as the stage of stag-head formation. Apron @ 0.2% as seed-dresser along with 3 sprays of Dithane M-45 @ 0.2% or Dithane Z-78 or Captafol or Copper oxychloride is recommended for reducing white rust and the simultaneous occurrence of downy mildew (*Peronospora parasitica*) incidence at the seedling stage. Sprays of Dithane M-45 or Zineb also help in minimising the severity of all these diseases on the leaves and pods. If Apron is not available, the spray of Dithane M-45 should be continued.

When there is an attack of powdery mildew (*Erysiphe cruciferarum*), which is characterised by a white powdery mass on the stems and leaves at the later stages of plant growth, it should be sprayed with Wettable sulphur (0.3%).

It is necessary to ensure that there is no delay in sowing. This would protect the crop from various diseases. To avoid the spread of club-root *(Plasmodiophora* sp.), crop rotation with cereals should be adopted and growing yellow sarson should be avoided.

Stem-rot or blight

The disease caused by *Sclerotinia sclerotirum* is characterised by elongated, water-soaked lesions on the stem with the cottony mycelial growth of fungus, wilting and drying of plants, stunting and premature ripening of the plant and the shredding of the stem. The disease can be controlled by collecting and burning the disease debris and refuse of the plants. Crop rotation may be followed using non-susceptible hosts to reduce the sclerotial population in the soil.

Harvesting and yield

Based on the varietal durations of mustard, it matures between 80-120 days. Usually, the mustard crop is harvested as soon as 75% of the pods turn yellowish and moisture content of the seed is around 40%. At this stage, the majority of the seeds will be firm when pressed between the fingers. The oil content in the seed is the maximum at this stage. To minimise shattering losses, the crop should be harvested preferably in the morning when the pods are slightly damp with the night dew. Harvesting too soon, when the moisture content of the seed is more than 45% (green watery stage), drastically reduces the yield, oil content and seed viability.

Bundles of the harvested plants are stacked and dried in the sun for a few days. Threshing is done by the usual method of treading by bullocks or by running a tractor over the dried plants spread on the threshing yard. The seeds are separated by winnowing. The moisture content of the seeds must be less than 8% at storage time. The yield per hectare ranges from 600-900 kg of seeds in rain-fed crops and from 900-1300 kg in irrigated crops.

Major steps to be taken for increased yield

i) Conservation of soil moisture by deep summer ploughings; (ii) use of improved variety; (iii) timely sowing; (iv) maintaining optimum plant population; (v) application of recommended dosage of fertilisers and, if possible, irrigation; (vi) timely plant-protection measures; (vii) harvesting the crop when 75% of the pods turn yellowish in colour.

Packing and storage

The seeds that are cleaned by winnowing are packed in gunny bags and stored in a cool, dry place.

Value-added products

Dried/Dehydrated mustard greens (leaves), powdered mustard or mustard flour, ground mustard (mustard meal), compounded mustard or mustard compound, mayonnaise, mustard cake, etc. are some value-added products of mustard.

Nutmeg

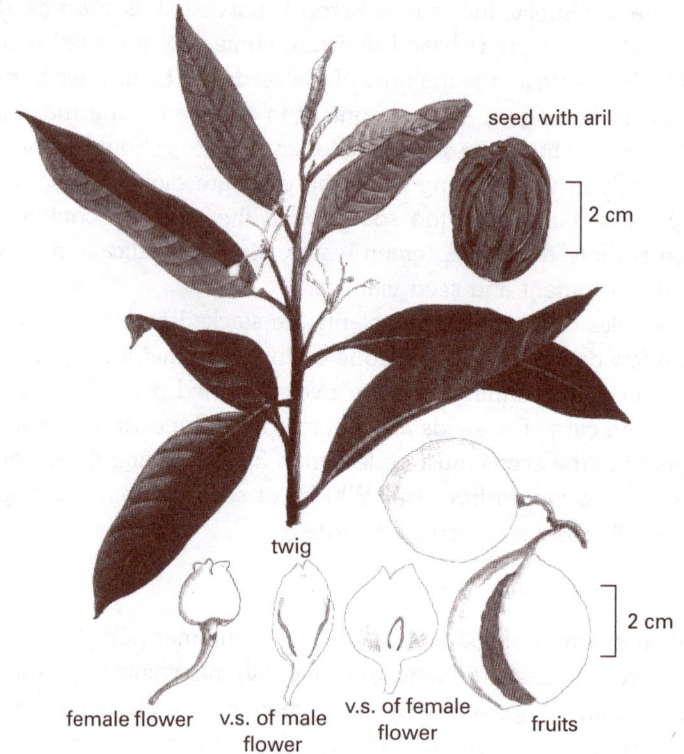

Plant Profile

Family	:	*Myristicaceae*
English name	:	Nutmeg
Indian name	:	*Jaiphal* (Hindi), *Jayikai* (Kannada), *Jathikka* (Malayalam), *Jatiphala* (Sanskrit), *Jathikai* (Tamil), *Jajikai* (Telugu).
Species	:	*Myristica fragrans* Hout., *M. malabarica,* M. beddomei, M. argentea, M. otaba,
Distribution	:	Indonasia, East Indies, Malaysia, India, Sri Lanka, Tanzania, Mauritius, Reunione, Trinidad & Tobago, China.
Uses	:	Culinary spice, flavouring

utmeg (*Myristica fragrans* Houtt.) is a spice crop belonging to the family *Myristicaceae*. Commercial nutmeg is the kernel, which is hard and brown, enclosed in a thin brittle shell. Surrounding this shell is the aril, which is scarlet in colour and is the mace of commerce.

The composition of dried nutmeg and mace are as follows.

Content (%)	Nutmeg	Mace
Moisture	14.3	15.9
Protein	7.5	6.5
Ether extraction (fat)	13.4	24.4
Carbohydrate	28.5	47.8
Fibre	11.6	3.8
Minerals	1.7	1.6
Calcium	0.12	0.18
Phosphorus	0.14	0.1
Iron	0.046	0.26
Vitamin B1	0.33 mg/100 g	0.35 mg/100 g
Vitamin B2	0.01 mg/100 g	0.42 mg/100 g
Niacin	1.40 mg/100 g	1.40 mg/100 g

Nutmeg also contains a volatile oil (6-16%), starch (14-25%), furfural (1.15%) and pectin (0.5-0.6%); and mace contains a volatile oil (4-15%) and amlodextrin (25%). The seeds and mace contain essential oils. The seed essential oil contains 80% pinene and camphene, 4% myristicin which is poisonous, p-cymene, d-linalool, terpineol, geraniol, safrole, eugenol and isoeugenol. The essential oil of mace is similar to nutmeg oil, but it is fresher than the seed oil.

Mace is used as a culinary spice and largely as a flavouring agent. The husk is used for pickling, when the fruit is at a tender stage. The fresh husks of the ripe fruit can be used for making jelly. While separating the seeds from the testa and aril, some are broken and such seeds are rejected. Medicinally, nutmeg acts as a stimulant, astringent, aphrodisiac and carminative. The essential oil has uses in medicine, toilet soap, dental pastes, flavouring chewing gums, chewing tobacco, baked goods, cakes, cookies, puddings and pickles. The aromatic oil has a butter-like consistency and is orange in colour. The oil has been recommended for the treatment of inflammations of the bladder and urinary tract. Nutmeg butter is used in the manufacture of scented oils, ointments, perfumes and soaps, and as a flavouring agent in cooking and confectionery.

Apart from this true nutmeg, inferior nutmegs are obtained from various species such as papua nutmeg and mace from *Myristica argentea*; Bombay nutmeg and mace from *M.malabarica* and Otaba nutmeg is obtained from *M.otaba*. These are used to adulterate the true nutmeg.

Origin and distribution

Nutmeg is a native of the Moluccas Islands of Indonesia, the East Indies or Spice Islands. It is now grown in most tropical countries of the world, but is cultivated on a large scale in the Malayan region. The British East India Company introduced this spice into India in 1800 AD. The countries producing nutmeg apart from Malaysia are Indonesia, Grenada, Sri Lanka, India, Tanzania, Mauritius, Reunion, Trinidad and Tobago and China. The major country importing nutmeg and mace from India is Russia.

Area and production

The annual world production is approximately 8,000 tonnes of nutmeg and 1,000 tonnes of mace, eighty-five per cent of which is produced in Indonesia and Grenada followed by Sri Lanka. Relatively small quantities are exported from India. In India, nutmeg is presently grown in Ernakulam, Cannanore, Kottayam, Thrissur, Idukki, Allepey, Mylapuram, Thiruvanathapuram, Palghat and Calicut districts of Kerala (3,502 ha), Dakshina Karnataka (110 ha) and also the Andaman and Nicobar Islands. The total Indian production of nutmeg and mace is 5,537 and 5,919 tonnes per year, respectively. Nutmeg is also grown near Courtallam in Tirunelveli district and around the Burliar zone of the eastern slopes of the Nilgiris, Anjarakandi near Coonoor of Tamil Nadu and Karnataka. The Araku valley in Andhra Pradesh and Wynad in Kerala are well suited for its cultivation.

Description of the plant

Nutmeg is an attractive, spreading, evergreen tree with dense foliage. Its somatic chromosome number is 2n=42. The trees are 10-20 m tall, sometimes reaching a height of 20 m or more. The branches are spreading with a dark gray bark. The leaves are shiny and oblong to oval in shape. The inflorescence appears in cymes. Each cyme has several branches bearing a number of flowers which hang down. The flowers are small, pale yellow and bell shaped and are slightly aromatic. The fruit is fleshy, globose in shape and lemon yellow to light brown in colour. Inside the fruit is the single, glossy brown seed with a brittle shell over which is the beautiful, brilliant, scarlet, net-like (reticulated) membrane or aril known as the mace, which is very fragile and aromatic. Nutmeg trees are dioecious in nature (the male and female flowers are borne on different trees). The monoecious condition is also reported to occur in aged trees at Burliar, in which case, double and triple nuts are produced and the yield from such trees is low. The quality of the spice is also poor.

The female and male trees can easily be identified by a trained eye. The male trees have erect branches and the leaves are generally smaller in size. The male trees are conspicuously less leafy than the female trees also; the shape of the tree is not as regular. The calcium oxalate content of the leaves is also taken as a criterion for the identification of sex in nutmeg.

Climate

Nutmeg thrives well in warm, humid conditions in locations with an annual rainfall of 150 cm to 250 cm and a temperature of 25-35°C. It grows well from sea level up to an elevation of 1300 m. Partial shade appears to be beneficial in the early stages of growth.

Soil

For the cultivation of nutmeg, river banks and hill valleys with clay loam, sandy loam and red laterite soils, which are rich in humus, are ideal. Both dry climate and water-logged conditions are not good for nutmeg. It can grow well even in comparatively poor types of soils provided the soil is not sandy and not too wet or dry. A certain amount of iron in the soil is said to be beneficial to nutmeg trees.

Varieties

There are two basic types of nutmeg valued in the world trade. These are the West India and East India types. The West Indian variety is grown in the Islands of Grenada and Trinidad, while the East Indian nutmegs are highly aromatic and superior compared the West Indian variety. The fruits of the East Indian nutmeg are ovoid, approximately 2.25 to 2.75 cm long, 1.75 to 2.25 cm in diameter, and longitudinally wrinkled. The colour is greyish brown, with a furrowed network of dark brown-veins, in which the volatile oil is found.

'Konkan Sugandha' is an improved variety of nutmeg released from the Konkan Krishi Vidya Peeth, Depoli, in Maharashtra.

Another high-yielding nutmeg line 'A 9/4', a clonal selection, has been released as IISR Vishwashree, with an average yield of 1000 fruits/tree in the eighth year @ 360/ha, with an average yield of 3122 kg dry nut and 480 kg mace/ha. This line has 7.3% oil in the nut and mace, 13% oleoresin in the mace and 2.5% oleoresin in the nut. This variety also contains 30.9% butter in the nut. The other promising accessions under evaluation at IISR, Calicut, are: A 11/29, A 11/70, A9/20, A 9/22, A 9/25, A 9/79, A 9/86, A 4/12, A 4/22 and A 4/52. The IISR, Calicut, has a germplasm collection of 421 in nutmeg.

Cultivation

Propagation

Nutmeg can be propagated by seeds as well as vegetatively by patch-budding and epicotyl grafting. The percentage of success in the vegetative methods of propagation is between 38-80%. The bio-technological method of *in vitro* synthesis of mace has been standardised.

Propagation by seeds

For raising seedlings, only well-matured, large seeds of uniform size, shape, light brown in colour with thick mace and low terpene content are collected. The seeds are dehusked and dried for a day before sowing. One kilogram contains about 200 small seeds or 90 big-sized seeds. The average number of small, medium and big seeds is 125 per kilogram. The seeds are collected from regular bearing and high-yielding trees, yielding more than 10,000 fruits per tree per year and having 30 g wet weight per fruit, 1 g wet weight of mace per fruit and 10 g wet weight of nuts per fruit. The seeds (7-9 g in weight) are sown three days after extraction and begin to germinate in four to six weeks. The sprouted seeds at the 2-leaf stage with a height of 15-20 cm are transferred to polythene bags or when the seedlings are six months old, they are potted and allowed to remain in the pots for about 12 to 18 months prior to planting in the main field. For raising the seedlings, healthy fruits harvested during June-July are used. The seeds soon lose their viability and should be sown immediately. Regular care regarding watering is necessary for its good germination. The seedling progeny will give about 50% of each sex, which is very difficult to distinguish until the trees flower, i.e., 4-6 years after planting.

Vegetative Propagation

Although nutmeg can be vegetatively propagated by patch-budding and epicotyl grafting, the latter has been reported to be more easy and successful. For epicotyl grafting, the selected root-stock should have a thick stem (diameter of 0.4 cm or more) with sufficient length to give a cut of 3 cm length. The scions with 2-3 leaves, collected from the high-yielding trees, can be used for grafting. The stock and scion should be of approximately the same diameter. A 'v'-shaped cut should be made in the stock and the tapered scion is fitted carefully into the cut. The root-stocks used for grafting or budding are *M. beddomei* and *M. malabarica*. It may be tied with 300 gauge polythene strips. The completed grafts should be planted in long polythene bags, containing coir dust as the medium, up to half of their length. After planting them in bags, the grafts should be covered with polyethylene bags to prevent drying of the scion and should be kept in a cool, shady place protected from direct sunlight. After one month, the bags can be opened and those grafts showing the sprouting of scions may be transplanted into bags containing a mixture of soil, sand and cowdung in the ratio of 1:1:1 and kept in the shade. The polythene bandage covering the grafted portion can be removed after three months of transplanting. During grafting, precautions should be taken to prevent wilting of the scions and grafting should be completed as soon as possible after the detachment of the shoots. Since the exudates from the cuts pose a problem by covering the cut surfaces in the grafting process hindering union, the root-stocks should be prepared earlier than the scion, so that by the time the scion is ready, the exudation in the root-stock would have ceased. Orthotropic shoots or straight growing shoots are used to get normal shaped nutmeg trees. If the lateral branches are used, it will result in a shrubby spreading plant, which is difficult to convert to a normal nutmeg tree.

Sex problem in nutmeg

Nutmeg is a dioecious plant and the sex of the trees is known only after they flower, i.e., about six or seven years after planting. Since the pistillate trees bear profusely, it is necessary to have a maximum number of them interplanted with a few male trees for pollination.

Land preparation and planting

The planting is done at the beginning of the rainy season. A spacing of 8 m × 8 m or 9 m × 9 m is required for nutmeg plantations. Pits of about 0.75 m^3 to 0.90 m^3 are dug and filled with organic manure and soil about 15 days prior to planting.

The plants should be shaded by planting banana or glyricidia to protect them from sun scorching in the early stages. Permanent shade trees should be planted when the site is on hilly slopes and nutmeg is grown as a monocrop. Nutmeg can also be grown as a mixed crop in the old coconut and arecanut gardens, where the light shade conditions are suitable. One nutmeg can be planted at the centre of 4 coconut plants; while in arecanut, nutmeg can be planted at every third row so that within the square formed by 4 nutmeg plants, there are 9 arecanut seedlings.

Manures and Fertilisers

Farm Yard Manure is applied at 10 kg per pit and gradually increased to 50 kg per plant for a 15 year old tree. The fertiliser doses vary with the age. In the first year after planting, a dose of 20 g N, 18 g P_2O_5 and 50 g K_2O is given per plant. The dosage for the second year is 40 g, 36 g and 100 g; and after five years it is 100 g, 90 g and 25 g; and after fifteen years a dose of 500 g, 250 g and 1000 g of N: P_2O_5 and K_2O, respectively, are applied.

The manure may be applied in shallow trenches dug sufficiently far away from the base of the tree. The manures and fertilisers are applied twice a year once in May-June and then again in September-October, depending upon the availability of moisture or rainfall.

Irrigation and inter-culture

Regular weeding and irrigation are required for its good growth, early bearing and higher yield. Irrigation in summer months is a must in dry areas. For 4-year-old plants, 20 litres of water per plant thrice a week is given and the quantity increases at the later stages of growth. Gramaxone can be used for chemical weed control. In lighter soils, the plant basins are mulched with heavy mulches of organic matter.

Pests and diseases

Pests

Scale insects (Parasaisettia nigra, Pseudaulacospis cockerelli)

Scale insects have been found to occasionally infect the tender leaves and shoots, generally in the nursery. The spraying of 0.05% Monocrotophos or Quinalphos will control the pest infestation.

Diseases

Die-back (Diplodia natalensis)

The drying up of mature and immature branches from the tip downwards is noticed in the affected plants. The infected branches should be cut and the cut ends should be pasted with Bordeaux mixture. Spraying trees with 1% Bordeaux mixture is also recommended.

Fruit-rot (Colletotrichum gloeosporoides and Botrydeplodia theobromae)

Immature fruit-rot and shedding are noticed in some trees without any other apparent signs of infection. In the case of fruit-rot, the infection starts from the pedicil end as dark lesions. It gradually spreads to the fruit causing a brownish discoloration, which results in the rotting of the fruit. In the advanced stages, the mace also rots emitting a foul smell.

Bordeaux mixture (1%), Diafoltan, Blue copper or Dithane M-45 at 0.5% concentration should be sprayed when the fruits are half mature, to control this disease.

Leaf blight caused by *Botrydiplodiea theobromae,* shot-hole by *Colletotrichum gloeosporiodes*, leaf-spot by *Alternaria citri*, algal leaf-spot by *Cephaleuros* sp. sooty mould caused by *Phragmocapinus betle* and thread blight (*Marasmius pulcherima, Marasmius equicrinus)* are the other common diseases of nutmeg, which can be controlled by repeatedly spraying 1% Bordeaux mixture.

Harvesting and yield

The female nutmeg tree starts fruiting from the sixth year onwards; but, the peak harvesting period is reached after 15 or 20 years and continues for 40 years or more. The fruits are ready for harvest in about 6-9 months after flowering. The flowering and harvesting continue throughout the year, but, June-August or December–May are the peak periods.

The fruits are ripe and ready for harvesting when the pericarp splits open. The harvesting is done by using a bill hook. The fruits are split open, the outer fleshy portion is removed, and the mace is manually separated from the nut. Then, the nut and mace are dried separately in the sun in a drying yard. The mace should be dried for 10-15 days and the nuts for 4-8 weeks till the kernel rattles within the shell. The scarlet-coloured mace gradually becomes yellowish-brown and brittle when the drying is completed. The fresh pericarp can be used for making pickles, jams and jellies.

Trees of 15 years of age and above will yield about 1000-2000 or more fruits and large trees, which are over 30 years of age, may yield about 3000-10000 fruits per year. The yield per hectare may vary from 1000-1500 kg of nutmeg and 200-500 kg of mace per annum. The mace to nutmeg ratio is about 3:20 on a dry weight basis. Generally, if the fruit weight is 60 g, 6-7 g nutmeg and 3-4 g mace can be recovered. The rest will be the pericarp. For every 100 kg of nutmeg, a tree produces only 4 kg of mace.

Grading

The dried nutmegs are graded by hand according to their weight, shape and colour. After grading, the nutmegs are fumigated with methyl bromide to protect them from storage pests.

The following classifications have been made in nutmeg trade.

1. Whole and sound nutmeg
   This is used in spice trade as: (a) large (b) medium and (c) small
2. Sound shrivels
   These are employed for grading, but are usually too expensive for oil distillation.
3. Rejections
   Considerably low-priced, this grade can be used for the distillation of oil.
4. Broken and warmy
   This grade is also suitable for oil distillation. The grades of mace are as follows;
   a) *Banda Mace* is considered to be the finest. It has a bright orange colour and a fine aroma
   b) *Jave Estate Mace* is golden yellow, interspersed with brilliant crimson streaks.
   c) *Siauw Mace* is of a lighter colour than *Banda mace* and contains less volatile oil
   d) *West Indian Mace,* often regarded as the fourth grade of East Indian mace, is derived from *M. argentea*. It contains less volatile oil with an undesirable turpentine-like aroma. It is unsuitable for distillation purposes. '*Banda*' and '*Penang*' maces are considered to be superior in quality by the trade the world over. Mace is available in the market as 'whole', 'broken' or 'ground'.

Adulterants

Nutmegs are sometimes adulterated with 'false' nutmegs (*M. malabarica*) or 'Bombay nutmegs', which are odourless and tasteless. The oil is sometimes adulterated by the addition of turpentine oil or pinene.

Mace is also sometimes adulterated with 'Wild Mace' (*M. malabarica*), which is of inferior quality.

Extraction of oil

The essential oil is extracted from the seed, mace, leaves and also the bark by steam distillation. For oil distillation, the economically viable and accepted materials are the rejections from spice trade. The oil yield ranges from 6 to 16% in nutmeg, 4 to 15% in mace, 0.14% in the bark and 0.4 to 0.6% in the leaves.

Nutmeg contains 25-40% fat, which can also be recovered using solvents or by mechanical pressing. It is highly aromatic and its major constituent is trymyristicin. The oleoresin is extracted with solvents. It may also have butter; about 7-16% nutmeg oil is found in it. Aromatic ethers, myristicin and elemicin are present in the oil and oleoresin.

Value-added products

The value-added products are nutmeg oil, mace oil, nutmeg oleoresin, mace oleoresin, myristicin, nutmeg butter, the volatile oil of the bark and flowers.

Onion

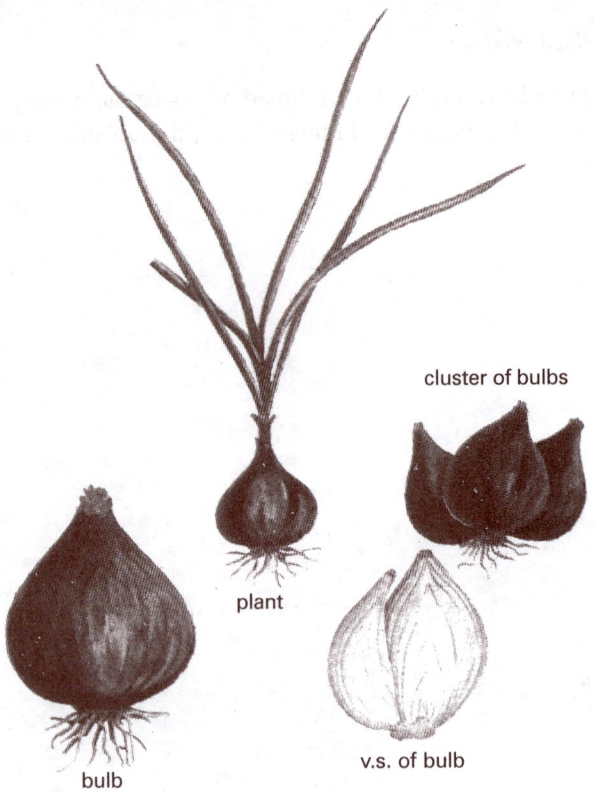

Plant Profile

Family	: *Liliaceae/Alliaceae*
English name	: Onion
Indian name	: *Pyaz* (Hindi), *Nirulli, Ullagadde* (Kannada), *Bawanj, Chuvannauli* (Malayalam), *Palandu* (Sanskrit), *Vengayam, Irulli* (Tamil), *Nirulli, Ulligadda* (Telugu).
Species	: *Allium cepa* var. *cepa* Linn., *A. sativum*
Distribution	: India, Afghanistan, Tajikistan, Uzbekistan, Western Asia, Mediterranean region
Uses	: Culinary spice, vegetable

The common Onion (*Allium cepa* var.*cepa* Linn.), belonging to the family *Alliaceae/Liliaceae,* produces underground bulbs which are edible. It is also called the Bellary onion or big onion. Onion is an important spice as well as commercial vegetable. The nutritive value of onion bulbs varies from variety to variety. However, the average composition per 100 g of bulb is as follows: moisture-86.0 g, carbohydrates-10.9 g, protein-1.4 g, fibre-0.8 mg, potassium-180.0 mg, sulphur-70.0 mg, phosphorus-44.0 mg, calcium-32.0 mg, chlorine-25.0 mg, magnesium-16 mg, sodium-7 mg, iron-9.7 mg, thiamine-0.06 mg, riboflavin-0.01 mg, niacin-0.4 mg, vitamin C-28 mg, calorific value-47 cal. The fresh herb and bulb yield an essential oil (0.005%), the chief constituent of the crude oil is allyl-propyl disulphide. Onions are cooked in several ways in all types of curries. They are baked, boiled, fried and used in the fresh, dehydrated or powder form in soups, pickles, sauces, etc. They are also eaten raw as in salads and relishes. Owing to the presence of several sulphur compounds, onion has antiseptic properties. Onion is said to possess stimulant, diuretic and expectorant properties and is considered useful in flatulence and dysentery. Freshly-expressed onion juice has moderate bactericidal properties.

Origin and distribution

The onion plant is said to have originated from the regions of North-west India, Afghanistan, the erstwhile Soviet Republic of Tajik and Uzbek; and western Tien shan, western Asia and the Mediterranean region are the secondary centres of development. The important onion growing countries are India, China, Tajikistan, Uzbekistan, Turkey, Brazil, USA, Vietnam, Pakistan, etc. India ranks second in terms of the area under onion cultivation and production, which contributes 12% of the world's average and 15.8% of the world's production. India has traditionally exported onions from 1950 onwards to countries like Malaysia, Singapore, West Asian Gulf countries, Sri Lanka, etc. Maharashtra is the leading state with regard to the area and production of onions. The other major onion producing states are Gujarat, Uttar Pradesh, Orissa, Tamil Nadu, Karnataka, Madhya Pradesh, Bihar and Andhra Pradesh. Onions are also cultivated on a limited scale in the other states of India.

Description of the plant

The genus *Allium* contains about 300 species of bulbous biennial or perennial herbs which emit the distinctive and pungent odour of onion or garlic when the tissues are crushed. The other main characteristic is the presence of bulbs, consisting of swollen leaf-bases on the underground part of the stem, which are like a flattened disc in shape. The leaves are hollow, cylindrical and tapering.

The flowers are borne in simple umbels at the apex of the floral stem, which is commonly hollow when mature, somewhat swollen at the middle, and white in colour. An inflorescence may continue opening for two weeks or more and a plant may be in bloom for more than 30 days. The stamens are 6 in number and the anthers are bilocular. The style is approximately 1 mm long, but it becomes receptive when it elongates to a length of 5 mm within a day or two after the anthers have all deshisced. Cross-pollination is common and is effected by honey-bees. Self-pollination also takes place as the insects visit many flowers in a single umbel.

Climate

Onions can be grown under a wide range of climatic conditions. However, most of the varieties are grown as winter season crops in the plains of India. A temperature of 12.5-25°C for the seed germination, 12.8°-21°C before bulb formation and 15.5°-25°C for bulb development is considered ideal for onions. Very low temperatures in the early stages favours bolting. A sudden rise in temperature favours early maturity of the *rabi* season crop, thereby resulting in smaller-sized bulbs. Relatively high temperature as well as a long photoperiod are essential for good bulb development in the commercial varieties grown in the *rabi* season. There are some varieties cultivated during the rainy season that require long days and high temperatures throughout the growing season.

In fact, the photoperiodic requirement in onion is a quantitative characteristic and each variety requires a minimum day length for bulb formation, which is known as the critical value. The bulb formation is influenced by the critical value in combination with temperature. The short-day varieties, when they pass the critical value limit, initiate bulb formation and bulb development is continued under long-day conditions. Conversely, a long-day variety cannot initiate bulb formation unless the longer critical value of its photoperiod is reached. It is, therefore, advisable to grow only those varieties which are acclimatised. Relatively high temperature as well as a long photoperiod are essential for bulb development in the commercial varieties grown in the *rabi* season.

Temperature is a more important factor than the length of the day for seed-stalk development. Most of the varieties that are transplanted early (before November in North India) form seed stalks.

In general, onion production is very good where the climate is mild. A *kharif* crop is also commonly sown in Maharashtra and other states where the climate is mild.

Soil

For successful onion production, friable soils rich in humus, which hold sufficient moisture, are desirable. It can, however, be grown on other types of soils such as sandy loam, silt loam and heavy clay soils. Sandy soils need more and frequent irrigation and favour early maturity; whereas, in heavy soils, the bulb development is restricted and the crop matures later. For a marketable yield and better quality production, the soils which are cool in nature are considered suitable. Onions are sensitive to high acidity and alkalinity. However, in the case of salt-tolerant varieties, the decrease in quality was not significant up to a salinity level of 7.5 m mhos/cm. It has been found that if the soil is highly alkaline and the sodium salts appear on the top of the soil, the growth of the plant is completely checked. In general, sandy loam to clay loam soils are recommended for growing onions with a pH range of 5.8 to 8.0.The electrical conductivity should be around 0.3 mm.

Types and varieties

The following are six different types of onions in India:

- Green onions–used in salads
- Mild onions–used for cooking as food or in salads
- Pungent varieties–used as a condiment for flavouring a number of foods
- Pearl onions or small onions–used in pickles, including vinegar pickles
- White and yellow onions–for dehydration purposes and for manufacturing onion powder
- Red and yellow onion–all purpose types for culinary use

The onion varieties which are commercially grown in India are mostly of three colours: brown to red, yellow and white. The brown to red varieties are the most pungent and have a better keeping quality in comparison to the white onion varieties. The yellow varieties of onion have less demand in the market.

The characteristics of some promising varieties which are under cultivation are described below.

Red onions

Pusa Red

This has been developed by selection from the local red variety. It is recommended for all parts of the country. The bulbs are bronze-red in colour, globular, moderately sweet and pungent. It has a good keeping quality and less bolting. The bulbs mature in about 160 days, yield about 250-300 q/ha with 13-14% TSS, and are recommended for the *rabi* season.

Pusa Ratnar

This is a selection from a hybrid variety introduced from the USA called Granex. This is recommended for the *rabi* crop for the North Indian regions. The bulbs are slightly oblate, flattish, and globular with an attractive light red colour. It is sweet or less pungent and suitable for salads. The bulbs mature after 145-150 days of sowing with a yield of 300-350 q/ha. It is also good for storage; the TSS is 11 to 13%.

Udaipur 101

This variety is recommended for *rabi* cultivation in Rajasthan and the adjoining states. The bulbs are deep red in colour, flattish, globular and sweet and less pungent. It is good for salads. It matures within 150-160 days, with yield of 200-300 q/ha and the TSS is 12 to 14%.

Udaipur 103

The bulbs in this variety are red but lighter than the Udaipur 101, oblate to globular in shape, sweet but slightly more pungent. It yields 250-300 q/ha. It matures within 150-165 days after sowing. The TSS is 10.5 to 13% and it is recommended for the *rabi* season.

Patna Red

This is most suited for Bihar and Uttar Pradesh. The bulbs are bronze-red in colour, slightly flat and globular in shape, very pungent and are therefore, suitable for spices. They mature within 150 to 160 days after sowing. The yield varies from 200 to 250 q/ha and the TSS is between 12 and 14%.

Punjab Selection

This is grown in Punjab and Harayana and has red, globular bulbs of medium size and moderate pungency. The bolting is less and it matures in 160 to 165 days after sowing. It yields about 300 q/ha and the TSS is 11.5 to 13.9%.

Hissar-2

This is also recommended for the *rabi* season in Haryana and Punjab. The bulbs are bronze-red in colour like the Pusa Red but are flattish globular, sweet but pungent. It matures 165 days after sowing. The bulbs yield 200-300 q/ha and the TSS is 11.5 to 13.9%.

N.53

It is a *kharif* season variety, recommended for Maharashtra and other states. The bulbs are light red like the Pusa Ratnar, globular in shape, moderately pungent and sweet. They mature in 140-150 days. The yield is about 225-300 q/ha.

Agrifound Dark Red

This is also recommended for cultivation in the *kharif* season in various parts of the country. The bulbs are dark red, globular in shape, with a tight skin and are moderately pungent. They mature in 140-146 days of sowing. The yield is 275-300 q/ha, with 12-13% TSS.

Agrifound Light Red

This is suitable for the *rabi* season in almost all the growing regions. It grows very well, specially in the Nasik region of Maharashtra. The bulbs are light red and globular. The bolting is less and it has a good keeping quality. They mature in about 160-165 days. The yield is 300-325 q/ha and the TSS is about 13%.

Kalyanpur Red Round

This is a popular variety of Uttar Pradesh. The bulbs are bronze, globular in shape, moderately sweet and pungent. The bulbs mature in around 130-150 days after sowing. The yield is between 250-300 q/ha. The TSS is about 13% and it is recommended for the *rabi* season.

Arka Pragati (Sel-11)

This is an improvement over the local collection (IIHR-149) from the Nasik area of Maharashtra. The variety has attractive, pink, globe-shaped bulbs of uniform size, with a thin neck, high pungency and early maturity. It can be grown successfully during both the *rabi* and *kharif* seasons and produces 200 q/ha in 140-145 days.

Arka Niketan (Sel-13)

This variety is also a selection from IIHR 153 from Maharashtra. The medium-sized, globe-shaped bulbs (4-6 cm across) are an attractive pink colour with outer scales, and tight, fleshy, single-centered inner scales. The bulbs mature within 150-160 days and have a high TSS and pungency with very good storage qualities. It produces about

250-340 q/ha and can be grown successfully both in the *rabi* and *kharif* seasons in Maharashtra.

Arka Kalyan (Sel-14)
This variety has deep pink, globe-shaped bulbs (4-6 cm across) with pungent, succulent internal scales. It is moderately resistant to the purple blotch disease and is preferred for cultivation during the *kharif* season. It produces 335 q/ha in 140-150 day of sowing.

Bangalore Rose
This is a popular variety grown near Bangalore exclusively for export. The bulbs are small, 2-2.5 cm in size, flattish and round in shape, with a deep red colour. The TSS is about 19-20%. The yield is about 180-200 q/ha and it matures in about 110-120 days.

Pusa Madhvi (Pusa Line 102-1)
This was developed at the IARI, New Delhi. It yields about 250-300 q/ha in 140 days. The bulbs are large, round and light red in colour, with a TSS of 12.5% brix. It is recommended for the *rabi* season in the Indo-Gangetic plains.

Arka Bindu
This variety was developed for the export market by mass selection from a local collection (IIHR 402) at IIHR, Bangalore. The bulbs are small in size (2.5 to 3.5 cm diameter), very deep red, highly pungent, are a flattish globe-like in shape and have a high TSS (14-16%). It yields 25 t/ha in 75-100 days of transplanting. It is recommended for the *rabi* season in Karnataka.

Arka Kirtiman
This is an F_1 hybrid of the CMS 65 x selection 13-1-1, developed at the IIHR, Bangalore and is suitable for the *kharif* and *rabi* seasons. The bulbs are medium-sized, dark red in colour and the scale leaves are compactly arranged. It can yield 45-60 t/ha in 130 days. The bulbs have a shelf-life of 3-4 months. They are suitable for export to the middle-Eastern countries.

VL Piaz 3
This has been evolved at the Vivekananda Pravatiya Krishi Anusandhan Shala, Almora, Uttar Pradesh, by hybridisation using an exotic line BYG 2207 A and a line derived from the local collection, followed by three cycles of mass selection. It is mainly suited for the hilly areas. It produces medium-sized, flat, round, tight-skinned, light red bulbs with closed necks. The bulbs are ready for harvest in 160-

170 days of transplanting. Its potential yield in the hills was found to be 40 t/ha. It can also be cultivated in the plains, where it registered an average yield of 33 tonnes of bulbs per hectare.

White onions

Pusa white flat

The bulbs are an attractive white colour and have a flattish-round shape. This variety produces 325-350 q/ha with 11-12% TSS. It is suitable for dehydration. It matures in 120-130 days. It is recommended as a *rabi* crop in zones IV, VI and VII.

Pusa white round

The bulbs are large-sized, flattish-round in shape and mature in around 140-150 days after sowing. The yield is 300-350 q/ha. It is good for storage with a TSS of 12-13%. It is suitable for dehydration. It is recommended for cultivation in zones IV, VI and VII.

Punjab 48

The bulbs are medium-sized, flattish in shape and the average TSS is 12 to 13% in this variety. The yield is 250-300 q/ha. It is good for dehydration.

Udaipur 102

The bulbs are medium- to large-sized. The average TSS is about 1.5% and the yield is around 200 to 250 q/ha.

Yellow onions

Early grano

This variety has been introduced from the USA. The bulbs are large, globular, less pungent and suitable for salads. The maturity period is between 135 and 140 days after sowing. It has less bolting, but has poor keeping quality. The yield is about 300-600 q/ha with 7-10% TSS.

Bermuda yellow

This yellow-skinned variety is from the Philippines. It is an excellent yielder and suitable for marketing freshly for eating it raw. Practically no bolting is observed, but its storage life is poor.

Arka pitamber (IIHR Yellow)

This is a short-day onion variety, having firm bulbs with an attractive yellow skin and a globular shape. The average bulb weight is 80 g, and it is moderately resistant to purple blotch. It has an yield potential of 35-38 t/ha. It is suitable for export to European countries.

Other varieties suitable for different regions

Agrifound rose

This is a pickling variety, suitable for Karnataka (Kolar and Bangalore) and Andhra Pradesh (Cuddapah). The bulbs are 2.5-3.5 cm across in diameter and mature in 95-110 days. The average yield is 190-200 q/ha.

Baswant 780

This is suitable for Maharashtra and matures in about 140-145 days. The average yield is between 200-250 q/ha and it is recommended for the *kharif* season.

Brown Spanish

This is suitable for the upper Himalayan zone. The bulb size is 2-2.5 cm, and it is a long-day type which matures in 130-140 days, with an average yield of 250-300 q/ha.

MDU-1

This variety, released for the northern parts of Tamil Nadu, produces 10-11 bulblets/plant. It matures in 60-75 days, with an average yield of 150 q/ha. It is tolerant to logging.

N-2-4-1

The bulb diameter of this variety is 4-6 cm, it matures in 140-145 days of transplanting and gives an average yield of 250-300 q/ha. It is recommended for the *rabi* season.

N-257-9-1

This variety matures in 140 days. The average yield is 300 q/ha and it is suitable for *rabi* cultivation and for dehydration purposes.

Punjab red round

This is well suited to the Punjab, and bears medium- to large-sized bulbs. It is an early-maturing variety, yields 300 q/ha on an average and is recommended for the *rabi* season.

Cultivation

Propagation

The Bellary onion/big onion is propagated by seeds by raising seedlings in the nursery and transplanting them into the field, either by planting the bulbs directly in the field, broadcasting or drilling the seeds directly and planting sets for the production of *kharif* onions. For an early *kharif* crop in South India, March-April is the optimum time for sowing. August-September is the most suitable time for a *kharif* crop. For a *rabi* crop in the plains, October-November is appropriate. For high-altitude hilly areas, March-April is recommended as the sowing time.

For raising a crop, the seeds are sown in nursery beds to raise seedlings for transplanting into the main field. The beds about 0.6 to 1 m in width, 3-6 m in length, which may vary according to the level of the soil, and 10-15 cm above the ground level are prepared in raised conditions to avoid water-logging in the flat beds too. About 8-10 kg seeds are sown in the nursery for raising seedlings for an area of 1 hectare. Around 50-55 beds of 3 × 0.6 m size are enough for 8-10 kg seeds. The seeds are treated with Thiram @ 2-3 g per kg of seeds to avoid damage from the damping-off disease. The soil in the nursery should also be treated with Thiram/Captan @ 4-5 g/ sq m. The seeds are sown in lines at a spacing of 5-7 cm on well-prepared beds and are covered with soil or powdered FYM or compost and then with dry straw to maintain the required temperature and also to preserve moisture. The beds are irrigated regularly with the help of watering cans till germination is completed. After this, drenching with Thiram/Captan @ 2 g per litre of water should be done at intervals of 7-10 days. The irrigation can be continued either by watering cans or by the flooding method if the area is large and the beds are flat, but the force of water should be regulated to avoid damage. The seedlings are considered ready when the plants are 8-9 weeks old, 20-25 cm in height and 0.6-0.9 cm thick. It has been seen that if younger seedlings are transplanted, the establishment is poor, resulting in a low yield and if over-aged seedling are planted bolting problems occur and the new growth takes longer to commence.

Planting by bulbs

This is practised to meet the demand of green onions for salads in early winter. The bulbs are dibbled at 15 cm on the sides of 45 cm-wide ridges or in beds. Another practice is to dibble bulbs in furrows made with a country plough. The field is irrigated after forming the beds and channels. To plant one hectare, 750 kg of medium-sized bulbs are required. Larger-sized bulbs tend to flower early, resulting in a low yield. In Northern India, the bulbs from the previous harvest are planted in August or September. The secondary bulbs developing from the mother bulbs find a ready market as green onions early in the season (October-December).

Direct sowing

Onions can also be grown by direct seeding. The soil is thoroughly pulverized and made free of clods. Varieties with bigger-sized bulbs are grown in rows 30 cm apart. Seedlings that are 6-8 weeks old may be thinned. For smaller-sized bulbs, sowing can be done by broadcasting and later thinned at 5 cm each way. A light irrigation is given immediately after sowing. The inter-culture and irrigation may be repeated after every 10 days.

Planting by sets

Sets are small-sized onions produced by the seedlings, and left to mature in the nursery beds instead of transplanting them. These are used to get an early crop of onion in some parts of Gujarat, Maharashtra and Rajasthan. Mahesena and Talaza in Gujarat are the areas where sets are raised for local uses as well as for supply to other states.

For planting one hectare, 5-8 kg of seed is sufficient to raise an adequate number of sets in a 200 m^2 area. The nursery is prepared as usual. Mid-January to February is the ideal time for sowing. The plants remain in the nursery till their leaves fall (till April). By this time, the plants form small sets due to the lesser spacing. Later on, the plants are dug out to separate the sets. About 1.5-2 cm-sized, disease-free sets are selected. Very small sets result in more splits, doubles and bolters, while larger-sized sets increase the cost of production. The graded sets are stored in hessian cloth bags, in shallow baskets or in racks, under well-ventilated conditions.

Generally, 10 q of sets of 1.5-2.0 cm diameter are required for planting one hectare. The sets are planted 10 cm apart in rows on both sides of ridges, spaced 35-45 cm apart. The sets are normally planted by mid-August to get an early crop by mid-November. The planting, however, can be continued upto mid-September, so as to stagger the availability of the produce up to mid-December. A light irrigation is given immediately after planting the sets

Land preparation

The land selected for growing should be cleared of vegetation, the field is ploughed 4-5 times, and 25 tonnes of FYM is applied at the time of the last ploughing. Along with this, a quantity of 50 kg of Zinc sulphate or Ferrous sulphate is applied and ploughed. Then, seed beds and channels of convenient sizes (2-3 m wide and 3-5 m in length) or ridges and furrows are formed at specified spacings before transplanting.

Transplanting

In Maharashtra, the early *kharif* crop is transplanted by the end of July end or the beginning of August, whereas the usual *kharif* onion is transplanted by mid-October. Red onion is transplanted in the last week of December or in the beginning of January.

In certain areas, pruning of the top to the extent of 25% is done to obtain a higher yield. The onion seedlings are transplanted at a spacing of 15 × 10 cm in Indian conditions. The sets and multiplier onion are planted at a spacing of 45 × 10 cm on both sides of the ridges for better bulb development. The field should be irrigated immediately after transplantation. In certain places, transplanting follows irrigation in order to avoid mortality.

Manures and fertilisers

The requirement of nutrients depends on the variety, soil type, region of growing and removal of major nutrients. It has been reported that a yield of 30 t/ha of onions requires 85 kg of N, 42 kg of P and 130 kg of K.

The recommended fertiliser dose (kg/ha) for different regions and varieties is given in the table below.

N	P	K	Variety	Region
150	80	0	Pusa Red (*rabi*)	Karnal, Haryana
150	40	50	Agrifound Dark Red (variety)	Karnal, Haryana
150	50	60	Pusa Red (*rabi*)	Jabalpur, Madhya Pradesh
100	50	50	Punjab selection (*rabi*)	Ludhiana, Punjab
80	0	0	Patna Red (*rabi*)	Sabour, Bihar
100	80	50	Arka Kalyan (*kharif*)	Bangalore, Karnataka
100	80	50	Arka Pragati and Arka Niketan (*rabi*)	Bangalore, Karnataka
125	50	125	Bangalore Rose (*rabi*)	Karnataka

In addition to the chemical fertilisers, some other nutrients are also useful for improving the quality of onions. The application of zinc (1-3 ppm) increases the yield as well as improves the quality of the crop.

Micro-nutrients

It has been reported that the application of micro-nutrients improved the quality and pungency of the onion bulbs. Some of them cause reduction of bolting in onions.

Onions grown in peat soil exhibit abnormal copper deficiency, with thin scales of a pale-yellow colour. The application of Copper sulphate, at the rate of 125 to 155 kg/ha, increases the thickness of the onion scales and changes the colour from pale-yellow to a brilliant brown. Copper-deficient onions lack firmness and solidity.

Sometimes, Boron deficiency is also observed in onions. It is characterised by transverse cracks on the upper sides of the basal leaves. The leaves become stiff and brittle. The leaf colour varies from dark grey to a deep blue green. The youngest leaves develop a conspicuous yellow-and-green mottling. The deficiency of boron can be corrected by the foliar application of Borax @ 18 to 20 kg/ha.

Magnesium deficiency symptoms in onion are exhibited by the appearance of irregular, elliptical-shaped areas, almost white in colour, near the ends of the leaves. Magnesium deficiency in the soil can be corrected by the annual application of 60 to 110 kg of Magnesium sulphate depending upon the soil conditions.

Irrigation

Weeding

Onion is a closely-planted and a shallow-rooted crop, thus hand weeding is difficult and may also damage the crop. Manual weeding has also becoming expensive. It has been observed that 2.5 kg of Teneran/ha 3-5 weeks after transplanting gives very good results. Basalin @ 1 litre/ha three weeks after transplanting also proved effective. The application of Tok E-25 @ of 4-6 litres and Basalin @ 0.5-1 litre/ha with one hand weeding controls weeds effectively. Nitrofen at 1 kg/ha applied before planting also controls the weeds, but Nitrofen treatment with 2 hand weedings is found to be better. The critical period for weed competition in onion is up to 40 days after transplanting.

Pests and diseases

Pests

Onion thrips

This is a major pest of onion and garlic. The insect *Thrips tabaci* is very minute and yellow to dark brown in colour. The life-cycle of these insects is completed in about 8-10 days. Thrips multiply in large numbers during March and April both on bulbs and seed onions in the northern parts of India, whereas, in the Nasik area, thrips incidence is observed even in plants in the onion fields and also over the bulbs in winter. It is found in the axil of green leaves where it sucks the juice of the early emerging leaves and it leaves behind numerous silver-white spots. The infested plants show stunted growth and, the plants die away slowly.

It can be effectively controlled by the application of Phorate @ 10 kg/ha in the soil before transplanting. Spraying with Quinalphos (0.04%) or Monocrotophos, Phosphomidon, Dimethoate (0.05%) or neem formulations (2-3 ml/l) also effectively control the pest.

Onion fly

This is also called the onion maggot. The flies, *Delia antiqua,* lay their eggs on old leaves or on the soil and the larvae enter into the soil and damage the disc portion of the onion bulbs. The injury caused by maggots also encourages basal-rot disease by pathogenic organisms. Aldrin dust @ 25 kg/ha should be incorporated into the soil before transplanting. The application of phorate (0.5 kg/ha) or neem cake (500 kg/ha) to the soil is also recommended for control.

Head borer

The gram pod borer, *Helicoverpa armigera,* is a ployphagous pest of the tropics, subtropics and warmer regions. The larvae cuts the pedicil of the flower and feeds on the stalk which, on maturation, pupates inside the scale in the soil. Spraying Edosulfan (0.2-0.3%) has been recommended to control the pest.

Mites

The tiny insects, *Aceria tulipae,* suck the sap, turning the plants pale with a sickly appearance. The pest also infests the stored onions and induces sprouting. The mite is reddish-green in colour and dry weather is conducive for infestation. For the management of mites, Dimenthoate/Ethion (0.05%) should be applied at the initial stage of infestation.

Tobacco caterpillar

The larvae of the insect *Spodoptera litura* enter the hollow leaves and eat them from the inside, leaving a papery epidermis. The small and gregarious larvae on the leaves should be collected and destroyed. Malathion (0.01%) or Dichlorvas (0.05%) can be sprayed at intervals of 15 days to reduce the larval population.

Cut worms

The larvae of this insect *Agrotis ipsilon* are seen in the nursery beds. The tender plants are found damaged at ground level during the night. The moth has a dull brownish colour, while the larvae is dark brown with a red head. An application of Carbofuran 1 kg a.i/ha or

Chloropyriphos (0.1%) to the soil at the time of planting may be used as a preventive measure. The clods in the field should be broken to destroy any hiding places for the larvae.

Leaf miner (Chromatomyia horticola)

The greenish larvae, measuring 1-2 mm in length, cause white serpentine mines on the leaves of the onion plants. The adult females lay their eggs beneath the epidermis of the onion leaves. To control the leaf miner, Malathion (0.1%) should be sprayed.

Beetle

The grubs of *Anthernus acenicus, A. jordanicus* and *Alpbitiobius laevigatus* beetles feed on the stored onion bulbs. The beetle damage leads to the attack of pathogenic micro-organisms, resulting in rotting. The storage houses should be fumigated with Carbon disulphide as a preventive measure before storing the bulbs.

Ground nut earwig (Euboreltea annulipes)

The caterpillars bore into the onion bulbs. For control, the application of Chlorophyriphos (0.1%) to the soil is recommended as and when the damage is observed.

Diseases

Damping-off

The disease is caused by *Fusarium oxyporum.* In pre-emergence, damping-off, the fungus kills the radicle and it attacks before emergence from soil; while in post-emergence, the pathogen attacks the collar region of seedlings on the surface of the soil. The collar portion rots and ultimately the seedlings topple down and die. The collar-rot is controlled by seed treatment with Thiram @ 2.5 g/kg of seed before sowing, or the top soil of the nursery should be treated with Thiram @ 5 g/sq m area. The drenching of the soil should be done with Thiram @ 2.5 g/litre of water at fortnightly intervals.

Purple blotch disease

This is a widespread fungal disease caused by *Alternaria porri*. The common symptoms occur on the leaves and flowerstalks as small, sunken, whitish flecks with purple-coloured centres. The incidence is more in closely-spaced plants than in those that are widely spaced. Raising the nitrogen rates from 40 to 120 kg N/ha decreases the incidence of the disease. The early-sown crop (1st November) is always more susceptible to the disease than the late-sown crop (1st December).

Stemphylium blight disease (Stemphylium vesicarium)

The infection appears as small yellow to pale orange streaks in the middle of the leaves/flowerstalks on one side. Later, the inner portion of the leaf turns brown from the tip of the base and ultimately drying starts from the tip. The disease spreads quickly and within a fortnight, the whole field gets infested. A fortnightly spray of Dithane M-45 at 0.25% along with sticky Trition can control the disease.

Downy mildew (Peronospora destructor)

This disease is prevalent in locations high humidity. A violet growth is seen on the surface of leaves or flowerstalk and as a result the leaves or seed stalks collapse. The onion bulbs grown for seed crops should be exposed to the sun for 12 days or heated to 41°C for 4 hours. The spraying of 0.2% Zineb controls the disease.

Onion smut (Urocystis cepulae)

The disease is observed in areas where the temperature remains around 30°C, like in the Mysore region. The disease appears on the cotyledon of the young plants soon after they emerges. Brown to black coloured, elongated blisters from the base of the seedling within the scale and on the leaves are the main symptoms of the disease. These blisters often break open, exposing the black powdery masses of spores of the pathogen. The disease can be controlled by treating the seeds with Thiram @ 2 g/kg of seeds.

Neck rot or bulb rot (Botrytis allii)

This fungus is primarily a parasite on the bulb scales. It may cause small, white, necrotic lesions on leaves and seed stalk. This may induce sufficient damage, causing blasting or reduction in the quality of the seeds. The bulbs should be harvested at full maturity, cured properly and should be heated at 40°C for 8 hours before storage.

White rot (Sclerotium rolfsii)

This pathogen is soil-borne and invades the roots and the basal parts of the bulb scales. The signs are the yellowing and wilting of the leaves, followed by a total collapse of the tops. Spraying fungicides such as Botran or Benomyl @ 0.1% controls the disease.

Pink rot (Pyronochoetia terrestris)

The affected roots turn yellow, shrivel and die, and change to a distinctly pink colour. Even the new roots growing from the infected plants get infected immediately and become functionless. The affected plants are generally not killed, but the development of bulbs is retarded.

The crop should be sown late to avoid the heavy incidence of this disease. Telone @ of 50 l/ha should be applied to control the disease.

Bacterial soft-rot

Soft-rot in onion is caused by *Pseudomonas allicola* and *Erwinia carotovora*. The rot begins at the neck of the bulb and later it emits an offensive smell through the neck when it is squeezed. The bulbs are generally attacked by this disease during the active growth period. The outer scales of fully grown bulbs usually and sometimes the fourth to seventh scales are also affected. The early varieties should be cultivated and the freshly-harvested bulbs should be cured for at least 10-12 days in the field.

Black mould (Aspergillus niger)

This disease is characterised by the black powdery mass of spores that are borne on the exterior of the scales and can be rubbed off easily. The disease is not only confined to the outer scales, but is also present on the inner scales. As a protective measure, the bulb should be thoroughly cured after the harvest and before storage. Dusting the plants with $CaCO_3$ at 12 kg/ha during harvesting helps in controlling the disease. The rotting of onion bulbs in storage is caused by *Fusarium solani,* which could be reduced to 58% by a pre-inoculation treatment with 2,4-D, Potassium Metabisulphite, Diphenylamine and Acetic acid.

Onion yellow dwarf virus (OYD)

The characteristic symptoms of this disease are severe stunting of the plants, dwarfing and twisting of the flower-stalk. The affected leaves and stems change from their normal green colour to various degrees of yellow. The leaves tend to flatten and crinkle and as a result fold over. The bulbs of the infested plants are malformed and difficult to store. The virus is transmitted both mechanically as well as by insect vectors. Spraying Dimethoate or Malathion (0.1%) is effective in controlling the vector.

Aster yellow

The virus-affected plants show yellowing and stunting symptoms. Such symptoms are also found in storage with sprouted onion bulbs. The diseased plants should be uprooted and burnt. Malathion @ 0.01% may be sprayed to control the insect vectors. Antibiotics such as Chlorotetracycline spray, suppress the symptoms of this disease.

Harvesting and yield

The onion bulbs should be harvested at the appropriate stage of maturity, since it is a deciding factor in the storage life and quality of onions. An early harvest results in the sprouting of the bulbs and a late harvest gives rise to the formation of secondary roots during storage. The development of a red pigment and the characteristic pungency of the variety are also important harvest indices of onions. In onions, neck fall is another common indication of maturity.

The time of harvesting also depends on several factors like the planting season, variety, market price, etc. The harvesting starts when 25-50% of the neck falls. It is recommended to fell the other tops manually, before harvesting the crop. This also results in better TSS and a higher bulb yield. The bulbs are harvested by pulling them out by hand or using hand implements like the modified potato digger on a large scale. After harvesting, the onions along with the tops are placed in windows in the field for 2-3 days to sun-scald. The tops are cut off either in the field or in the shade if the sun is strong. In case of the *kharif* crop, forced toppling should be done to stop the growth 25 days before harvesting. The roots are then removed and bulbs are sent to the market.

In the *rabi* season, the common onion gives a yield of about 200-300 q/ha depending upon the variety, manuring, irrigation and other cultural operations. However, in the *kharif* season, the yield is comparatively lower.

Cold storage

The onions are kept in cold storage during the summer. The best temperature for storing onions is 1-2°C, with a relative humidity of 70-75%

Inhibitors for sprouting of onions during storage

The sprouting of onions while in storage could be prevented by spraying Maleic Hydrazide in the concentration of 2000 ppm, one week before the harvest of the onion crop. This spray checks sprouting, rooting, rotting and reduces the total loss in the weight of the onion bulbs, even when they are stored at at room temperature.

Value-added products

Dehydrated onion, onion powder, onion flakes, kibbled onions, and onion salt are some of the value-added products of onion, of which the first five are used for flavouring ketchups, sauces, etc.

For dehydration purposes and for the manufacture of onion powder, white onions of the desired quality are preferred. The desirable traits needed in onions for dehydration are: (i) white-coloured scales, (ii) full globe to tall globular shape of bulbs with 5-6 cm diameter,

(iii) solids content around 15%, preferably 20%, (iv) high degree of pungency, (v) high yield, (vi) good keeping quality (at least 2-3 months), (vii) freedom from joints, and (viii) disease-free bulbs.

Onion powder is prepared by grinding the dehydrated onion slices in hammer mills to a suitable mesh. It is highly hygroscopic and, hence, the important precaution to be followed regarding its storage is to keep it in air-tight containers in a cool, dark and dry place.

Onion salt is prepared by mixing 19-20% onion powder with 78% free-flowing pulverised, refined table salt and 1-2%, of an anticaking agent, which prevents water absorption and caking.

Multiplier (aggregatum) onion

The multiplier or Aggregatum onion (*Allium cepa* L. var. *aggregaum* Don.), of the family *Alliaceae* produces a number of edible underground bulbs from a single bulblet. It is also called the country onion or potato onion.

Climate

It grows very well under tropical conditions with sufficient soil moisture. Heavy rains during the germination as well as the bulb-formation period are deleterious to the crop growth.

Soil

Well-drained loamy soils are the best suited to this crop, while in clayey soils the bulb formation is hindered. It grows very well in soils with a pH range of 6-7.

Varieties

The following are a few promising varieties of the multiplier onion, which are being cultivated in the country.

Co-1: this is a clonal selection from CS 450, an onion type introduced from Manachanallur, suitable for cultivation in all the onion-growing districts of Tamil Nadu. It has a yield potential of 9-10 t/ha in a crop duration of 85 days. The bulbs are large sized, pink in colour and pungent, with a good keeping quality. It has 4-8 bulbs/clump, with a weight of about 25 to 50 g/clump.

Co-2: The variety has been selected from the germplasm type CS 911. It has a potential yield of 12 t/ha in a crop duration of 65-70 days. The bulbs are bigger in size and have a crimson colour. The number of bulbs/clump ranges from 6-10 and the weight of bulbs ranges from 30 to 60 g/clump. This is another variety that is suitable for Tamil Nadu.

Co-3: This is a clonal selection from the open-pollinated progenies of CS 450. The crop duration is 65-70 days and it yields 15-16 t/ha. The number of bulbs/clump ranges from 6 to 10 and they have a full red colour. The bulb size is bigger than that of Co-2.

Co-4: This is a hybrid derivative of the cross AC 863 × Co-3. The crop duration is 65 days and the yield is 10 t/ha. Each clump weighs, on an average 90 g and the bulbs/clump range from 8 to 11. The bulbs are bold in size with an attractive pink colour. The bulbs store well for more than 150 days without sprouting in well-ventilated store-houses.

Agri-found red: This variety has been developed by the NHRDF, Dindigul. Each clump weighs, on an average, 65 g. The bulbs are an attractive light red colour. It yields 18-20 t/ha in 65 days with a TSS of 16° Brix.

Cultivation

Propagation

The aggregatum onion is propagated through the bulblets produced in the clump.

Land preparation

After clearing of the land, it is ploughed two to three times and the soil is brought into fine tilth. At the time of the last ploughing, a quantity of 25 tonnes of FYM is incorporated into the soil. The ridges and furrows are formed at a spacing of 45 cm. A basal dose of inorganic fertilisers is also applied and thoroughly mixed with the soil before planting commences.

Planting

A quantity of 1.5 tonnes of seed bulbs (medium-sized bulbs) is required to plant an area of 1 hectare. The bulbs are planted at a spacing of 10 cm between plants on both sides of the ridges. The bulbs start sprouting in 4-6 days of planting.

Manures and fertilisers

Inorganic fertilisers are necessary for better yields, in addition to the organic manure. A dose of 30 kg of N, 60 kg of P and 30 kg of K per hectare are applied as a band near the ridges (on both the sides of the ridges) and thoroughly mixed with the soil. Another 30 kg of N should be applied after 2 months of planting as a top dressing.

Irrigation

The field should be irrigated immediately after planting and again on the third day. Later on, irrigations are given depending on the weather and the soil moisture conditions. Irrigating the crop at 1:2 IW/CPE ratio to a depth of 5 cm has been found to be the optimum.

Weeding

The critical period for weed competition is up to 40 days. Therefore, the crop should be kept free of weeds either by hand weeding or by using herbicides. The application of 2.0 kg of Teneran/ha, 3-5 weeks after transplanting, is effective in controlling the weeds.

Pests and diseases

Pests

Thrips cause considerable damage to the crop. To control this pest, Metasystox spray (0.1%) is recommended.

Diseases

The purple blotch, caused by *Alternaria porri,* is the major disease. Here, the collar portion rots and, ultimately, the seedlings topple down and die. Spraying Indofil M-45 (0.2%) is effective in controlling this disease.

Harvesting and yield

Fifteen days before harvesting, the plants are sprayed with 500 ppm of Maleic Hydrazide to enhance the shelf-life of the onions in the storage house. At the time of maturity, more than 75% of the leaves will dry up. The last irrigation is given one week before the harvest. After the harvest, the tops are removed and the bulbs are shade-dried and stored in a well-ventilated room.

The yield varies with the variety, season and location.

Opium Poppy

Plant Profile

Family	:	*Papavaraceae*
English name	:	Opium Poppy
Indian name	:	*Ahiphenam, Aphukam* (Sanskrit), *Kas-Kas Aphim* (Hindi), *Asagase* (Kannada), *Kasakasa, Posttakkai* (Tamil)
Species	:	*Papaver somniferum*
Distribution	:	India, Argentina, Bulgaria, China, Czecholovakia, Egypt, Germany, Greece, Holland, Hungary, Iran, Japan, Poland, Portugal, Spain, Turkey, Russia, Yugoslavia
Uses	:	Confectionery, culinary items, drugs, oil

Papaver somniferum (2n=22), commonly called the opium poppy, is an annual herb belonging to the family *Papavaraceae*. It is mainly cultivated either for the manufacture of opium or for the seeds. The var. *album* of poppy with white seeds has been cultivated for many years for the production of seeds in India. Poppy is an important medicinal plant, being the source of over 40 alkaloids, including psychoactive agents, and is a great boon to psychiatry for the treatment of mental and nervous diseases and medical research. The commercial product 'Opium' is an addictive narcotic obtained from the seed capsules of the opium poppy, the source of a number of very valuable alkaloids like morphine, codeine, narcotine, papaverine and thebain. Other minor alkaloids include aporeine, codamine, cryptopine, guoscopine, hydrocotarnine, laudanine, narcotoline, neopine, oxynarcotine and papaveramine. The analysis of the seeds from 5 types of Indian poppy gave the following ranges of values—moisture-4.3-5.2%, protein-22.3-24.4%, ether extract-46.5-49.1%, N-free extract-11.7-14.0%,crude fibre-4.8-5.8%, ash-5.6-6.0%, calcium-1.03-1.45%, phosphorus-0.79-0.89%, iron-8.5-11.1 mg/100 g, thiamine-740-1,181, riboflavin-756-1,203, and nicotinic acid-800-1,200 µg/100g, carotene-absent. The minor minerals in the seed are iodine (6 µg/kg), manganese (29 mg/kg), copper (22.9 mg/kg), magnesium (15.6 g/kg), sodium (0.3 g/kg), potassium (5.25 g/kg) and zinc (130 mg/kg). The seeds also have lecithin (2.80%), oxalic acid (1.62%), pentosans (3.0-3.6%), traces of narcotine and an amorphous alkaloid, the enzyme diastase, emulsion lipase and nuclease. The seeds have a high protein content, the major component being a globulin, which accounts for 55% of the total nitrogen. The seeds contain a high percentage of linoleic acid, which lowers blood cholestrol in the human system. The alkaloids morphine and codeine are widely used as sedatives, to relieve pain, and induce sleep, in addition to their use in treating coughs. Opium is a very valuable but a dangerous drug. It should be used in only limited quantities and under strict supervision of a physician.

In India, this plant is mainly cultivated for its latex (opium) and the seeds come as a by-product. These seeds are quite a rich source of fatty oil and protein and, in many countries of Europe, they are employed as a major source of cooking oil. The seed called *khus khus* is also an important culinary item in India. It is extensively used in the preparation of confectionery, pasteries, cookies, curries, cakes, sweets and bread, etc. The seeds are demulcent and are used in the form of emulsion, as an emollient and especially obstinate constipation and in the treatment of catarrh of the bladder. The white seeds are sometimes used in the manufacture of pharmaceuticals. Poppy seeds or seed-meal also find use in the production of lecithin (0.67-0.91%). In some places, the young plants are also consumed as a leafy vegetable. This is one of the medicinal plants which is highly lucrative. A crop raised properly, in an area of one hectare, fetches about 1 lakh rupees. However, its cultivation has to be done under the strict supervision of the Central Excise Department and it cannot be cultivated everywhere, except in the areas specified by the Government of India.

Origin and distribution

The centre of origin is reported to be the Western Mediterrannean region and from there it has spread through the Balkan peninsula to Asia Minor and India. It has traditionally been cultivated in Italy, Greece and Asia Minor. It was during the 15th century that the herb was introduced into India. First, it was cultivated along the coastal areas and it later spread into the interior of the peninsula.

Area and production

The 1953 United Nation's Opium Conference Protocol (still in effect), for limiting and regulating the cultivation of opium poppy plants, asserts that Bulgaria, Greece, India, Iran, Turkey the erstwhile USSR, Egypt, Czechoslovakia, Poland, Denmark, Sweden, Germany, Holland, China, Japan, Argentina, Spain, Hungary, Portugal and Yugoslavia are the countries that may legally produce opium. Other countries may grow if for other purposes. The only country where substantial amounts of opium are now produced for export is India; this amounts to 1,465 tonnes annually and contributes to over 90% of the world's production. In India, all the opium of commerce is now grown mainly in the states of Uttar Pradesh, Madhya Pradesh, Kashmir, and Rajasthan, covering an area of about 18,000 hectares.

The control and regulation of all the aspects of poppy cultivation and its wholesale trade have been under Government of India since 1773. After the partition of the Indian sub-continent in 1947, the essentials of the present system of narcotics control, which derived from the system instituted by the British in the 1857 Opium Act, were retained with few changes.

The Government of India is able to effectively licence farmers to grow opium. This is highly organised, so as to allow the elimination of those growers whose yield of opium is poor.

Description of the plant

Opium is a small, erect, scarcely branched (towards the top) herb, growing up to 120 cm in height, with large, serrated leaves and attractive bright coloured (white, pink, purple, red and variegated) flowers. The main shoot and branches terminate into large, oblong to globose capsules, filled with small, white, flat seeds. The seeds are small, kidney-shaped, whitish, yellow, grey brown, reddish-brown or black and are rich in oil.

Climate

In general, the crop needs long spells of cold weather (20°C) with adequate sunshine in the early stages for healthy vegetative growth. Heavy rains after sowing cause losses in seed germination. Warm, dry weather with a temperature of 32-38°C is needed during the reproductive period. Cloudy weather, frost, hailstorms and high gusty winds, particularly during the lancing operations, cause immense damage to the growing crop. Dry, warm weather conditions in February-March favour the good flow of latex and results in higher yields.

Soil

The opium crop needs deep clay loam, highly fertile and well-drained soils, with a pH range of 6.0 to 7.5. Such soils containing adequate organic matter retain moisture and there is no need of irrigation during lancing. However, with adequate manuring and use of fertilisers, even light loam to sandy-loam lateritic soils can give high yields under good management. Heavy clay or fine sandy soils generally pose a problem, since they remain wet during the rains and are too difficult to cultivate during the dry periods.

Varieties

The most popular varieties grown in the country remain in the field from 140 to 160 days. The following are some of the important varieties, of which Talia, Ranghatak and Dhola Chota Gothia are the popular varieties recommended for heavy black soils.

(i) Talia

This is sown early and it remains in the field for 140 days. Its flowers reflect a pink shade on their large petals. The capsule is oblong to ovate in shape, light green in colour and shining (waxy).

(ii) Ranghatak

This is a medium-tall variety, maturing for lancing in 125-130 days after sowing. It bears white and light pink flowers. It produces medium-sized capsules (7.6 cm × 5.0 cm), which are slightly flattened on the top. It yields opium of a comparatively thin consistency that changes to dark brown colour on exposure.

(iii) Dhola Chota Gothia

This is a dwarf cultivar (85-90 cm), bearing pure white flowers and light green capsules with an oblong-ovate shape. It is ready for lancing after 105-115 days of sowing and matures for the seeds in 140 days.

iv) MOP-3

This variety has been developed at the Jawaharlal Nehru Krishi Vishwa Vidyalaya, Mandsaur. It bears pinkish-white flowers, made up of large non-serrated petals. Its capsules are ready for lancing at 120 days after sowing and the variety is recommended where adequate irrigation facility exists in the later part of the season.

(v) MOP-16

This is another promising selection made at the Jawaharalal Nehru Krishi Vishwa Vidyalay, Mandsaur. The plant bears white flowers with serrated petals and round capsules with a flat top. This is comparatively drought-tolerant and is ready for lancing in 105-110 days after sowing. It is recommended wherever an early-maturing crop is preferred.

(vi) Shama

This variety was released by the Central Institute of Medicinal and Aromatic Plants (CIMAP), Lucknow, during the year 1983. The main alkaloids like morphine (14.51 - 16.75%), codeine (2.05-3/24%), thebaine (1.84-2.16%), papaverine (0.82%) and narcotine (5.89-6.32%) in this variety are reported to be on the higher side compared to the existing commercially cultivated variety. It yields 39.5 kg of latex and 8.8 kg seeds per hectare.

(vii) Shweta

This variety was also released by the CIMAP, Lucknow, along with Shama. However, it is reported to be superior to Shama in terms of the content of the main alkaloids—morphine (15.75-22.38%), codeine (2.15-2.76%), thebaine (2.04-2.5%), papaverine (0.94-1.1%) and narcotine (5.94-6.5%). It gives an average yield of 42.5 kg of latex and 7.8 kg of seeds per hectare.

(viii) BROP 1 (Botanical Research Opium Poppy-1) (NBRI-3)

This is a synthetic variety developed at the National Botanical Research Institute, Lucknow, by crossing selections from Kali Dandi, Suyapankhi and Safaid Dandi. This variety is highly adaptable to varied agro-climatic conditions and gives a higher yield than the national checks. It is moderately resistant to diseases and to lodging. It yields about 54 kg opium/ha and 10-13 q seeds/ha. The morphine content is 13% and above.

(ix) Kirtiman (NOP-4)

This was developed at the Narendra Dev University of Agriculture and Technology, Kumarganj, Faizabad, through selection from the local races. The variety is moderately resistant to the downy mildew disease. It yields 35-45 kg/ha of latex and 9-10 q/ha of seeds. The morphine content is up to 12%.

(x) Chetak (U.O.285)

This variety was developed at the Rajasthan Agriculture University, Udaipur. It is moderately resistant to diseases and lodging. The opium yield is up to 54 kg and the seed yield is 10-12 q/ha and it contains up to 12% morphine.

(xi) Trishna (IC-42)

This variety was developed at the National Bureau of Plant Genetic Resources, Delhi, through selection and selfing. It yields 49 to 53 kg latex/ha and 12 to 14.78% morphine. The variety is resistant to frost and root-rot and is tolerant to the downy mildew disease.

(xii) Jawahar Aphim 16 (JA-16)

This is a pure-line selection of local land races developed at the Jawaharalal Nehru Krishi Vishwa Vidyalay College of Agriculture, Mandsaur (Madhya Pradesh). It is moderately resistant to the downy mildew disease. It gives 45-54 kg of latex/ha, 8-10 q of seeds/ha and contains up to 12% morphine.

Recently, another three varieties, 'NBRI-3' for opium and 'Sujatha' an opium-free poppy for the production of oil and seed and 'Shubhra' for high morphine and seed yield, have been released from the National Botanical Research Institute, Lucknow; Regional Research Laboratory, Jammu; and the Central Institute of Medicinal and Aromatic Plants, Lucknow, respectively.

'Vivek' is another variety with a high latex and morphine yield; whereas 'Suchitra' is for high morphine content in the straw, meant for the solvent extraction of capsules, eliminating lancing in the field.

Land preparation and sowing

The crop is propagated through seeds. Poppy seeds should be sown in very well-prepared soil. The field should be given 5-6 cross-ploughings followed by planking, and the land should be divided into small plots to facilitate irrigation. The seed rate is around 8 kg/ha when sown by the broadcast method but when sown in lines at 30 cm spacing, about 5-6 kg of seeds per hectare is sufficient. In addition, this method is reported to result in higher yields, besides promoting inter-culture and lancing operations. Experiments conducted at Mandsaur showed that a plant population of 300,000/ha is conducive to higher yields. Treating 1 kg of seed with 4-5 g of Thiram or Brassicol before sowing protects the seeds from soil-borne diseases. The seeds are sown between late October and mid-November in different parts of the country. After sowing, the seeds are covered by a thin layer of soil, followed by a light irrigation.

Manures and fertilisers

Usually farmers maintain a rich top layer of humus by adding a large amount of Farm Yard Manure to the opium field. It is found that the application of 10-20 tonnes of FYM is enough to maintain the soil in good physical condition and to support a good initial growth. In addition, the field is applied with 30-50 kg P_2O_5 and wherever potash is deficient K_2O should be applied at the rate of 25-30 kg/ha. Both these fertilisers are placed at a depth of 5-6 cm in the rows before sowing. The amount of nitrogen varies from 60 (10 t of FYM) to 90 kg (20 t of FYM/ha). It is applied in three equal split doses. The first dose is given at sowing and the subsequent doses at 30-40 and 60-65 days after sowing. For correcting zinc deficiency, 12-30 kg/ha of Zinc sulphate should be added.

Irrigation

The first irrigation is given immediately after sowing, if there is not enough moisture available in the soil. For subsequent irrigation, a 7-10 days' irrigation schedule is optimum, depending upon weather and soil conditions. A total of 10 to 15 irrigations are required by the crop.

Inter-culture and weeding

All cultivation operations are done by hand. When the plants are 5 to 6 cm high with 3-4 leaves, i.e., after 3-4 weeks of sowing, the first weeding is done. The intervals of the other weedings have to be adjusted locally. In all, about 4-5 weedings may be required during the whole cropping season.

Crop rotation

The maize–opium, chillies–opium or sunhemp–opium rotations are generally followed.

Flowering and fruit-set

After about 90 to 100 days of sowing, the plants which are waist-high begin to flower; i.e., flowering will take place during the first week of March if the crop was sown during the second fortnight of November. Usually after 3 days of flowering, the petals fall off and after another 10-14 days the capsules are ready for lancing.

Pest and diseases

Pests

Cutworms (*Agrotis sulfusa*) cause great harm to plants. They remain burrowed in the soil during the day and during the nights they cut and eat the leaves. This insect can be controlled by flooding the field with water and dusting the crop with 2% Carbaryl.

Weevils (*Stenocarus fuliginosus*-root weevil and *Ceutorhynchus macula-alba*-capsule weevil) attack the young seedlings and feed on the young leaves. As the plant matures, the insect eats its way up the stem and ultimately bores its way into the poppy heads and proceeds to destroy the capsule.

A capsule borer (*Helicoverpa armigera*) and aphid (*Myzus* sp.) are also reported to damage the capsules and shoots, respectively.

Diseases

The downy mildew (*Peranospora arborescens)* disease may appear any time from the seedling to the maturity stage. In the primary infection, the plants develop chlorosis, curling, abnormal plant forms with stunted growth or the plant may not even survive. If the plants are attacked in the later stage, the capsules are affected. The disease can be controlled by the application of Diathan Z-78 (0.4%).

Powdery mildew (*Erysiphe polygoni*) appears after 90-100 days of sowing as white mycelial growth on the stem and lower side of the leaves. It later grows into large patches of white powdery masses turning to black. The diseased plants in the early stage should be uprooted and burnt. At a later stage, the crop should be sprayed with Sulfex or Microsulf (1 litre in 100 litres of water). Spraying Diathane Z.78 or Diathane M-45 (2 g/lit.) beginning at the flowering stage and repeating it 2-3 times at intervals of 15 days also prevents the spread of the disease.

The other diseases reported on this crop include root-rot caused by *Fusarium semitectum, Sclerotinia sclerotiorun, Macrophomina phaseolina* and *Rhizoctonia sp,* leaf blight caused by *Helminthosporium papaveris;* capsule infection (*Macrosporium papavaris, Alternaria brassicae* var. *somniferum, A.alternata* and *Dendryphion penicillatum*); and virus diseases (cabbage ring-spot virus, beet yellow virus, turnip mosaic virus and bean yellow mosaic virus). Broomrape (*Orobanche papavaris*) is also reported to infest the crop.

Lancing and latex collection

The lancing operation is performed by skilled labour (an average of 6 to a plot), usually on bright sunny days between noon and 4 p.m. During the hottest part of the day, the pedicil is said to form on the surface of the freshly-exuded latex due to the hot sun, resulting in a greater degree of evaporation and quicker thickening. This also helps to prevent the latex from falling off the capsule. The lancing operation begins at the edge of the field and works backward to avoid contact with the exuding latex. The hand is quickly passed over a capsule and a subjective decision is made as to whether it is ready for incision or not.

The lancing instrument called '*Nastar*' or '*Naka*' comprises of four lines about the dimension of ordinary needles spaced at 1.5 to 2 mm distance and affixed to a holder around 18 cm long. The '*nastar*' is held carefully as one holds a pencil while writing and the incision is made by a swift downward stroke, starting just below the stigmatic rays. The depth of the incision is controlled by the affixation of the lines to the holder; if the incisions are too deep the latex is exuded into the interior of the capsule and is thus lost. If the cut is too shallow the yield of the latex will be low. Usually, an incision with a depth of 0.4 cm is considered ideal. About 150-200 capsules can be lanced per hour by an experienced worker.

Immediately after lancing, the latex exudes. It is intitally milky and gets accumulated in the outer wall of the capsule. It quickly darkens and dries during the course of the day; it is generally collected the next day before 10 a.m. by scraping it off with a trowel called a '*Seetoah*'. The collection may also be delayed for one or two days, depending upon the 'appearance' of the capules and vigour of the latex flow. In other words, the interval is subjectively determined by the collector. The lancing process is ordinarily repeated twice, making a total of three series of vertical incisions. The spacing of the incisions on the capsule is generally even. If the capsules are exceptionally good, the lancing can be done 4-5 times. The air-dried latex, which has now become blackish in colour, is scraped from the capsules into small earthernware pots that are lined with polythene sheets. The scraping is carried out by grasping the capsule between the thumb and forefinger of the left hand and inclining it gently, and the scraper is then drawn upwards. The capsule is finally 'cleaned' with the thumb.

The semi-dry, blackish latex is then transferred to wooden trays and further dried. Upon arrival at the factory, appropriate samples are drawn from each grower's produce and the quality is determined in the laboratory of the Chief Opium Chemist. The material is graded according to morphine content as follows:

A = With morphine content more than 12%,
D_1 = With morphine between 11 and 12%,
B_2 = With morphine between 10 and 11%,
B_3 = With morphine between 8 and 10%.

Processing

The crude opium is transferred to the appropriate storage vats capable of holding 3 to 30 tonnes. The opium of variable consistency contained in these tubs is removed as required and is carefully air-dried to 70% consistency. The drying generally takes 21 days and is carried out in wooden trays, by turning the latex 3-4 times a day. The sun-drying is considered as a critical process, since artificial drying cause a substantial loss in the alkaloid content. Each processed tray comprises of 36 kg opium with a 70° consistency. The opium is then packed in polythene bags, each containing 5 kg of the product, which are then shipped in wooden chests (60 kg/chest) for air freight and 50 kg/chest for surface shipment.

In addition to export, a small amount of 90° consistency opium is prepared for domestic, quasi-medical purposes. This is called 'Government Excise opium' and is available in cubical cakes. These are wrapped in 'butter paper', tied, stamped and shipped to various states where the opium is taxed and sold to those who are duly registered as addicts and consumers.

Harvesting of seeds

After the lancing operation and collection of the opium latex, the capsules are allowed to dry in the plant itself. The drying process takes about 15 days after the lancing is completed. In India, the capules are plucked by hand and the seeds are separated after breaking the capsules. A bold, dry capsule weighs about 7 g and contains 11 to 12,000 seeds weighing about 3.5 to 4 g.

Yield of crude opium and seeds

On an average, 25-30 kg/ha of crude opium and 400-500 kg/ha of seeds are obtained in India. The Indian produce of opium latex, on an average, contains 20% of total alkaloids by weight. Of this, 9-14% is morphine, 1.25-3.75% is codeine, 5-7%, is narcotine, 0.4-1.0% is papavarine and 0.2-0.5% is thebaine.

Storage of seeds

The seeds are packed in gunny bags or on a smaller scale in friction-top tins.

Chemical Evaluation

i) Extraction of alkaloids from the poppy husk

For this purpose, 28 of poppy husk is treated with 20% NaCl solution (10 ml) and it is stirred for an hour. Then, this is further treated with a solvent mixture of methanol and benzene (1:1) (25 ml). The mixture is shaken for half-an-hour. Later on, this is filtered under suction. The residue is washed three times with solvent mixture, taking 5 ml each time. The filtrate thus obtained is treated with 20 ml of HCl (0.5 N) and is kept in a separating funnel. The pH of the aqueous layer is maintained between 4-6. Now, 5 ml of HCl (1 N) is added and it is made up to 100 ml. This solution is tested by TLC for the presence of the major opium alkaloids and used for the estimation of morphine.

ii) Determination of morphine in a poppy capsule

Grind a dry poppy capsule to a fine powder with the help of a pestle and mortar. Weigh 5 gm of this dry powder into a 100 ml beaker and treat it with 25 ml 10% W/V aquoeus sodium carbonate solution and keep it for an hour. The stirred mixture is then treated with 5 ml of 20% W/V aqueous sodium carbonate solution and shaken for an hour with a solvent mixture (45 ml) of 3:1 volumes of chloroform and isopropanol. This proportion of the organic solvent minimises emulsion formation. The resulting mass is suction filtered with 150 ml sintered Buchner Funnel and washed with 5 ml of solvent mixture four times. The combined filtrate and washings are extracted with 1N HCl (20 ml) and then with distilled water (2 quantities each of 10 ml). The aqueous layer is filtered, and then the pH is adjusted to pH 5 by treating with 1N HCL (45 ml) and it is diluted to 200 ml.

Then, take 20 ml aliquot into a volumetric flask and add 5 ml 1N HCl and make up the volume to 100 ml.

Take three 10 ml aliquots into three 25 ml volumetric flasks. To one, add 5 ml of 0.1 N HCl, and to the other two add 2 ml Iodic acid each. Exactly after 2 minutes, 10 ml ammonium bicarbonate-nickel chloride complex is added and the volume is made up to 25 ml with distilled water. After 90 minutes, the Optical Density (red filter) in balanced cell colorimeter-102 is measured and the morphine content is computed from the calibration curve.

iii) Extraction procedure for opium by colorimetric analysis

The sample is dried at 65 to 70°C for 24 hours and is ground to a fine powder with the help of a pestle and mortar. Of this, 0.5 g (based on availability) of the sample is weighed and transferred to a 15 ml centrifuge tube, containing 5 ml of water. The sample is warmed and stirred with a glass rod and kept overnight at 45°C, so that all the suspension is dissolved and it becomes a homogeneous mixture.

To this mixture, 400 mg of calcium hydroxide is added and mixed well till a homogeneous paste is formed. It is kept for half-an-hour with intermittent stirring. Further, it is centrifuged for 10 minutes and the supernatant liquid is taken in a 100 ml volumetric flask. Ten ml of distilled water is added to the residue in the volumetric flask, stirred well again with a glass rod and centrifuged for 10 minutes. The same procedure is repeated 6 times, which finally gives 60 ml of the combined filtrate.

Then, 3 to 4 ml of 1N HCl is added to this to make the pH 4 to 5 and then distilled water is added to make up the volume to 100 ml.

Two ml aliquot (in case of a 1 g opium sample) and 4 ml aliquot (in case of a 5 gm opium sample) is taken in a 25 ml volumetric flask containing 5 ml 1N HCl (pH 1.6) and the volume is made up to 10 ml with water.

Further, colour development is done similarly by adding Iodic acid and Ammonium bicarbonate-nickle chloride complex and the reading is taken at 670 nm in the visible spectrophotometer after 90 minutes.

Paprika

Hungarian Cherry Paprika

Flower–*C.Annum*

Plant Profile

Family	:	*Solanaceae*
English name	:	Paprika, Hungarian paprika, Spanish pimento
Indian name	:	*Paprika*
Species	:	*Capsicum annum* Linn.
Distribution	:	Peru, Central and South America, Spain, Hungary, Mexico, Bulgaria, Morocco, Kenya, India
Uses	:	Food colouring, flavouring

aprika (*Capsicum annum* Linn.), belonging to the family *Solanaceae*, also called Hungarian paprika, sweet pepper or Spanish pimento, is the mild or non-pungent (sweet) variety of chilli or capsicum, which has a great commercial potential the world over. The dried red paprikas are valued chiefly for their brilliant red colour and mild flavour.

Paprika is one of the richest sources of Vitamin-C (ascorbic acid). The nutritional composition of paprika (per 100 g) is given the table below.

Composition	USDA Hand Book 8-2[1]	ASTA[2]
Water (g)	9.54	7
Food energy (K, cal)	289	390
Protein (g)	14.76	14
Fat (g)	12.95	10.4
Carbohydrate (g)	55.74	60.3
Ash (g)	8.6	7.02
Calcium (g)	0.177	0.2
Phosphorus (mg)	345	300
Sodium (mg)	34	20
Potassium (mg)	2344	2400
Iron (mg)	23.59	23.1
Thiamine (mg)	0.645	0.6
Riboflavine (mg)	1.743	1.36
Niacin (mg)	15.32	15.3
Ascorbic acid (mg)	71.12	59
Vitamin-A activity (RE)	6060	5800

Source: 1. Composition of foods: Spices and Herbs, USDA Agricultural Hand Book 8-2, January 1977.

2. The nutritional composition of spices, ASTA Research Committee, February, 1977.

The pungent nature of paprika has been found to be due to a mixture of seven homologous-branched chain alkylvinillyl amides, named capsaicinoids. The seven capsaicinoids are given in the following table.

Structural formula	Name
$(CH_3)_2$. CH.CH = CH. $(CH_2)_4$ – CO-R	Capsaicin
$(CH_3)_2$. CH. $(CH_2)_6$ – CO-R	Dihydrocapsaicin
$(CH_3)_2$. CH. $(CH_2)_9$ – CO-R	Nordihydrocapsaicin
$(CH_3)_2$. CH. $(CH_2)_9$ – CO-R	Homodihydrocapsaicin
$(CH_3)_2$. CH.CH = CH. $(CH_2)_5$ – CO-R	Homocapsaicin
CH_3. $(CH_2)_7$ – CO-R	Nonanoic acid vanillylamide
CH_3. CH. $(CH_2)_8$ – CO-R	Decanoic acid vanillylamide

Where R is — NH — CH_2 —(benzene ring)— OH, with OCH_3

The distribution of the pungent principles of the fruit is uneven and is the greatest in the placenta. Generally, paprika contains less than 0.1% of capsaicinoids. The pungency is measured in SH (Scoville Heat) units which normally ranges from 0 to 700 for non-pungent paprika. The highly pungent red pepper powder has >8000 SH units.

The colour of paprika powder is the principal criterion for assessing its quality. The pigment content of paprika can range from 0.1 to 0.8%. The major colouring pigments in paprika are capsanthin and capsorubin, comprising 60% of the total carotenoids. The other pigments are β-carotene, zeaxanthin, violaxanthin, neoxanthin and lutein. The extractable colour is measured using a spectrophotometer and expressed in terms of ASTA (American Spice Trade Association) Units. The ASTA values are positively correlated with colour. The more the ASTA values, the more the brightness of the product. The fruits of paprika have a relatively low fixed oil and a relatively low volatile oil (0.1 to 2.6%). The fixed oil comprises mainly of triglycerides (about 60%) in which linoleic and other unsaturated acids predominate. The characteristic aroma and flavour of the fresh fruit used as a vegetable is imparted by the volatile oil. The volatile oil contains 24 components. One of the major components, 2-methoxy-isobutyl pyrazine, possesses an aromatic characteristic of the fresh fruit and dominates the organoleptic profile.

Paprika is used for its colouring and flavoring properties. It is used largely as a garnish for light-coloured foods, such as eggs (poultry), fish, potatoes, pasta, salads and salad dressings. As flavour is important in food manufacturing, this spice becomes a vital ingredient of sausages, soups, salad dressings and many ready-to-eat foods.

Origin and distribution

The sweet peppers found in the pre-historic remains in Peru, were widely grown in Central and South America in pre-Colombian times. The pepper seeds were carried to Spain in 1493 and from there spread rapidly throughout Europe.

Paprika is now produced in Spain, Hungary, the USA, Mexico, Bulgaria, Morocco, Kenya, Yugoslavia, India, etc. In the USA, it is mainly cultivated in California and the annual production is about 10,000 Mt. In India, paprika has been successfully grown in IARI, New Delhi, and CFTRI, Mysore, and in some other places. Very small quantities of Indian paprika are being exported to the US. There is a tremendous export potential for paprika powder and oleoresin. However, in India, not much work has been done to develop pepper cultivars suitable for paprika making. Although some institutes have developed some varieties for this purpose, their cultivation is limited. Since environmental conditions are favourable for growing paprika spice, it can be cultivated on a large scale. Plant breeders, should thus focus more on developing suitable paprika varieties aimed at consumers, growers and processing industries.

Climate

Paprika requires a long, warm and frost-free (4-5 months) growing season. It can be grown in the plains of the east, west and southern parts of the country both as a *kharif* and *rabi* crop (May-June and September-November). But in hills, it is grown only during summer (April-May). The optimum temperature range for seed germination is 18-24°C and for plant growth it is 24-27°C, with a maximum of 32°C±1 or 2°C.

Soil

Paprika can be grown in many types of soils, ranging from sandy to heavy clay. A well-drained, fairly light-fertile loam, with a fair moisture, holding capacity is ideal. Poor and medium quality land produces good early crop, if managed properly. Heavy soil is not suitable for raising the rainy season crop, as water stagnation for more than 24 hours leads to the death of the plants. The light soils produce better quality fruits than the heavy soils. This crop prefers a pH ranging from 6-7. In acidic soils, liming is beneficial.

Types of paprika

The paprika of commerce comes from different producing areas. The major characteristics of paprika existing in the world market according to the American Spice Trade Association (ASTA) are described below.

Hungarian paprika

This type has a great demand in Europe. Hungarian paprika is produced in eight grades and three qualities, ranging in colour, power and pungency. It is made from longer, more conical and pointed fruit types, which can be pungent or non-pungent. Hungarian paprika has a distinctive flavour, which is very much favoured in Europe and is used as a seasoning and colouring agent. The flavour is due to the presence of some aromatic substances. The compound which imparts flavour is 2-isobutyl-3-methoxyprazine.

Spanish paprika

Spain produces a sweet paprika in a wide range of colour values. It is made from round fruits. Here, the paprika is ground according to the customer's specifications which are normally fixed by the Spanish Government. Bola, Agridulce, Jaranda, Jariza, Dattler, Ocal, Numex, Conquistador, Belrubi, Datoca and Larguillo are some cultivars of Spanish papkrika.

Moroccan paprika

This is similar to Spanish paprika and produced as medium- and high-colour paprika.

Bulgarian paprika

Bulgaria is the only East European source that predominantly produces mild paprika. It is mainly used for food manufacturing purposes. It is made after removing the seeds from the paprika fruit. A round fruit about the size of a peach or an apricot is used for making this type of paprika.

American paprika

This paprika is grown and produced in Southern California from the NuMexican or the Anaheim type of fruits, where the pods are long, slender and tapering to a point, with medium to dark green fruits turning red, which are low or non-pungent with a brilliant red colour. Here, the paprika is mechanically dried using a scientific method for the retention of the maximum colour extraction value. Some of the commonly grown paprika cultivars for the preparation of American paprika include NuMex Sweet, NuMex Conquistedor, NuMex R Naky, Anaheim TMR, Anaheim F-6, Papri Queen, Doux des Landes, Sandia, etc. California has become a large supplier of paprika, larger than any other individual country.

Yugoslavian paprika

This is quite similar to the Hungarian variety. It is normally finely-ground and contains slight heat or pungency.

Czechoslovakian and Chilean

These are sweet to mildly pungent paprikas.

Romanian, Turkish and Greek

These are slightly pungent to pungent types of paprika.

Portunguese

This is a sweet paprika with medium to high colour strength.

Varieties

Many paprika varieties exist in the world. Bydagi, Warrangal chilli, Arka Abhir, Kt-Pl 8, Kt-Pl 18, Kt–Pl 19, Kt–01, Kt–02, Kt–03, Kt–04, Agni (F_1) are some of the varieties of paprika grown in different parts of India. Of these, Kt–Pl 19 is an idotype with the desired traits and Byadagi chilli, a popular local type, is a sweet spice paprika. The description of the three important varieties is as follows.

Kt-Pl 19

The plants of Kt-Pl-19 are upright (height 50-100 cm), and spreading (38-40 cm) with profuse branching (2-3 primary and 3-5 secondary branches). The plants flower in 30-45 days after transplanting. The fruits with a pendant bearing habit (55-59 in numbers per plant) are mostly flattish bell-shaped (used in the USA), 2-celled, dark green when immature, turning to dark red at maturity. They are conical, 13.5-16.8 cm long and 2.2-3.1 cm in diameter. The flesh thickness is less than 0.2 cm. The fruits mostly taper in a slightly curved manner with pointed tips. The average weight of the fresh fruit will be 25-30 g (pericarp weight of 22.5-27.3 g and seed weight of 1.9-2.7 g/fruit). There are 145-175 in number per fruit. The ascorbic acid content and the total extractable colour will be 113-127 mg/100 g and 173-213, respectively. It is a high-yielding variety, providing 25-35 t/ha of fresh fruit with a dry recovery of 15-17.6% (68.7% skin, 25.0% seed and 6.3% pedicle).

Byadagi chilli

This variety is suitable for cultivation mainly under rain-fed conditions and also under intensive irrigated conditions.

The special characteristics of this variety are that the fruits are deep red in colour, wrinkled, the pungency is low, and the keeping quality is good. Besides, there exists a lot of variability in this variety for fruit shape, size, colour, pungency, yield, etc. Several strains like Byadagi kaddi, Byadagi dabbi, Dyavanur kaddi, Delux and Anthur Benthur local exist in this variety.

Arka Abhir

This variety has been developed at the Indian Institute of Horticultural Research Institute, Bangalore. The fruits are less pungent and the oleoresin content is 4.3-7.21%.

Cultivation

The cultural practices for paprika are almost similar to those of capsicum and chilli, and the same could be also adopted for the commercial production of paprika. However, in the Kt-Pl-19 variety, it is always beneficial to remove the first flower bud from the first dichotomous branching of the plant, followed by pinching of two or more buds if possible from the nodes of the first-stage branching, as it helps produce a higher yield.

Considerable research work has been carried out at the University of Agricultural Sciences, Dharwad, Karnataka, on the production technology of 'Kt-Pl-19' and 'Byadagi chilli' varieties and the highlights of the research achievements on these varieties is given below.

Kt-Pl 19

The variety can be grown as a *kharif* crop.

- A spacing of 60 × 30 cm gave the highest fruit yield (12.08 t/ha)
- The application of vermicompost (2 t/ha), instead of FYM (25 t/ha), *Azospirillum* treatment of the roots of seedlings and 150 kg of inorganic nitrogen recorded a significantly high yield, quality and higher net returns as well as a high benefit: cost ratio.
- For seed production, the fruits could be harvested at 60 days after flowering and could be dried either by mechanical air-drying at 40°C or shade-dried to obtain better seed quality.

Byadagi chilli

Crop improvement

The efforts on studies on variability, genetics and breeding have yielded a good accumulation of basic knowledge. Resistance breeding has led to the release of KDC-1, which is tolerant to leaf-curl.

Production technology

- A raised bed seedling nursery is recommended.
- Hardening of the seedlings before transplanting is recommended.
- A spacing of 90 × 90 cm for rain-fed cultivation and 75 × 45 cm for irrigated conditions is recommended.
- A fertiliser application of 150:75:75 kg NPK/ha and 100:75: 50 kg NPK/ha for irrigated and rain-fed cultivation, respectively, along with 25 of FYM/ha.
- A fertiliser-combination of DAP with CAN has increased the yield and quality of chilli compared to urea application.
- Taking groundnut as an inter-crop in between chilli rows is found to be profitable and better.
- Seed treatment: Emidachloroprid 70 WS @ 10 g/kg seed controls thrips, mites and leaf-eating caterpillars for 3-4 weeks.
- Seedling treatment: dipping the seedlings in Emidachloroprid 70 WS solution (2.5 g/5 litre water) is benefical.
- Bio-fertilisers: VAM cultures like *Glomus microcarpum,* and *G.vesiculatum* are found effective when they are inoculated into the nursery beds.
- Azospirillum culture 'ACD-15' seedling inoculation reduced the nitrogen requirement of this crop.
- Spraying NAA (150 ppm) during the flowering and fruiting stages will reduce the flower and fruit drop significantly.

Plant protection

- Seed treatment with Bavistin (3 g/kg seed); drenching the seed beds with Metalaxyl MZ (1.5/l); raised bed nursery practices will control damping-off
- The use of Acephate + Dicofol + Monocrotophos for the control of chilli leaf curl complex.
- Fenzaquin 10 EC @ 150 g ai/ha is effective against mites and thrips and therefore, suppresses leaf curl in chilli.
- The use of Sorghum leaf extract @ 5.0% + Monocrotophos (2 ml/l) + Dicofol (2.5 ml/l) for control of chilli leaf curl complex.
- The application of CAN recorded a significantly lower % of leaf-curl index.
- Spraying Difenoconazole @ 0.1% or Kitazin 0.15 and 0.20% or Carbendizim (1 g/l) or Hexaconozol (1 ml/l) at the fruit ripening stage will control anthranose.

- The application of Penconazole @ 0.1% or 1 ml/l or Carbendaizim (1 g/l) or Wettable sulphur (3 gl/) will control the powdery mildew disease.
- Spraying Carbendaizim (1gl/) controls *Cercospora* leaf-spot.
- Spraying Mancozeb (2 g/l) controls *Alternaria* leaf-spot.

Biocontrol

- The release of *Crysopherla* parasitoid @ 10,000 larva or 20,000 eggs/ha at 30-40 days after transplanting controls fruit borer larvae, jassids, thrips and mites.
- Six releases of *Trichogramma* @ 20,000 insects each time.
- The use of NPV controls the fruit borer (*Helicoverpa armigera*) and the leaf-eating caterpillar (*Spodoptera litura*). NPV 2000 EL + 0.1% Teepol + 0.5% Jaggery solution + 200-300 litre water.
- The use of NPV + Methomyl @ 2 ml/l controls the fruit borer.

Harvesting and processing

Generally, the fully-ripe, red fruits are ready for picking 60-80 days after transplanting. The fruits should be harvested 3-4 times, so that the fruits which are set in the later stages get proper nourishment for their growth. The ripe fruits are stored in open places, but preferably in the shade to avoid the direct sun. Then, these fruits are spread out thinly to dry naturally or dehydrated in specially constructed tunnels, depending on the producing area. If a cut is made at the centre of the fruit, the moisture evaporates quickly and drying can be increased up to 21%. It is a unique fact in the spice realm that paprika is always referred to as a ground product. Hence, the dried fruits, after removing the seeds and pedicles, are ground into paprika powder.

Storage

Since paprika is considered as a semi-perishable product, it is recommended to store it in a cool, dry and dark place; with temperatures ranging from 4°C to 16°C and a relative humidity of 50%. Under these conditions, paprika can be protected against infestation and it also helps in proper flavour and colour retention. Exposure to light, high temperature and oxygen bleaches or decomposes their colour, for which they are valued the world over. The high moisture content in paprika leads to mould growth. Frozen paprika can be stored for longer periods.

Parsley

Plant Profile

Family	: *Umbelliferae/Apiaceae*
English name	: Parsley
Indian name	: *Ajmood* (Hindi), *Achu Mooda (*Kannada), *Kothamberlari* (Malayalam).
Species	: *Petroselinum hortense* (P. Miller), Nymann ex.A.W.Hill (Syn *P. cirspum* Mill.)
Distribution	: Mediterranean region, USA, Dominion Republic, Mexico, Canada, Germany, France, Hungary, Belgium, Italy, Spain, Yugoslavia, India
Uses	: Vegetable, condiments, medicine, soups, spice

Parsley [*Petroselinum hortense* (P.Miller)] Nymann ex.A.W.Hill. (Syn *P.cirspum* Mill), also called *Ajmood*, is a hardy, biennial herb of the family *Umbelliferae/Apiaceae*. The leaves, young stems and seeds are the economic parts.

The leaves contain moisture-68.4%, protein-5.9%, fat-1.0%, carbohydrates-19.7%, fibre-1.8%, mineral matter-3.2%, calcium-390 mg, phosphorus-200 mg/100 g, iron-17.9%, carotene (as vit A) -3200 I.U, thiamine-0.04 mg, nicotinic acid-0.5 mg, vit C (ascorbic acid)-281 mg/100 g. The vitamin A content ranges as high as 8,230 I.U/100 g; riboflavin and biotin are also present.

The leaves, stems and fruits contain a glucoside apiin, which on hydrohysis yields apigenin, glucose and a sugar—apiose. A second glucoside, consisting of luteolin, glucose and apiose, has also been reported. All the parts of the plant contain the essential oil, which is responsible for the characteristic aroma and flavour of the spice. The oil is mainly used for flavouring food products. The oil (0.06%) obtained from the flowering tops is of the finest quality. The whole herb yields 0.25% and the fruits about 7% volatile oil. The fruit (seed) oil contains apiol (Parsley camphor) and α-pinene, with small amounts of myristicin, aldehydes, ketones and phenols. The herb oil is also reported to contain apiol. Besides, the fruits also yield about 20% of a greenish, fatty oil, which contains petroselinic acid (upto 76%) and is used for making plastics, synthetic rubber, lubricating oil, additives and protective coatings.

The fresh leaves mask even strong culinary odours and are commonly used for garnishing and seasoning. They are eaten fresh, incorporated in salads and as an ingredient in soups, stews and sauces. They are also used in meat and poultry seasonings. The leaves are also employed to make a sort of tea, which is considered to possess anti-scorbutic properties since it is a very rich source of vitamin C. The dried herb is available as whole, rubbed, or in the ground form.

The roots are used as a vegetable in soups, while the dried leaves and roots are used as condiments. The herb is reported to possess diuretic, carminative, emmenagogue and antipyretic properties and has long been used for uterine troubles. The juice of the fresh leaves is used as an insecticide, and the bruised leaves are applied to the bites of parasites. The extracts and infusions of the leaves and roots on subcutaneous administration in mice produced a depressing effect on the nervous system.

In a series of experiments conducted on the hearts of frogs, cats and rabbits, it was noted that a small dose (2 drops) of 1% aqueous infusion of leaves increased the amplitude and slowed cardiac contraction; a large dose (5-10 drops) of 10% leaf infusion was followed by the depression of cardiac activity. The intravenous administration of the infusion as well as the extract caused a fall in blood-pressure and dilation of the blood vessels. Rabbits and guinea pigs, when treated with increasing oral doses of apiole during the gestation period, usually abort and die after 26-28 days. Apiole it also known to cause poisoning and sometimes fatal lesions in the liver and kidney.

Origin and distribution

Parsley is a native of Sardinia and is extensively cultivated in the Mediterranean region and the USA. The USA is the largest commercial producer of parsley in the world. Parsley produced in the dehydrated form is mostly consumed within the country, since it is one of the most popular herbs known there. Owing to the increased demand, it imports parsely from Mexico and Dominion Republic and to some extent from Canada, West Germany, Haiti, France, Hungary, Belgium, Italy, Spain and Yugoslavia. In India, the var.*crispum* is grown in limited areas in Himachal Pradesh, Punjab, Haryana, Uttar Pradesh and in the high-altitude areas of Southern India.

Description of the plant

Parsley is a hardy, aromatic, biennial, umbelliferous, much-branched green herb, sometimes living up to 4 years, producing a rosette of finely-divided radical leaves in the first year and a flowering stalk, up to 100 cm high in the second year. The plant has a white taproot and a branched, striated cylindrical stem. It has rich green compound leaves, 2 or 3 pinnate. The flowers are yellow to yellowish-green, borne in compound umbels. The fruits (commonly known as seeds) are 2-3 mm long, crescent-shaped, conspicuously ridged and, consisting of two mericarps. The colour of the dried herb is green; its aroma is pleasant, characteristic, fragrant and spicy due to the presence of volatile oil.

Climate

It is a hardy cool season crop, which flourishes both in temperate as well as tropical climates in the country, but it grows better at higher altitudes. It is grown occasionally in gardens. Sowing is done in the months of March-May in the hills and August to November in the plains. It can also be produced under glasshouse or polyhouse conditions.

Soil

Parsley thrives best in loamy as well as rich moist soil with good drainage, amenable to deep cultivation.

Types/varieties

There are two horticultural types; those cultivated for the leaves (var. *crispum*) and those grown for their turnip-like roots (var. *radicosum* Danert). There are three types of leafy varieties: plain leaved, double curled and moss curled of which, only the former type is cultivated in India. In the moss-curled type, the plants are dwarf, compact and bushy. The leaves are dark green with exceedingly fine cuts, serrated and deeply curled. These are used for garnishing and decorative purposes. The fleshy-rooted parsley has plain celery like leaves.

In the plain-leaved variety, Hamburgh, the plants are tall and coarse, with tough stems. The leaves are deeply cut, resembling those of celery shoots. They are fleshy and edible.

Cultivation

Propagation

Parsley is mainly propagated by seeds. The seedlings are grown in raised nursery beds and then transplanted into the field.

Land preparation

The main field selected for crop production is ploughed 2-3 times, cleared of all the weeds, the clods are broken and the field is brought to a fine tilth. Farm Yard Manure @ 15 t/ha is applied at the time of the last ploughing and mixed well with the soil. Then, the field is divided into convenient blocks to irrigate the crop easily. Ridges and furrows are made at a 60 cm spacing.

Nursery raising and planting

The crop is raised by the direct sowing method (6-8 kg seed/ha) and also by raising the seedlings in well-prepared nursery beds. Abut 250-300 g of seed is sufficient for one hectare. It takes 18-24 days for the seeds to germinate. Soaking the seeds in hot water for 24 hours along with radish slightly reduces the period required for germination. The seeds are then sown in beds. As the radish seeds germinate first, they are rougued out while the thinning is done. Two-month-old seedlings are ready for transplanting. These should be transplanted at spacings of 15-20 cm × 30-40 cm or 60 cm × 15 cm.

Manures and fertilisers

The application of 65 kg of N, 40 kg of P and 25 kg of K is recommended for optimum productivity. The nitrogen should be applied in split doses, while phosphorus and potassium should applied as a basal dose.

Irrigation

The crop should be irrigated immediately after transplanting. Later on, irrigation is provided at 15-20 days' intervals, depending on the weather and moisture level of the soil.

Weeding/Inter-culture

Shallow and clean cultivation is recommended for parsley. Hoeing and weeding should be done as and when required to keep the crop weed free.

The flowering stem arising in the second season should be removed for more vegetative growth.

Pests and diseases

The major pests of the crop are carrot fly and celery fly. Leaf-yellowing, due to nematodes, has also been noticed. Among the diseases, septoriosis (*Septoria petroselini*) has been reported ot affect this plant.

Harvesting and post-harvest management

The leaves are ready for harvesting in about three months. Usually, two to three cuttings of leaves are made annually. The harvested leaves are sorted out and tied in bundles or small bunches and packed in baskets and then sent to the market for disposal. Placing ice-blocks in the baskets prevents the heating and decay of the leaves during transit. The characteristic flavour and green colour can be retained if the leaves are dried rapidly at 12°C in an oven.

Yield

The yield varies with the variety and the location. In commercial plantations, 2,500 dozen bunches are obtained per hectare at each cutting.

The roots of the turnip-rooted variety of parsley are usually dug out after the fruits have been harvested. The thick roots are cut longitudinally to facilitate drying. The fruits are collected for the extraction of the parsley oil of commerce. The fruit yield generally varies from 800-1500 kg/ha.

Value-added products

The value-added products of parsley are dehydrated parsley, volatile oil (oil of parsley) and fatty oil.

Adulterants

The commercial apiol is frequency adulterated with trio-o-cresyl phosphate, which may have toxic effects.

Rosemary

Plant Profile

Family	:	*Lamiaceae; Labiatae*
English name	:	Rosemary
Indian name	:	*Rusmari* (Hindi); *Rosemary* (Tamil)
Species	:	*Rosmarinus officinalis* L.
Distribution	:	India, Algeria, France, Italy, Morocco, Spain, Tunisia, Portugal
Uses	:	Cosmetics, culinary purposes, flavouring, perfumery

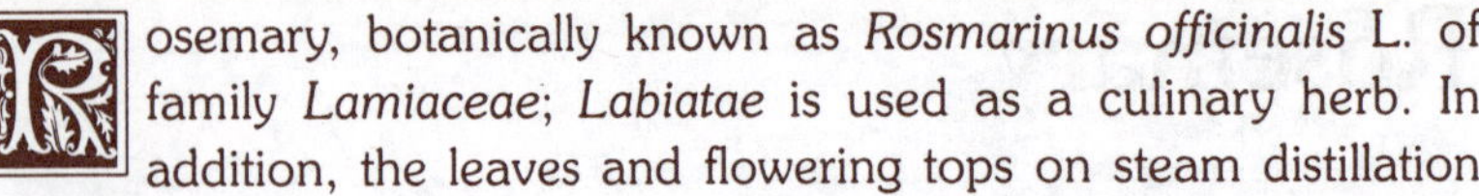

Rosemary, botanically known as *Rosmarinus officinalis* L. of family *Lamiaceae*; *Labiatae* is used as a culinary herb. In addition, the leaves and flowering tops on steam distillation yield the essential oil. Dried rosemary leaves contain–moisture-5.7%, protein-4.5%, fat-17.4%, crude fibre-19%, carbohydrates-47.4%, ash-6.0%, calcium-1.5%, phosphorus-0.70%, iron-0.03%, sodium-0.004%, potassium-1.0%, vitamin-A-175 IU/100 g, vitamin-B-0.51 mg/100 g, vitamin-B_2-0.04, niacin-1.0, vitamin-C-61.3 mg/100 g, calorific value 440 calories/100 g. Rosemary oil has 1, 8 cineole (20-50%), borneol (20%), camphor, linalool, α-pinene, camphene, β-pinene, sabinene, myrcene, α-phellandrene, α-terpinene, limonene, α-terpinene, p-cymene, terpinolene, thujone, copaene, terpinen-4-ol, caryophyllene, methyl chavicol, terpineol, thymol, carvacrol, etc.

The fresh tender tops are used for garnishing and for flavouring cold drinks, pickles, soups and other foods. The leaves are employed as a condiment. They are dried and powdered, and added to cooked meats, fish, poultry soups, stews, sauces, dressing, preserves and jams. They are mixed with sage in pork and veal stuffings and sometime, are added to biscuits. The oil is valued for its use in culinary, medicine, perfumery and cosmetic industries. It is an excellent fixative material and the oil also contributes to a strong fresh odour, which blends well with various other oil odours and also serves to mask the unpleasant smell of certain other ingredients.

Rosemary oil is known to have anti-microbial properties against certain gram-positive and gram-negative organisms. It is also used in the formulations of compounded oils for flavouring meat, sauces, condiments and other food products. The leaves are used in cooking. A distilled water is obtained from the flowers, which is used as an eye wash.

Origin, distribution and production

Rosemary is a native of the Mediterranean regions of Europe, Asia Minor and North Africa. It is grown in Spain, Italy, France, Algeria, Morocco and Portaugal for its essential oil. Spain has traditionally been the largest supplier of the oil, but appears to be rapidly losing prominence to Tunisia. The annual world production of the oil has been increasing gradually over the years and now 200-300 tonnes of oil is being produced against the negligible quantity produced in India. In India, rosemary is cultivated to a limited extent in the Nilgiris in South India. Its cultivation in the plains is of recent origin and now it is being cultivated in and around Bangalore on a small scale. The oil is comparable to the Spanish oil in quality and has been well received commercially.

Description of the plant

The plant is a hardy, dense, evergreen shrub, growing up to 1 m in height. It has an erect stem divided into numerous, long, slender branches bearing many sessile opposite leaves, which are smooth and green, woody, whitish and glandular beneath. The leaves are 2 to 4 cm long and cylindrical, leathery and are green on top, white, hairy below and sticky to the touch. The flowers are situated in little clusters towards the end of the branches. The calyx is two lipped; the upper with a single broad oval lobe, and the lower with two-segmented triangular lobes. The corolla is also two lipped with two violet stamens and a long style projecting from it. The fruit is an oval, four-sectioned cremocarp.

Climate

Rosemary prefers the Mediterranean type of climate with low humidity, a warm winter and a mild summer for its successful growth. However, any place where frost occurs frequently should be avoided as the plant is susceptible to it. The climate of the Nilgiris and Bangalore in India has been found suitable for its cultivation.

Soil

It is a very hardy plant and found growing on the rocky terrain in the temperate parts of the world. In India, the plant grows well in the light loamy soils of the Nilgiris and the sandy loam soils of Bangalore. The crop requires a soil pH ranging from 6.5-7.0 for its successful growth.

Types and varieties

There are two types of rosemary under cultivation. They are the 'French rosemary' and 'Italian rosemary'. The French type produces white-coloured flowers and its oil is of a more superior quality than the Italian type, where the flowers are of a purple colour. Of late, the CIMAP has developed a variety which bears flowers (French) and produces oil of a superior quality.

Cultivation

Propagation

Vegetative propagation

It is best propagated vegetatively by means of stem-cuttings. Cuttings of 10-15 cm length are taken from healthy mother plants. About half the length of the leaves from the bottom should be removed. The cuttings are then planted in nursery beds made up of sandy soil, under partial shade, to a depth of about 6.0 to 10.0 cm. Thereafter regular watering and weeding is provided to the nursery for about a month. After about 6-8 weeks, the cuttings are ready for transplanting into the main field. They can also be raised in small polybags or seed-pans; this helps in easy transportation of the rooted cuttings.

Seed propagation

The crop can also be propagated by seeds. The ideal season for raising the nursery is during September-November. The seeds are very small and about 0.2 to 2.5 g seeds are required to cover 1 sq m area and are sown at a depth of 1-2 cm. After they are sown in well-prepared nursery beds, regular watering and weeding of the nursery is continued. The seeds germinate best at a soil temperature ranging from 14-15°C. When the seedlings are about 8 to 10 weeks old, they are ready for transplanting into the main field.

Land preparation

The land is well prepared by repeated ploughing and harrowing. About 15-20 t of well-rotten FYM is incorporated into the soil at the time of the final ploughing.

Planting

Eight to ten-weeks-old rooted cuttings or seedlings are planted in the main field at a spacing of 45 cm × 120 cm. About 20,000 plants per hectare give the highest yield of the oil. It is reported from the Central Institute of Medicinal and Aromatic Plants (CIMAP), Lucknow, that plant spacing of 45 × 45 cm is optimum for obtaining the highest yield.

Manures and Fertilisers

Prior to transplanting, 15-20 t of FYM, along with 40 kg of P_2O_5, 40 kg of K_2O and 20 kg N is applied to the soil as a basal dose. After each harvest, 80 kg/ha of N is applied in 4 equal split doses as a side-dressing to promote vegetative growth. For obtaining the highest yield, CIMAP, Lucknow, has recommended the application of 300 kg N/ha/year.

Irrigation

Initially, the crop is irrigated twice a week till the plants are established. Afterwards, only a weekly irrigation is enough. By nature, this crop is drought-resistant and can survive longer drought periods.

Inter-cultivation

About 5-6 cultivations between the rows and an equal number of weedings within the row are required to keep the weeds under control.

Pruning

After 2 to 3 years, the bushes are cut frequently to keep them from becoming too leggy and to promote the formation of numerous shoots, which can be harvested for obtaining the oil.

Pests and diseases

Among the pests, mites have been reported to cause considerable damage to the crop. Ethion (0.05%) could be sprayed to control the pest.

When the crop was introduced, there were no major pest and disease problems for this plant. Recently, however, blight caused by *Rhizoctonia* sp. a soil-borne fungus, *Phytocoris rosmarini* and *Orthotylus ribesi* have been reported. Maneb (1.0%) can be sprayed on the crop and drenching may be done at 8-10 day intervals to keep the diseases under control.

Harvesting

During the first year, the crop is ready for harvesting 8 months after planting and only 2 harvests are obtained. In the subsequent years, 3 to 4 harvests at 100 to 120 day intervals can be taken.

Depending upon the exposure of the plantation, the plants start flowering earlier in warmer and low altitude areas and later at high slopes.

Harvesting should begin at the 50% blossoming stage and should continue till 75-90% inflorescence emergence and should end when the flowers have finished blossoming. The shoots are cut for distillation when they have reached their maximum size, but have not become woody. The hardwood should not be distilled as it imparts an odour of turpentine.

Distillation

The essential oil is obtained by steam distillation of the freshly-harvested herbage. The most modern method uses steam directly or indirectly, produced at 2-3 atmospheric pressure. The herbage can also be shade-dried, stored and distilled at convenience without any loss of oil. The time required for distillation of one charge is 3 hours. Recently, CIMAP, Bangalore, has recommended the distillation of rosemary for 2 hours for recovery of the maximum oil.

Oil content and yield

In the laboratory, fresh rosemary leaves yield 1% oil and shade-dried leaves yield 3% oil. However, in field distillation units, an yield of

0.7% is considered satisfactory. About 12 to 15 t of herbage per hectare per year, yielding about 85 to 100 kg of oil per hectare per year, is obtained.

Rejuvenation of the old plantation

A 10-12-year-old plantation starts ageing and its yield is reduced. Hence, its regeneration is necessary. This is achieved when the plants are in a period of rest.

For this purpose, the plants are cut at 4-5 cm above the ground. The quick regeneration of a plantation requires abundant fertilisation. A normal yield from an adult plantation is obtained during the second year after regeneration. Regeneration restores the production capacity of old plantations, which should otherwise have been uprooted and replanted.

Saffron

Plant Profile

Family : *Iridaceae*
English name : Saffron
Indian name : *Keshara, Kunkuma, Aruna, Assa, Asrika* (Sanskrit) *Zaffran, Kesar* (Hindi), *Kungumapu* (Tamil) *Kunkumapuva* (Telugu), *Kunkuma Kesari* (Kannada) *Keshar* (Gujarathi), *Kesara, Keshar* (Marathi), *Jafran* (Bengali)
Species : *Crocus sativus* Linn.
Distribution : Spain, France, Italy, Austria, Greece, Iran, Turkey, India and China
Uses : Spice, Drugs

affron (*kesar*) consists of the dried, tri-lobed stigmas of *Crocus sativus* Linn., a bulbous perennial of the family *Iridaceae*. Saffron is one of the oldest and certainly among the world's most expensive spices. The true saffron must not be confused with either meadow saffron (*Colchicum automnale* L., family *Liliaceae*) or safflower, or bastard saffron (*Carthamus tinctorius* L., family *Compositae*), which are occasionally used as adulterants. One pound of saffron consists of about 2,25,000 to 500,000 dried stigmas, and requires the picking (by hand) of 75,000 flowers. The colour of saffron is bright yellow-red, the aroma is powerful, somewhat bitter, peculiar and exotic. The average composition of commercial saffron is as follows: water-15.6%, starch and sugars-13.35%, essential oil-0.6%, fixed oil-5.63%, total N-free extract-43.64%, crude fibre-4.48%, and ash-4.27%. The essential oil and fixed oil yields are as high as 1.37% and 13.4%, respectively. The ash is rich in potassium and phosphorus and contains traces of boron. Kashmir saffron contains 8.5-10.2% moisture and 5.9-13.3% of total ash.

The principal colouring agent of saffron is the glucoside 'crocin' and the bitter substance is the glucoside 'picrocrocin', besides which crocetin and safranal are the other main constituents. Saffron is very popular for its medicinal, flavouring and colouring properties. It is used abroad in exotic dishes, particularly in Spanish rice specialities and French fish preparations. It is also used in fine bread in many countries, viz., in Scandinavia as well as the Balkans.

In medicine, it is used occasionally in exanthematous diseases to promote eruptions. It is used in fevers, melancholia and enlargement of the liver and spleen. It also has stimulant and stomachic properties and is considered to be a remedy for catarrhal affections in children. In modern phamacopoeias, it is employed only to colour other medicines or as a cordial adjunct. Saffron has been employed as an abortifacient and several fatal cases have been recorded. Saffron bulbs are toxic to young animals, and the stigmas in large doses act as a narcotic.

Saffron is an important ingredient of both the *Ayurvedic* and *Unani* systems of medicine in India. It is popularly known as a stimulant, warm and dry in action, helping in urinary, digestive and uterine troubles. It is mixed with other drugs to help in normal menstruation. If soaked overnight in water and used with honey, it facilitates micturation. Its oil is used as an external application in uterine sores. When pounded with *ghee*, it is reported to be very effective in controlling diabetes. It is also reported to strengthen the heart and brain, but only when administered in large doses. Saffron is also being investigated as an anti-cancer agent.

Origin and distribution

Saffron is a native of Southern Europe. It is cultivated in Spain, France, Italy, Austria, Greece, Iran, Turkey, Persia, England, India and China. The La-Macha belt of Spain is the largest producer of saffron in the world contributing to about 80-90% of the world's saffron production. In India, it is mostly cultivated in the Pampore and Kistwar areas of Jammu and Kashmir. Attempts have also been made to cultivate it in the Kinnaur area of Himachal Pradesh, the Kumaon and Garhwal regions of Uttar Pradesh and Arunachal Pradesh and the North-eastern region. Approximately 27% of the saffron produced in India is exported to European and Middle East countries. India faces stiff competition from Iran, Spain, Greece and Morocco in the export trade.

Description of the plant

Saffron is a low-growing, autumn flowering perennial herb (15-25 cm high) with an underground globular corm about 1 ¼" across. The leaves are numerous, narrowly linear, usually 1-½' long after blooming, ciliate, grey-green, with a proper spathe which is 2-valved. The flowers are lilac, reddish-purple or white, the throat is of the same colour, pubescent and fragrant. The perianth segments are 1 ½ - 2" long and obtuse. The anthers are yellow, exceeding the filaments. The style is trifid or tri-lobed stigma, the branches are blood-red, cuneate and toothed drooping or lying on the perianth, much exceeding the stamens and caps were not produced. Saffron is a triploid ($2n = 3x = 24$). The plant is hysteranthus, with flowers arising directly from the corms.

Climate

Saffron is a temperate season crop and it thrives well in cold regions. It can tolerate frost and snowfall easily; however, during October-November the cultivated areas should be free from snowfall. In the Kashmir valley, saffron thrives well in a sub-temperate climate at an elevation ranging from 1500-2400 m. Sunny days during flowering are favourable for a good yield. An annual rainfall of 300 to 400 mm and snow during the winter are suitable for its commercial cultivation. Spring rains are favourable for the promotion of new corms, while a second spell of rains at the end of summer or at the beginning of autumn encourages profuse flowering. The number of flowers and time of blooming are dependent on the temperature prevalent in spring and autumn, and also upon the amount of rainfall. A warm spring and long autumn are conducive to early flowering. An optimum day temperature of 15 to 20°C is required during the flowering period with an optimum night temperature of 6 to 8°C.

Soil

It can be grown on different types of soils ranging from sandy loam to clay loam. However, saffron prefers well-drained soil. In water-logged soils, the corms rot and so a proper drainage system is an important requirement for this crop. A medium-light soil, neutral to slightly alkaline, is suited for its cultivation. The clay-loam soil of the Kare area of the Kashmir valley, locally known as *grut,* is most suited for its successful cultivation.

Varieties

SKU-S-86-KIS, SKU-S-86 WAT, SKU-S-86-QUIL, SKU-S-86-Sona, SKU-S-86 Lath, SKU-S-86 Chand, SKU-S-86 Chad are some of the indigenous and EC-37328, EC-221039, EC-217010, SKUAST, Nag C8708 and Badi C8606 are some of the exotic varieties available in this crop. The latter two exotic varieties have been found to give higher yield when compared with all the other local varieties.

Cultivation

Propagation

Saffron is propagated vegetatively by using the underground corms. The corms selected for planting should be 2 to 5 cm in diameter. Each mother corm produces about 2 to 6 cormlets. Saffron undergoes deep dormancy from mid-June to August. The new cormlets produced annually remain attached to the mother corms till the beginning of the dormancy period.

Land preparation

The land is well prepared by ploughing 4-5 times to a depth of 30-35 cm during March-April. The last ploughing is done in May. At the time of the last ploughing, FYM @ 15-20 t/ha is incorporated into the soil. Strips of 2 m × 8 m, with 20 cm deep inter-drainage channels, are made. The channels are made around the beds to drain out the excess water. It is desirable to make raised beds in loamy soils.

Planting

The planting season differs with the location. In the Kashmir valley, planting is done from mid-August to mid-September, whereas in Himachal Pradesh, planting is done from mid-July to mid-August. Only healthy, well developed corms free from cuts, wounds or diseases are selected for planting. The mean diameter of the seed corm should be 4-5 cm and the corms should be treated with copper fungicide before planting to protect them from corm-rot.

The corms are planted in furrows at the rate of 30-40 q (500,000 corms) per hectare, at a depth of 7.5 to 10 cm, with a spacing of 5 to 7.5 cm between plants and 7.5 to 10 cm between rows. The corms are also planted at 12 and 18 cm or 15 and 20 cm apart in rows and at a distance of 10 cm in a row.

Manures and fertilisers

It is suggested to annually apply 40 kg nitrogen and 50 kg phosphorus or NPK at 50-100:10:50 kg/ha for the commercial production of saffron. The usual practice in Himachal Pradesh is to apply 80-90 kg of nitrogen per hectare in 4-5 split doses at an interval of 20-25 days. Besides, it has been reported that 3-3.5% urea spray is effective in enhancing the total dry weight of saffron, but higher doses adversely affect the total dry yield of this crop.

Irrigation

In the absence of rainfall, irrigation is required during August-September. However, saffron should not be irrigated during the flowering season. Saffron is grown as an irrigated crop in Spain, while it is cultivated as a rain-fed crop in India.

Interculture

The saffron fields should be kept weed-free for the proper growth and production of flowers. Weeding is done two the three times in a year to check the weed growth and also to avoid nutrient loss by weeds. Hoeing is also an important operation in the cultivation of saffron. The crusty soil around each bulb is loosened sometime in the spring itself to facilitate the emergence of the flower-stalk. The first hoeing should be done in the month of June, the second in the month of September and the third some time later. Incorporating sawdust mulch in to the crop beds is reported to increase the yields of the spice significantly.

Pests and diseases

Pests

Rodents (rats and moles) cause heavy damage to the crop by eating away the developing corms. As a control measure, the burrows should be fumigated with Phostaxin or Phosphene or poison-baited using Zinc phosphide and the treated cereal grains should be placed in the freshly-opened burrows.

Diseases

Corm-rot is an important disease of saffron. To protect the corms from rotting, they are treated with 5% solution of Copper sulphate or any other copper fungicide before planting.

Picking of flowers for obtaining saffron

The flowering period starts in middle or late October and lasts only until the first or second week of November. The flowers must be picked each morning before the sun gets too hot. The flowers are cleaned during the day and the style and stigmas are separated from the perianth.

The separation of the stigmas from the perianth has to be carried out every day, otherwise the flowers wilt. If the flowers are left for a couple of days, operations become very difficult. The collected flowers are picked up individually. An employee for this job holds a flower in her hand and, with the thumb-nail of the same hand, removes the pistils below the perianth, at the same time tearing off the stigmas with the fingers of the right hand and depositing these in the containers held for this purpose. Following a rather sluggish period in the first two years, the growth and production of saffron flowers reaches its peak by the third year and, after about six years, the bulbs are uprooted for subsequent planting elsewhere leaving the land fallow for about five years. Interestingly, the bulbs continue the cycle for 10-15 years, new bulbs being produced annually while the old ones rot.

Drying or Toasting of the stigmas of obtain saffron

The value of saffron depends mainly on the method by which the stigmas are dried. The techniques followed in Kashmir and Spain are described below.

In Kashmir, the stigmas picked from the flowers are dried and are arranged to constitute the first grade (*Shahi saffron*). The flowers are dried in the sun for three to five days, then lightly beaten with sticks and passed through coarse sieves. The material which passes through is thrown into water. Those parts of the flowers which float are discarded and the parts which sink to the bottom are collected and further dried, constituting the second grade (*Mongra saffron*). The discarded parts of the flowers are again subjected to the beating process and the process of throwing the entire pounded mass in water is repeated. The product which sinks is collected, and is very much inferior in value, constituting the third grade (*Lachha saffron*).

Curing

The value of saffron depends mainly on the method by which the stigmas are dried. The techniques of curing or drying of flowers followed in the preparation of different grades of saffron are described below.

Shahi saffron

In the Kashmir valley, the finest quality saffron '*Shahi saffron*' is obtained from the red tips or the stigmas borne on the styles. The stigmas are sun-dried or dried over low charcoal fires. After this, the product is packed immediately, preferably in containers.

Mongra saffron

The flowers are dried in the sun for three to five days, then lightly beaten with sticks. The entire material is then passed through coarse sieves into water. The heavier stigma products which sink to the bottom are collected and when dried subsequently constitute the best quality *Mongra saffron* or second-grade saffon.

Lachha saffron

The discarded parts of the flowers are again subjected to the beating process and the process of throwing the entire pounded mass into the water is repeated. The product which sinks, is collected and is quite inferior in value, constituting the 3rd grade which is locally known as *Lachha saffron* or *Kong* in the Kashmir valley.

Toasting

In Spain, the curing process is called toasting. The stigmas are placed in sieves, in layers 2 to 3 cm thick, and the sieves are placed 15 cm above fire. By stacking them and changing their order and position, the product is carefully dried. Special stoves for the purpose of drying are also utilised. During drying, the saffron has to be kept protected from dampness and light, because light bleaches it to a dull yellow.

In general, saffron is dried to a moisture content of 8-10% and, after grading and packing, is stored in moisture-proof containers. Saffron is usually sold in 1 g to 10 g packs.

Yield

About 1,50,000 flowers are required to produce a kilogram of good-quality dried saffron. At the maximum stage of production, i.e., from the fourth year onwards, a farmer can get a yield of 2 to 2.5 kg/ha of *Mongra*-grade saffron in Jammu and Kashmir. However, in some countries the yield is as high as 12.5 kg/ha, particularly in Spain, depending on the crop management. Yields as high as 24.3 kg/ha have been reported from New Zealand. Generally, the yield will be highest between the third and the sixth year.

The yields of saffron, both in Kashmir and Jammu, are much lower than in other countries, where yields are 4-5 kg per acre. About 5 kg of fresh stigmas yield 1 kg of saffron.

The final product as sold in the *bazaar* is a loosely-matted mass of dark, reddish-brown flattened stigmas with a characteristic aromatic odour and bitter taste. When fresh and dry they are glossy and unctuous (soapy or greasy) to the touch, but after keeping, they become dull and brittle.

Adultration

Because of high cost, saffron is frequently adulterated with the styles, anthers and parts of the saffron corolla. Exhausted saffron, flowers and floral parts of some plants belonging to the *compositae* family, like *Calendula* sp. and *Carthamus tinctorius*, corn silk and various materials coloured with coal-tar dyes are also used as adulterants. Water, oil or glycerine are added to increase the weight. The cake saffron of commerce often contains safflower florets with adhesive sugary substances.

Sage

Plant Profile

Family	:	*Lamiaceae*
English name	:	Sage
Indian name	:	*Salvia, Sefakus* (Hindi)
Species	:	*Salvia officinalis* Linn.; *S. sclara*
Distribution	:	Yugoslavia, Italy, Turkey, Albania, Portugal, Spain, Cyprus, England, Canada, USA
Uses	:	Spice, drugs

Sage is the dried leaf of *Saliva officinalis* Linn., of the mint family *Lamiaceae*, and is the most popular herb that is used in culinary preparations in the West. Of late, *S. sclara* (clary sage) is replacing *S. officinalis* because of its superior and more expensive oil.

The herb contains moisture—5.75%, protein-10.2%, fat-14.1%, crude fibre-16.0%, carbohydrates-46.3%, total ash-7.7%, calcium-1.8%, phosphorus-0.09%, iron-0.03%, sodium-0.01%, potassium-1.0%, vitamin-A-2395 I.U/100 g, vitamin-B_1-0.75 mg/100 g, vitamin-B_2-0.34, vitamin-C-29.8, niacin-5.7 mg/100 g, calorific value-415 calories/100 g. The dry sage leaves, on steam distillation, yield an essential oil (1.3-2.6%). The essential oil contains α-pinene (1.8%), cineole, linalyl acetate (10.1%), thujone (44.45%), borneol, bornyl acetate, farnesol and camphor. The quality of sage oil is determined on the basis of its thujone content; the higher the thujone content the better the oil.

In addition to the essential oil, the leaves contain tannin: 3%, fumaric, malic and ursolic acids, a bitter principle, picrosalin, saponin, pentoses, a wax, and potassium nitrate. The seeds contain 18% protein and yield a drying oil.

As a food flavourant, sage is used in pork and liver sausage, meat, fish dishes, poultry stuffings and baked loaf. The colour of the dried herb is grey, tinged with green; the aroma is strong, fragrant and spicy. The taste is fragrant, spicy, warm, astringent, and a little bitter. The dried and powdered leaves are mixed with cooked vegetables and sprinkled on cheese dishes, cooked meats and other similar preparations. The young leaves are pickled and are also used for making tea.

Sage is used as a mild tonic, astringent, diaphoretic, antipyretic, dentifrice, as a carminative and for treatment of sore-throat. Sage is reported to cure feminine disorders and estrogenic substances are said to have been extracted from the dried leafy tops. The dried leaves are used as a fumitory. The herb has been used in toothpastes and mouth-washes, gargles, poultices, tooth powders, hair tonics and hair dressings.

Sage oil is used in perfumes as a deodorant, in insecticidal preparations, for the treatment of thrush and gingivitis and as a carminative. The oil is also used as a convulsant. Sage oil is sometimes used as an adulterant for rosemary and lavender oil. Sage and sage oil exhibit antioxidant properties.

Hot water extracts of the residue, after steam distillation, on concentration yield an oleoresin, which is used in conjunction with the oil for flavouring foods. The drying oil of the seeds is used as a bonding agent in the production of oil paints.

Origin and distribution

Sage is a native of the Mediterranean area and is often cultivated as a spice, for medicinal purposes, and also for ornamental purposes, mainly in Yugoslavia, Italy, Turkey, Albania, Portugal, Spain, Cyprus, England, Canada and the United States. Sage is found growing wild in the Dalmatian region of Yugoslavia. In India, the plants are reported to do well in Jammu.

Description of the plant

It is a hardy, variable sub-shrub. The stems are shrubby, white and woody, 30-60 cm tall. The leaves are grayish-green, aromatic, entire, petiolate, oblong 7-8 cm long, and the base narrowed or spear shaped or round. On drying, the leaves turn silvery grey. They have a soft, velvety texture. The flowers are blue, purple or white, borne in simple racemes.

Major types of sage in the world trade

Dalmatian sage

This is the term for the product of Yugoslavia, derived from the Dalmatian coast. This has traditionally been the prime source of sage and its product commands the highest price. It is highly aromatic, with a pleasing particularly mellow flavour, its mellowness being its hallmark among the varieties of sage. It is shipped in tightly compressed bales, each weighing approximately 110 pounds.

Albanian sage

Albania is the other source of *Saliva officinalis* Linn. This is also recognised as an excellent variety of sage, but does not typically draw the same demand as Dalmatian sage in the US. While the two are comparable in terms of essential oil levels, there are certain differences in flavour characteristics, due to the variance in the components of the oil. These components give the Dalmatian its smoother, more fully rounded flavour notes. Albanian sage is shipped in loose semi-pressed bales.

Other types

In any given year, there are several other types of sage, known by their national origins, which come into the US market. Turkish, Italian, Spanish and Cretan are typical, but there are other sources and they vary in volume from year to year. In total, the poundage of all these other types is relatively small when compared to the shipments of the

Dalmatian and Albanian products. A number of botanical varieties of sage are represented in the exports of these other countries, but none is considered to be at par in flavour with either the Albanian or Dalmatian sage. The other varieties contain flavour notes which are harsher than those of *Saliva officinalis* Linn. sage.

Different forms of sage

For the industrial user, sage leaf is available in the market in a variety of forms, viz., (i) whole (ii) cut (iii) rubbed and (iv) further ground to specified granulations.

Whole-leaf sage

This is the term for the product which has been cleaned and prepared for sale, but not cut or otherwise processed.

Cut sage

This is known variously as: (a) cut (b) cracked (c) sliced (d) chopped (e) butcher's chop, etc. Whatever the term, it refers to leaves that have been cut (rather than ground) into smaller pieces, usually to the buyer's specifications. Particle sizes of 0.3 to 0.6 cm are common, but the size is determined by the buyer. Cut sage is chosen when the manufacturer wants the sage to be apparent in the end product.

Rubbed sage

This is a traditional term for sage which has been put through only a minimum grinding and a coarse sieve. The result, due to the soft, velvety texture of the sage leaves, is a fluffy, almost cotton-like product, which is unique among ground herbs. Many sausage-makers prefer this minimal processing for sage because it preserves flavour longer and at the same time makes it suitable for easy blending into a product. More sage is sold in the rubbed form than any other. However, the buyer should be aware that '*rubbed*' is a traditional term, not a standardised one. The consistency can vary according to the processing procedure of the supplier.

Ground sage

This is often called '*finely ground*' to differentiate it from '*rubbed*', which also goes through a grinding process. In general, this simply means that the sage has been further pulverised and sifted through finer sieves. The final granulation (particle size) is up to the buyer and this, in turn, depends on the end-use. Basically, the finer the powder, the more completely the sage will blend into the food, and lesser are its chances of being visible in the finished product.

Climate and soil

Sage is tolerant to cold and drought and is adaptable to a wide variety of situations. However, a hot and dry climate is not suitable for its cultivation, since it produces an inferior crop. Higher altitudes with ample sunshine and a few good showers in spring result in a good yield of superior quality oil. The crop can be grown as an annual or perennial.

The plant is generally grown on poor soils. Slightly acidic, soils with a of pH 4.0-5.5 are better. The crop can thrive on soil that grows cotton, but a better quality product is produced on rich clayey and loamy soils.

Varieties

There are no named varieties available in this crop in India. Many high-yielding, improved varieties are being grown in the traditional sage-growing countries.

Cultivation

Propagation

It is propagated by seeds. The seeds can be sown directly in the field or the seedlings can be transplanted. The seed rate is 3-4 kg/ha for transplanting. The seeds germinate in 10-15 days after sowing in the nursery. Since the crop raised from seeds is likely to be a mixture of different types, it is preferable to grow the crop from rooted stem-cuttings. The cuttings are obtained from broad-leaved plants that have the least tendency to flower.

Land preparation

The land should be well prepared by ploughing it once with a soil-turning plough, followed by one or two ploughings with the cultivator. The field should be levelled properly so that the irrigation water is uniformly distributed. Farm Yard Manure is applied @ 10-12 t/ha before the last ploughing and mixed well with the soil.

Sowing/transplanting

The sowing time has significant effect on the productivity of sage. The ideal sowing/planting time is March-April.

The seeds are broadcasted over well prepared land in the broadcast method of sowing. However, transplanting of seedlings is preferred over the broadcast method, as the latter gives better yield and helps in better management. The seedlings are transplanted when they are

30-35 days old, at a 1 m row spacing, while the rooted stem-cuttings are planted 30-45 cm apart in rows that are 90 cm apart.

Fertiliser application

Sage crop requires 100-120 kg N and 30 kg each of P_2O_5 and K_2O per hectare. Phosphorous and potash should be applied as a basal dose. Nitrogen should be applied in four equal split doses, ¼ before sowing/planting and the remaining ¾ after a 30 day interval.

Irrigation/inter-culture

The crop may be irrigated once or twice in case of a drought situation. Sage suffers severely from weed infestation, mainly during the early stages of crop growth. Depending on intensity of the weed infestation, one or two hand weedings are required. The first weeding should be done before the first top-dressing and the second one at the time of second top-dressing of N. Alternatively, chemical weed control can be used with Fluormeturon or Diuron at 2 kg/ha (pre-emergence application) and post-emergence application of Preforan or Introchlor at 3 kg/ha. In addition, 2-3 hoeings should be done before the flowering season and one hoeing is given after the harvest.

Pests and diseases

Pests

The aphid (*Acyrtosiphon salviae*) is found to infest sage and this can be controlled by contact insecticide.

Root-knot nematode (*Meloidogyne incognita*) infects the plant heavily. It can be controlled by the application of Carbofuran granules @ 1 kg a.i. per hectare.

Diseases

The fungus *Rhizoctonia solani* causes root-rot disease. Drenching the soil with Copper oxychloride or Bordeaux mixture is recommended for control.

Harvesting

The flowering tops and leaves are harvested twice a year during July and September. However, it is preferable to harvest the plants which have not reached the flowering stage, as at this stage they possess the finest aroma and contain the highest amount of oil. Excessive stalk growth is removed as it contains no significant amount of oil. Harvesting is done by hand or cutting the tops with a mower. Sometimes, the whole plants are cut by small sickles. The leaves could be shade-dried by spreading them in a thin layer and turning them occasionally. The plants remain productive for 5-6 years and a new plantation is started in a different location.

Drying

Sage is dried in the shade in order to retain as much of the natural colour and flavour as possible. The leaves and small tops are tied into small bundles or spread on screens and dried in a well-ventilated, warm room, away from direct sunlight. If the leaves are dusty or gritty, they are washed in cold water before drying. The dried bunches can be sold without further treatment or the leaves may be cut, rubbed or pulverised for the packeted herb trade.

Yield

The first year crop is usually light, but after the plants are established, in the second and subsequent years, a herbage yield of 1,750 to 2,250 kg/ha may be obtained.

Storage

Sage should be stored under cool, dry conditions. It can be stored for 9-12 months at room temperature.

Processing

For obtaining sage oil, the harvested herb is distilled immediately to avoid loss of essential oil by evaporation. Distillation is carried out for a period of 2-3 hours. An oil recovery of 0.15% is obtained on poor soils, whereas 0.2-0.3% is achieved with improved varieties and good management when the yield of oil will be 40-50 kg/ha.

Adulterants

Spanish sage and Greek sage derived from other species of *Saliva* are used either as substitute or adulterant of the genuine sage. Sage oil is sometimes adulterated with American cedar leaf oil which contains thujone.

Small Cardamom

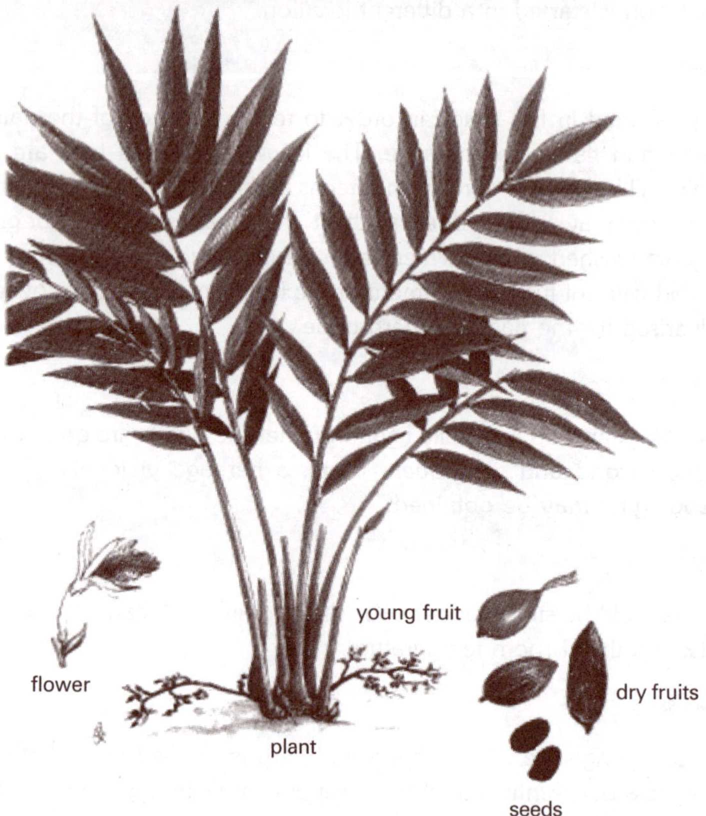

Plant Profile

Family	: *Zingiberaceae*
English name	: Small cardamom, Lesser cardamom, Malabar cardamom.
Indian name	: *Chhoti Elaichi* (Hindi), *Yelakki* (Kannada), *Elathari* (Malayalam), *Ela* (Sanskrit), *Yelakkai, Elakkai* (Tamil), *Yealakkayulu, Elakkayi* (Telugu)
Species	: *Elettaria cardamomum* Maton.
Distribution	: India, Sri Lanka, Guatemala, Thailand, Cambodia, Laos, Vietnam, Costa Rica, El Salvador, Tanzania.
Uses	: Flavouring agent in cooking, confectionery.

The true 'Cardamom' of commerce is the dried capsules obtained from the perennial plant *Elettaria cardamomum* Maton of the family *Zingiberaceae*. It is commonly called the lesser cardamom, Malabar Cardamom or *Chhota Elachi*. It is one of the first few spices, along with cinnamon, which entered the spice market of the West. It is the second-most important 'National Spice' of India and is also rightly known as the 'Queen of Spices' being next to black pepper, the 'King of Spices', which incidentally is also the world's most important spice as far as international trade is concerned. The dried capsules yield an essential oil and oleoresin. The aroma and therapeutic value of the spice are due to the essential oil (3-8%), whose main constituents are cineole, terpineole and limonene. Cardamoms are largely used as a masticatory, a flavouring agent in cooking, in confectionery and for medicinal purposes. In the Scandinavian countries it is used in flavouring cakes and pastries. In the Arabian countries, a beverage made from coffee and cardamom is offered to guests as a gesture of goodwill and hospitality. This cardamom coffee is called "*Gahwa*". In India, cardamom is used for flavouring sweets, pastries, cakes, other bakery products, gingerbread, *kheer*, meat, curries, sausages, various kinds of foods, liquors, beverages, condiments, culinary preparations and tea, which is called '*Elachi tea*'.

In medicine, it is used as a powerful aromatic stimulant, carminative, stomachic and diuretic and it is also reported to be a cardiac stimulant. A good nasal application is prepared by using extracts of cardamom, neem and myrobalan along with animal fat and camphor. Cardamom seeds are chewed to prevent a bad smell in mouth, indigestion, nausea and vomiting due to morning sickness, pyrosis, etc. Gargling with an infusion of cardamom and cinnamon cures pharyngitis, sore throat, and hoarseness during infective stage of flu.

The powdered seeds of cardamom boiled with tea and water impart a very pleasant aroma to tea, and the same can be used as a medicine for scanty urination, diarrhoea, dysentery, palpitation of the heart, and exhaustion due to overwork, depression, etc.

Eating cardamom once every day, with a tablespoon of honey improves the eyesight, strengthens the nervous system and keeps the body healthy. It is also believed by some people that excessive use of cardamom causes impotency.

Cheaper grade materials like thrips-damaged capsules and splits, which otherwise do not fetch a good price in the market, are used for the production of the essential oil. Of the three varieties of cardamom, viz. Coorg greens, Saklespur bleached and Alleppey greens, the

Alleppey greens produce a higher percentage of oil and is mild with a sweet odour due to the presence of linalool and linelyl acetate to the extent of about 10%, while, in the oil of Coorg greens these components are less and it has a harsher camphoraceous odour. The oil content of the Indian Cardamom is higher than that of the Sri Lanka Cardamom. The essential oil is used for flavouring certain bitters and liquors and in the manufacture of perfumes. The oleoresin has similar applications to essential oil, but it is less used. The oil and the oleoresin also find use in the preparation of aromatic, stimulant, stomachic and diuretic tinctures.

Origin and distribution

The plant is indigenous to the evergreen forests of South India and Sri Lanka. India once held the virtual world monopoly for cardamom, supplying 90-95% of the world off take of cardamom; but of late it is facing a stiff competition from Guatemala and other countries. The other major cardamom-producing countries besides India are Guatemala, Sri Lanka, Thailand and Cambodia. Cardamom is also produced on a smaller scale in Laos, Vietnam, Costa Rica, El Salvador and Tanzania.

Area and production

India accounts for 90% of the world's area under this crop. In India, small cardamom cultivation is restricted to the Western Ghat region of Kerala, Karnataka and Tamil Nadu. The Indian cardamom has higher oil content and aroma. Kerala occupies the first place, contributing to about 60% of the total area and Karnataka covers about 31% of the area, followed by Tamil Nadu (9%), with a production of 58.5, 34 and 7.5%, respectively. Coorg and Hassan in Karnataka; Kottayam in Kerala; and Madurai, Tirunelveli, Coimbatore, Salem and the Nilgiris in Tamil Nadu are the important cardamom-growing districts in the country. These areas are well suited for cardamom cultivation since they receive heavy rainfall and are characterised by a deep, natural mulch of fallen and decaying leaves which favour good growth.

The average annual production of cardamom is reported to be around 4,750 tonnes obtained from 1,05,000 hectares. India still meets about 60-65% of the world's demand through the export of cardamom to about 50 countries. The important countries importing cardamom from India are Kuwait, Qatar, Saudi Arabia, the erstwhile USSR and Japan.

Description of the plant

Cardamom is a tall growing, cross-pollinated, herbaceous perennial, with branched subterranean rhizomes. The real stem is the rhizome, which is subterranean in habit. Cardamom is a shallow-rooted plant. The leaves are 35 cm long and 1-10 cm wide, distichous, linear and lanceolate in shape, with short petioles. The inflorescence or the long panicles with racemose clusters arise from the underground rhizome. The flowers open in succession from the base to the top and develop into fruits. The calyx is cylindrical and persistent and the corolla tube is shortly exerted. The fruit is a trilocular capsule; the capsules are globose and rounded. The seeds are black when fully ripe in a capsule and embedded or covered with a white mucilagenous coat. There are 15-20 seeds in each capsule with a hard seed-coat.

Climate

The crop thrives well in evergreen tropical forests at an altitude of 600 m to 1500 m. above MSL. They occur in the preclimax stage of the forest. Cardamom is found to grow well even in areas where the annual rainfall ranges from 150-400 cm, provided the rainfall is well distributed. The better distribution of rain rather than high annual rainfall is the most important factor for a good crop. It grows well in a temperature range of 10° to 35°C. Valleys and gentle slopes facing the North or North-eastern direction, protected from strong winds, form an ideal situation for cardamom plantation. Cardamom also requires overhead shade.

Soil

Well-drained soils rich in organic matter, forest loams rich in phosphorus and potassium and deep light-textured lateritic soils with a pH range of 4.6 to 6.5 and plenty of humus are good for the crop. It is mostly grown as a pure crop in the lower portion of undulating valleys.

Varieties

Based on the size of the fruit, two varieties viz, *Elettaria cardamom* Maton variety *major*, made up of the wild indigenous types and the variety *minor* Watt (Syn.var.*miniscula* Burkill) comprising of the cultivated types are recognised. Among the various cultivated types, three major races viz., "Mysore" "Malabar" and "Vazhukka" are most popular and can be identified on the basis of certain specific morphological charactereristics.

Malabar or prostrate type

This is the most common race cultivated on the Travancore hills and the southern Western Ghats. The plants are of medium size and grow 2-3 m in height. The leaves are either pubescent on the dorsal side, or sometimes glabrous. The panicles are prostrate and the fruit is roundish. This type is suited to the 600-1200 m elevational zone. The plants are more productive than those of the Mysore type, but more drought resistant. The Malabar cardamom is considered to be more tolerant to the attack of thrips. This variety is mostly suitable for lower elevations.

Mysore or erect type

The plants are robust and grow to 3 to 4 m in height. The leaves are laceolate, glabrous on both surfaces, large and dark green. The panicles are erect, the capsules bold, and more productive. This variety is suited to high elevations and withstands exposure to wind better.

Vazhukka or semi-erect type

This is a hybrid of the "Mysore" and "Malabar" types and consequently the plants exhibit various characteristics which are intermediate between the two types. They are also adaptable to a fairly wide range of environmental conditions. The plants are robust like the "Mysore" types and have deep green leaves, semi-erect panicles and bold capsules.

In India, 336 germplasm accessions have been collected, conserved and characterised under the AICRPs. The commercially important released varieties of small cardamom are described below.

Mudigere-1

This is a high-yielding clonal selection from the Malabar type developed at the Regional Research Station, University of Agricultural Sciences, Mudigere, and was released as a variety in 1984. It is a pubscent type and bears oval, medium-sized capsules. It yields 250 to 300 kg of capsules per hectare.

CL-37

This is clonal selection from the open-pollinated seedlings of the Malabar type released by NRCS, Calicut. This is a high-yielding clone having a productive sucker and a long panicle with a bold capsule. It gives a yield about 410 kg of capsules per hectare.

MCC-49

This is also a selection from the Malabar type developed by the Spices Board, Myladumpara (ICRI). It is a high-yielding, prostrate type cultivar, suited for both irrigated and rain-fed conditions. This variety has bold, dark green capsules and is recommended for medium elevations. It has yield potential of 325 kg of capsules per hectare.

MCC-61

This is a clonal selection from the Mysore type developed by the Spices Board, Myladumpara (ICRI). The cultivar is an erect type and a high yielder. It bears long bold, parrot-green coloured capsules and is recommended for high elevations. It is tolerant to the azhukal disease and yields about 375 kg of capsules per hectare.

PV-1

This is a Malabar type variety, a selection from the Walayar collection, developed at the Cardamom Research Station, Pamadumpara (KAU). The capsules mature in 110 days. The average plant height is 269 cm and it produces 39 tillers and 26 panicles per clump. Each panicle bears 79 flowers and 49 capsules of 1.3 cc. The capsules are long and possess 17 seeds per capsule. Each clump yields 1310 g of green capsules. The average yield of dry capsules per hectare is 260 kg (potential yield 500 kg) and the dry recovery is 19.9%. The seeds contain 6.8% of essential oil, which contains 46 and 33% α-terpenyl acetate and 1, 8-cineole, respectively.

Coorg cardamom selectin-1 (C.C.S–1)

This is a Malabar type, a selection from the open-pollinated progeny (No.872) of CL-37 made at the Cardamom Research Centre, Appangala (ICRI). The capsules require 110 days to mature. The average plant height is 129.6 cm, and each clump produces 13 tillers and 8 panicles. Each panicle has 129 flowers and bears 60 capsules. The capsule is oblong, 0.9 cc in size and weighs 0.24 g (dry). There are 21.4 seeds in each capsule. The average yield and potential yield per hectare are 409 kg and 1156 kg of dry capsules, respectively. The dry recovery is 22%. The seeds contain 8.7% oil with α-terpenyl acetate and 1-8-cineole content of 37 and 42%, respectively.

ICRI-1

This plant belongs to the Malabar type. This variety is a selection (MCC-49) from the Chakkupallam collections made at the ICRI (Spices Board), Maladumpara. The capsules mature in 117 days after flowering. The mean plant height is 299.4 cm; each clump produces 84 tillers and 70 panicles. There are 330 flowers in an inflorescence, which bears 313 capsules. The capsules are round, 2.2 cc in size and weigh 1.1 g (green). Each capsule contains 18.5 seeds and a clump produces 2900 g of green capsules. The variety can yield 325 kg of dry capsules per hectare (potential yield, 650 kg/ha) and the dry recovery is 22.9%. The seeds contain 8.3, 37.6 and 28.8% essential oil, α-terpenyl acetate and 1, 8 cineol, respectively.

ICRI-2

This variety belongs to the Mysore race and is a clonal selection (MCC-61) from the Pampadumpa collection made at the ICRI (Spice Board), Myladumpara. The capsules mature within 127 days of flowering. The plants grow up to 275 cm and produce 76 tillers per clump, each clump producing 55 panicles. There are about 220 flowers in a panicle and 175 capsules are produced per panicle. The capsule is oblong, 2.8 cc in size and weighs 1.26 g (green). The capsule contains 16.9 seeds and each clump can yield 3450 g of capsules (green). The average and potential dry capsule yield per hectare would be around 375 kg and 766 kg, respectively, and the dry recovery is 22.5%. The seeds contain 9% oil, 36% α-terpenyl acetate and 29.3% 1, 8-cineole.

RR-1

This is a rhizome-rot resistant, high-yielding, variety developed by selection at the Cardamom Research Centre, Appangala. It is a clonal selection from the open-pollinated progency of early bearing plants of the 4th generation of clone-37, a collection from the RRS, Mudigere. The selection is an intermediate type between the Malabar and Vazukka type. The capsules are oval to elongated and 59% are bold. The essential oil content of the capsules is 6.7%. The oil contains 34.6% α-terpenyl acetate and 30.4% 1, 8-cineol. The variety can yield 238-848 kg of dry capsules per hectare. The variety is also less susceptible to foliar diseases.

Mudigere-3, SKP-14, ICRI-3 and ICRI-4 are the other varieties. The variety Njallani Green Gold is a farmer's selection. The lines P_3 and P_6 are drought-tolerant/resistant.

Cultivation

Propagation

Cardamom is generally propagated by vegetative means, using rhizomes from a clump but usually seedlings are also used for planting due to the lack of adequate clonal material. A micro propagation protocol for rapid multiplication and the synseed technology have been developed.

Vegetative propagation

In this case, rhizome cuttings are obtained from two and two-and-a-half years old clumps. The plants thus propagated are noted for their high yields, greater uniformity and earlier bearing than seedlings. However, a serious disadvantage is that the cardamom mosaic disease is spread through rhizomes.

Propagation by the trench method has been standardised. The multiplication rate in this method is 1:30-40 from one sucker/year. Tissue culture technology of cardamom clones is adopted by the organised industrial sector. Tissue cultured planting material is available in large numbers.

Seed propagation

Propagation by seeds prevents the spread of the '*Katte*' disease. Thus, it is the most common and widely prevalent method among planters. The main disadvantage, however, is that the progeny is highly variable with no uniformity in yields. The seeds also do not remain viable for a long time.

Selection of seeds

The seeds from the fully ripe capsules of robust, disease-free, high-yielding plants producing compact panicles, are collected and dried. A few days before sowing, the capsules are immersed in water and pressed gently to eject the seeds. The seeds are then washed in cold water for about 4-6 hours to remove the mucilaginous covering. After washing, the seeds are mixed with ash and dried under the shade for 2 to 3 days. To obtain good uniform germination, the seeds should be sown immediately after harvest. Shaking the seeds with coarse sand and scratching the seed-coat or seed scarification with nitric acid 2.5% for 10 minutes gives a quicker and higher percentage of germination than the untreated seeds. The seeds start germinating within a few weeks and the germination is complete within six weeks.

Seed rate and season

To plant a hectare, about 600 g of seeds are sufficient. The sowing season varies with the regions but the ideal sowing season is during the dry months. In Karnataka it is between September-October, while in Tamil Nadu and Kerala it is between November-January.

Raising of seedlings

Nursery sites are generally selected near a water source. The soil should be loamy and rich in humus. The selected site is cleared of all the vegetation like stumps, roots, etc., and the soil is worked to a depth of about 30-45 cm to a fine tilth. Raised beds, of 1 × 6 m size, are prepared. The seeds are sown in lines of 15 cm using 50 to 60 seeds per line in the beds, followed by a good stirring to cover the seeds. The beds are covered with straw, dry leaves or other material like polythene sheets during the winter months. Mulching the seed beds with dry leaves is found to give a high percentage of germination. It is better to have short beds along the *pandal* to provide shade. A temperature of 30°C would be ideal for seed germination.

Transplanting

First transplanting

When the seedlings are about 25-30 cm height and 8-9 months old, they are forked out and transplanted 15-20 cm apart in raised nursery beds. The first transplanting is done just before the onset of the monsoon.

Second transplanting

In Kerala and Tamil Nadu, the seedlings are transplanted into the secondary nursery beds. They are planted at the rate of 1:10; i.e., for every primary nursery bed of seedlings planted in 10 secondary beds, at a 20 cm spacing in 3 to 4 rows. Farm Yard Manure is applied before planting and mulching is also done. To provide shade, a *pandal* is erected. Regular watering is necessary.

After one year, the seedlings are ready for planting in the main field. The seedlings are planted with the onset of the monsoon during June-July, in pits of 45 × 45 × 30 cm or 60 × 60 × 35 cm in size, filled with compost and top soil at a spacing of 1.8 × 1.8 m. or 2 × 2 m. The spacing varies with the variety and fertility status of the soil. A spacing of 2 × 1 m is advisable along the contours. While planting, the seedlings are uprooted from the nursery, the clumps are separated and the overgrown roots are trimmed. They are planted in pits in such a way that the tallest tiller stands in the centre with the smaller ones on the sides. The plants are also provided with stakes for support.

Provision of shade

A light intensity of 30-40 k.lux is ideal for cardamom. To provide shade, about three months prior to planting, shade regulation may be carried out in August-September. The shading should be so regulated as to provide a canopy of about 8-10 m. Cardamom can be grown under the shade of *Gravellia robusta, Albezzia lebbeck* and *Artocarpus heterophyllus*. All the drying shoots and leaves are removed once a year in May-June after the commencement of the monsoon.

Manures and fertilisers

The manure are applied in 2 doses, one in June-July and the other in September-October. In the first year for plantations with irrigation facility, a dose of 37.5 kg N, 37.5 kg P_2O_5 and 75 kg K_2O per hectare is applied in two equal splits. The quantity of fertilisers to be applied is doubled from the second year onwards (75:75:150 kg NPK/ha). A fertiliser dose of 120:120:240 kg NPK per hectare for high density (5000 plants/ha) planting is recommended for Karnataka. In addition, the application of well-decomposed FYM or leaf mould @ 10 kg and neem cake @ 1 kg per plant, prior to the application of fertiliser, during May is recommended. The fertiliser recommendation for rain-fed plantations is 30:60:30 kg NPK per hectare. The fertilisers are applied in a circular band of 20 cm width and 30-40 cm away from the base of the clumps and incorporated into the soil.

During the first year of planting, one kilogram of lime is also mixed with soil at a depth of 30 cm per pit.

Irrigation

Cardamom is usually cultivated as a rain-fed crop, but supplementary irrigation is provided in certain pockets to overcome the dry spells during summer. Irrigation may be provided by the overhead or drip method. The flow of water through the drippers is so adjusted that 2-3 litres of water is discharged per hour. The provision of 10-15 litres of water/ plant/ day is sufficient for cardamom. The varieties P_3 and P_6 are resistant/tolerant to drought.

After care

In established plantations, the annual operation consists of weeding, digging, removing the old and dried stems (trashing, lopping of shade trees, manuring and dusting of the plantations).

Generally, 2-3 weedings are necessary in young plantations. The first weeding is done in May-June and the second in September before the fertiliser application and the third in December-January. Spraying Paraquat (625 ml in 500 litres of water/ha) should be done only in the inter-spaces between rows leaving 60 cm around the base in the pre- and post-monsoon periods. Besides, mulching is also practised to conserve moisture, reduce weed growth and to overcome dry situations. Forking is necessary in hard soils.

Trashing is carried out during June-July with the commencement of the monsoon, to prevent the spread of diseases and to expose the panicles for pollination by honeybees. Shade regulation is essential to provide optimum shade. Red cedar (*Toona ciliata* Roem.) is an ideal shade tree, which sheds its leaves during the rainy season and thus provides natural shade regulation.

Pests

Cardamom Thrips (Sciothrips cardamomi Rank)

This is a very serious pest occurring in cardamom. It attacks the flower buds and tender capsules which are punctured and on maturation, then scabby warts and corky outgrowths appear on the capsules.

For the control of the pest, spray the crop with 0.025% Quinalphos or Monocrotophos (0.05%) or Phosalone (0.05%) alternatively during the months of March-September at monthly intervals.

Black hairy caterpillar (Eupterote canerica, E. cardomomi. E. fabia, Lenodera vittata and *E. testacea)*

The caterpillars eat up the leaves and completely defoliate the clumps, resulting in the shrivelling up of the pseudo-stem, retardation of growth and reduction in yield.

This pest can be controlled by dusting with 2% Methyl parathion dust or spraying Quinalphos (0.05%). In addition, the collection of the hairy caterpillars, which are found in large numbers, and destroying them, is recommended. *Apanteles tabrobanac* has been reported to parasitize caterpillars of *Attacus atlas* and *E.fabia*.

Shoot/panicle and capsule borer (Conogethes punctiferalis; Thamurgides cardamomi)

The grubs bore into the stem and also into the fruit capsules. The infected plants are cut and removed. Spraying the crop with 0.075% Monocrotophos or Quinalphos (0.03%) is effective in controlling shoot and capsule borer. *Eriborus trocharteratus* and *Xanthopinpla australis* are the bio-control agents of this pest.

Shoot fly (Formosina flavipes)

Spraying Quinalphos (0.03%) or Phosalone (0.05%) or Monocrotophos (0.05%) will control the pest.

Mites

The infested plants show the symptoms like the drying up of leaves, which also becomes yellow in colour. To control this pest, Parathion (0.2%) may be sprayed.

White fly (Kanagarajiella cardamomi)

Both the adult and nymphs suck the sap from the leaves and the leaves become yellow. The nymphs are seen on the lower surface of the leaves. To control the pest, yellow sticky traps (rectangular metal/ polythene sheets painted yellow and coated with a sticky substance such as castor oil or grease) are fixed vertically inside the cardamom plantation at 3'-5' height from the ground level. The simultaneous spraying of neem oil @ 0.3 to 0.5% (mix 300–500 ml each of neem oil and Triton or Sandovit in 100 litres of water) should be carried out at 20-25 days, interval.

Root grub (Basilepta fulvicorne)

Occasionally this pest also causes damage to the crop and this can be controlled by soil application of Carbofuran @ 10 kg/ha during July and December. The fungus *Metarhizium* sp. has been found to infest the root grubs.

Nematodes

The root-knot nematode, *Meloidogyne incognita* and *Radopholus similes*, infest cardamom plantations. The suggested control measure for this is the application of Phorate 10 G @ 30g/clump twice, during May-June and again in September. *Verticillium chlamydosporium* and ten isolates of fluorescent *pseudomonas* and VAM are the biocontrol agents of the nematodes.

The bonnet monkey (*Macaqua radiata*) is reported to damage the cardamom suckers. The soft scale *Coccus hespiridum* and its parasite (*Coccophagus ceroplastae*), the leaf beetle (*Lema fulvimana*) and *Brevipalpus californicus* have also been reported as pests.

Diseases

The major diseases that attack cardamom are the *'katte'* disease, leaf disease, rhizome-rot, damping-off, leaf rust and nursery leaf-rot.

Katte or Marble disease

This is a virus disease and is transmitted by the banana aphid *Pentalonia nigronervosa* f. *caladii*. The disease is characterised by general chlorisis, followed by slender interrupted streaks on the leaves. The subterranean rhizomes will be shrivelled leading to a reduction in size and productivity of the clump. The vector may be controlled by regular spraying with Parathion (0.2%) or with systemic insecticides like Metasystox, Rogor or Dimecron at 0.5% concentration or Bioneem (0.5%) and Margolin (1%). The affected plants have to be uprooted with the rhizomes intact and destroyed three days after spraying. Disease-free planting materials should be used for planting. It is also recommended that phytosanitation measures are adopted.

The green lacewing (*Chrysoperla carnea*) is highly predaceous to the aphid. The sources of resistance/tolerance for this disease are NKE 3 and NKE 73.

Leaf disease

The affected tissue becomes dry and brown and shredded into thin strips. The leaves dry up and cause the eventual death of the plants. Spraying 1% Bordeaux mixture will control the disease.

Rhizome-rot

This is caused by *Pythium* sp. and *Rhizoctonia solani*. The leaves become pale yellow and it causes the premature death of older leaves. The base of the shoots get brittle and break away. The affected plants have to be removed and destroyed. Drenching the soil with 1% Bordeuax mixture will control this disease. The variety RR-1 released from the Cardamom Research Station, Appangala, is resistant to this disease. *Trichoderma harzianum* using coffee husk as carrier and *Paecilomyces lilacinus* are the biocontrol agents of this disease.

Damping off/seedling rot/clump rot (Pythium vexans; Rhizoctonia solani)

This disease causes heavy mortality of seedlings in the nursery. To control this disease, nurseries have to be provided with proper drainage and drenching with Bordeaux mixture or Copper oxychloride (0.2%) during June, August and October has to be done.

Nursery leaf-rot (Coniothrium sp. and Phyllosticta elettariae)

On the leaves of the plants in the nursery, small water-soaked spots first appear, gradually extending into large areas in which the chlorophyll is completely destroyed. The affected tissues become flaccid and shrink up to the mid-rib. This may be controlled by spraying 0.2% Captafol or 1% Bordeaux mixture at 15-day intervals during the monsoon.

Bacterial wilt and nematodes are the other nursery diseases. The application of Streptocycline (200 ppm) against wilt during June, and Ablicarb granules @ 1 g a.i/sq.m. against nematodes has been suggested.

Capsule rot (Azhukal)

The causal organism for this disease is *Phytophthora meadii, phytophthora nicotianae* var. *nicotianae*. The control measure suggested for this disease is spraying and soil drenching with BM (1%) or Copper oxychloride (0.2%) or Aliette (0.3%) during August and October.

Kokkekandu (Vein-clearing virus)

This is a viral disease caused by the potty virus. The banana aphid is also the vector for this virus. This disease is characterised by vein clearing on leaves and the affected plants produce hook-like shoots.

Leaf mottling, shredding, quick decline and the death of the clumps result in poor yield. For control, the same disease management practices as that of *Katte* are followed. The clone 893 is resistant to this disease. The other viral diseases are Cirke disease (mosaic), Nilgiri necrosis virus (NNV) and Forkey (reported from Sikkim)

Seed rot (Fusarium sp.)

The rotting of cardamom seeds can be prevented by treating the seeds with Bavistin 2.5% or Captan (4 g/kg of seed).

Cropping

The flowering starts in April-May and continues actively for about 3 to 4 months, i.e., till August-September. The inflorescence comes up from the base of the stems of the previous year's growth. In fact, flowering can be observed almost throughout the year. Due to this fact, the ripening of capsules is irregular, which necessitates several pickings. The fruiting period (i.e., from the opening of the flower to the ripening of the fruit) is nearly 3-4 months. As many as 75 inflorescences with a number of flowers in each are borne by a single plant.

Harvesting and yield

Cardamom plants start yielding by the third year after planting, but good yields are expected from the 5th year onwards. It is recommended to spray NAA 25 ppm to reduce fruit drop and to increase fruit set. The peak harvesting period is September-February. Harvesting is continued at intervals of 30-40 days and will be completed in 5-6 months. Rearing four honeybee colonies per hectare in cardamom plantations helps in increasing the fruit set and capsule production. The average yield of dried capsules is 200-500 kg/ha year.

Processing

After the harvest, the produce is dried either in the sun or in specially-built curing houses. The harvested capsules are first washed with water and spread out thinly on wire-net trays in the curing chamber or heated chambers (kilns) and hot air is passed through the flue pipes in the furnaces. The drying process will be completed within 36-42 hours at 50-60°C in the curing chambers. By drying the cardamom in the curing chamber, the green colour is retained.

The dried capsules are polished by rubbing in trays fitted with wire mesh or rough coir matting to remove the ends, soil particles, etc., and winnowed. They are then sorted out into different grades according to their size (7 mm, 6.5 mm and 6 mm) and colour as ‘longs’, ‘medium’ and ‘shorts’. Since the green colour is an important quality determining factor in trade, it is essential to retain the colour as long as possible. It has been found that soaking the freshly-harvested green cardamom capsules in 2% washing soda solution for ten minutes before drying is effective in preserving the green colour. Similarly, bleaching is done by exposing the dried capsules to the action of sulphur dioxide, produced by burning sulphur or bleaching powder or hydrogen peroxide. The graded produce is stored in polyethylene-lined gunny bags to retain the colour and to avoid exposure to moisture. The output of dried capules is 20-25% of the harvested capsules.

Grading

The Government of India and the ISI have prescribed fairly well-defined grades, popularly known as ‘Agmark’ grades, on the basis of important quality factors like colour, weight per unit volume (litre), size and percentage of ‘empties’, malformed, shrivelled and immature capsules. Separate specifications (about 33) have been laid for different qualities or types of cardamom such as ‘Aleppey Green’, ‘Coorg Green’, ‘Bleached’, ‘Half Bleached’, ‘White’ and ‘Mixed’ cardamoms.

Essential oil

Cardamom oil is a greenish–yellow liquid with a spicy aromatic odour, somewhat pungent and faintly bitter at high concentrations. It is commercially produced by the steam distillation of the crushed capsules. The yield and organoleptic properties of the essential oil so obtained are dependent upon the type of spice used. In general, the spice from a recent harvest, which has not suffered excessive volatile-oil loss, should be employed in order to obtain a good yield. It has also been shown that distillation should be continued at least for 4 hours to produce the full ester content of the oil. The oil content is 3-8%; up to 11% oil is available in the seeds while it rarely exceeds 1% in the husks.

Oleoresin

Cardamom oleoresin is produced on a relatively small scale and little information is available on the methods of extraction employed by manufacturers. In solvent extractions, the yields are of the order of 10% and the content of volatile oil in the oleoresin is dependent upon the raw material and solvent used. Commercial cardamom oleoresins have been offered in the market with volatile oil contents ranging between 52 and 58%.

Sweet Flag

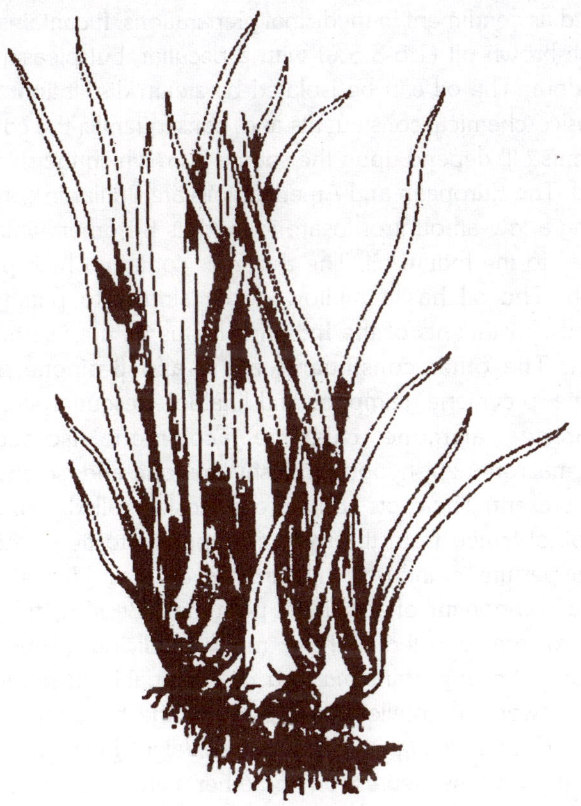

Plant Profile

Family	:	*Aeraceae*
English name	:	Sweet Flag
Indian name	:	*Ugragandha* (Sanskrit), *Bacc, Gorbacc* (Hindi), *Baje* (Kannada), *Vashampu* (Tamil)
Species	:	*Acorus calamus* L.; *A. gramineus* Soland
Distribution	:	India, North America, Asia, Central Europe, Japan, Romania, Russia
Uses	:	Aromatics, confectionery, drugs, flavouring liquors, beer, gin, vinegar, snuff and various other preparations, substitute for ginger, insecticide

Acorus calamus L., commonly called sweet flag or calamus, is a semi-aquatic rhizomatous perennial herb belonging to the family *Aeraceae*. The dried root (rhizome) has long been employed as condiment in medicinal preparations. It contains a volatile yellowish-brown oil (1.5-3.5%) with a peculiar, but pleasant, slightly sweet odour. The oil can be isolated by steam distillation. The yield and physico-chemical constituents and, consequently, the composition of 'calamus oil' depend upon the source from which the rhizomes are obtained. The European and American oils are similar in composition, consisting a low amount of 'asarone' and its β-isomer, which can be compared to the Indian oil. The rhizomes contain a high percentage of starch. The oil has a mellow odour similar to patchouli. The important constituents of the Indian calamus oil are asarone and its β-isomer. The other constituents are a and β-pinene, myrcene, camphene, p-cymene, camphor and linalool. Sesquiterpenic ketones like asarone, calamone, calacone, acolamone, iso-acolamone, acoragermacrone, epishyobunone, shyobunone and iso-shyabunone are also present. The roots contain a glucoside called acorin.

The oil obtained from the rhizomes appears to be of the greatest use in the perfumery industry in *agarbattis, dhup* and *havan samagries* and as a component of the finest perfumes. Besides, the rhizomes and the oil are reported to have many medicinal properties. The sweet flag is an important *madhya drug*, capable of improving the memory power and intellect. It is used for the treatment of chronic diarrhoea, delirium, cough, bronchitis, odontalgia, inflammations, gout, epilepsy, convulsions, depression and other mental disorders, tumours, dysentery, skin diseases, numbness and general debility. It possesses emetic and antispasmodic properties. The rhizome is an ingredient of preparations like *Vacaditalla, Ayaskrit, Kompancadi gulika, Valiya rasnadi kasaya*, etc.

The powdered rhizome also possesses insecticidal properties and is useful against bed-bugs, mosquitoes, moths, lice, etc. The rhizome as well as the oil are used in the manufacture of liquors, essences and bitters, and safe insecticides for the protection of foodgrains. In moderate doses, the oil produces an antispasmodial action on the involuntary muscle tissue, inhibiting excessive peristalitic movements of the intestine.

Origin and distribution

It grows wild on the edges of swamps on the banks of rivers and ponds in North America, Europe and Asia. The countries growing this crop commercially are the erstwhile USSR, the USA, Romania, Sri Lanka, India and Japan. In India, the leading states are Karnataka,

Kashmir, Manipur, Arunachal Pradesh, Meghalaya and it is also grown in the foot hills of Himalayas up to 1800 m. In Karnataka, it is being grown in the Koratagere Taluk of Tumkur district in an area of 100 acres and the annual production is 400 tonnes of dried rhizomes.

Description of the plant

Sweet flag is a monocot plant. The plant is a herb, with narrow, 0.75-1.5 m long leaves. The breadth of each leaf is 2-4 cm and they resemble the leaves of Iris. The leaves are bright green, thickened in the middle with wavy margins. The spathe is 6-30 cm long and encloses the spadix, which is slightly curved. The fruit is a berry, yellowish green in colour, angular, with 1-3 seeds.

The rhizomes are horizontal and jointed, with a spongy texture. The rhizomes are 1.5-2.5 cm thick and sometimes up to 2 m long.

Climate

Sweet flag is a hardy plant and is thus found growing from tropical to subtropical climates. However, it prefers areas with a good and well-distributed rainfall throughout the year.

Soil

Sweet flag is cultivated in almost the same way as rice and may be grown in any part of the country where suitable irrigation facilities are available. Although calamus in the wild state grows in water, it may be cultivated in almost any good, but fairly moist soil. It is usually grown in the areas where paddy can be grown. It is also grown in the clayey loams and light alluvial soils of river banks.

Species/varieties

Acorus is a small genus of herbs, comprising of 2 species.

***1. Acorus gramineus* Soland**

This is a Japanese species, occasionally found in Sikkim (up to 1800 m) and Khasia hills (1200-1400 m). It is used for the same medicinal purpose as *A. calamus*. Here, the mid-rib of the leaf is absent. It is a semi-aquatic, perennial herb, with creeping and much-branched aromatic rhizome with a spongy texture of 1.5-2.5 cm thickness and sometimes several metres long. The leaves are 10-22 cm in width and 0.60-1.8 m in length. The rhizome is brown outside and white and sponge-like inside.

2. *Acorus calamus L.*

This is a semi-aquatic perennial herb with a stout mid-rib, creeping and much branched aromatic rhizome. It is found growing wild or cultivated throughout India and Sri Lanka, ascending to 1800 m in the Himalayas. The plant thrives in marshy places and moist situations, like the edges of lakes and banks of streams and rivers. It is found plentifully in the marshy tracts of Kashmir and Sirmoor in Manipur and Naga Hills. It is also grown in North America, Europe and Asia. The variety *variegatus* has leaves with longitudinal yellow stripes.

There are no named varieties available in this crop.

Cultivation

Propagation

The planting material consists of the live ends or tops of the previous crop. At harvesting time, the mature portion of the rhizome is cut off which forms the marketable part and the tender portion formed by the growing end is used for replanting and propagation. This propagating material can be planted at once or stored for about a week by covering it with straw or dried leaves. If it has to be stored for a longer period, this can be done by keeping it in open pits.

Planting season

The best planting season is spring (March-April) but, the crop may be planted almost at any time of the year, taking into consideration the fact that plenty of sunshine must be available at harvest time to dry the rhizomes.

Land preparation

The field is laid out exactly as for rice. It is irrigated sufficiently and after two ploughings, it is again watered heavily and ploughed. The field is then left for a few days and ploughed to a puddle again. It is then levelled with the levelling board. Farm Yard Manure is incorporated into the soil before puddling.

Planting

The recommended spacing for sweet flag is 30 cm × 30 cm. The rhizome pieces are pressed into the mud at a depth of about 5 cm inside the soil. The rhizomes are planted in such a way that the plant in the second row comes in between the first row and not opposite to it.

Manures and fertilisers

The field should be manured with green manure (10 to 12 t) and compost (15-25 t/ha). Along with this, 100 kg ammonium sulphate,

300 kg superphosphate and 100 kg muriate of potash (25:50:60 kg NPK/ha) should be added. Otherwise, 125 kg NPK/ha is recommended to be applied in 3 split doses—a basal dose of 1/3 of N, P_2O_5 and K_2O are applied along with FYM during March-April and the remaining 2/3 of nutrients in two equal split doses at 4 months and 8 months after planting.

Irrigation

The field should be kept regularly irrigated. About 5 cm of water is left standing in the field in the beginning. Later, the water level is increased to 10 cm as the plant grows.

Weeding

The field is carefully and regularly weeded. In all, eight weedings are given and at each weeding, the growing plants are lightly pressed down into the soil.

Pests and diseases

Pests

The plant is attacked by shoot and root mealy bugs which suck the sap from the respective parts resulting in their drying up. The mealy bugs can be controlled by spraying the shoot and drenching the roots of the grown up plants with 0.1% Methyl parathion, 0.15% Oxydemeton methyl or 0.2% Quinalphos. Leaf-eating caterpillars are also reported to cause damage to the crop. A spray of 0.1% Monocrotophos will control the insect.

Diseases

Leaf-spot is a major disease of *Acorus* which can be controlled by spraying Carbendazim (0.1%). An aphid transmitted mosaic streak disease of wheat has also been reported on sweet flag.

Harvesting and Yield

After about a year, the crop is ready for harvesting. The field is allowed to dry partially so that sufficient moisture is retained to facilitate the necessary deep digging. The leaves start turning yellow and dry, which is indicative of maturity. The rhizomes will be found at a depth of 60 cm and are about 30-60 cm long. Therefore, harvesting should be done carefully. After harvest the rhizomes are cut into short lengths of 5-7.5 cm and all the fibrous roots are removed. The pieces are then washed thoroughly and dried in the sun. The dried material is put into gunny bags and rubbed to free them of leafy scales. The yield of rhiozmes is about 7-10 t/ha.

The fresh rhizomes yield about 1.2% and the dry rhiozmes 3-3.5% of essential oil and the essential oil from dried rhizomes is bitter. The fresh aerial parts yield 0.125% of oil. The rhizomes should not be peeled prior to drying, because peeling results in considerable loss of oil by evaporation and resinification. The highest yields of oil are obtained from dried unpeeled rhizomes that have been granulated immediately before they are charged into the still. The peeled and dried European calamus roots yield 1.5-1.8% oil, whereas the Japanese material yield up to 5% of oil. The Indian roots growing in the plains yield about 3.1% and those growing in the Kashmir valley yield not more than 1.4% of oil.

Adulteration

The dried rhizomes constitute the spice as well as the drug, i.e., the calamus of trade, and are sometimes substituted for the rhizomes of *Alpinia galanga* and *Aconitum* sp. The powdered drug is reported to be adulterated with siliceous earth, ground root of 'marsh mallow' (*Althaea officinalis*) and cereal flours.

Sweet Marjoram

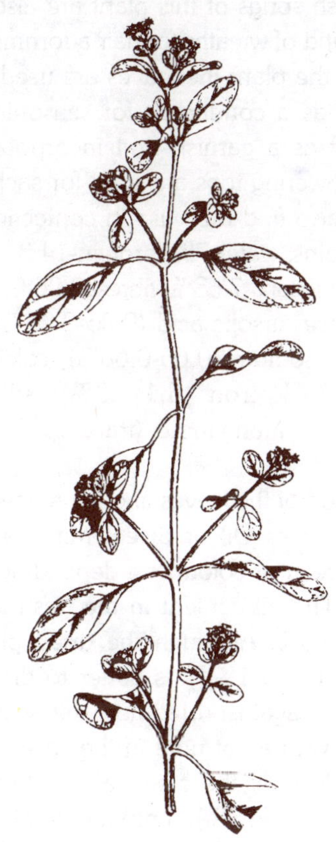

Plant Profile

Family	:	*Lamiaceae*
English name	:	Sweeṭ marjoram
Indian name	:	*Marwa* (Hindi), *Maruga*(Kannada), *Maruvemu* (Malayalam), *Magu, Sinshi, Murwo* (Sanskrit), *Maruvu, Marra* (Tamil)
Species	:	*Marjorana hortensis* Moench.
Distribution	:	Southern Europe, North Africa, Asia Minor, North and South America.
Uses	:	Food flavourant

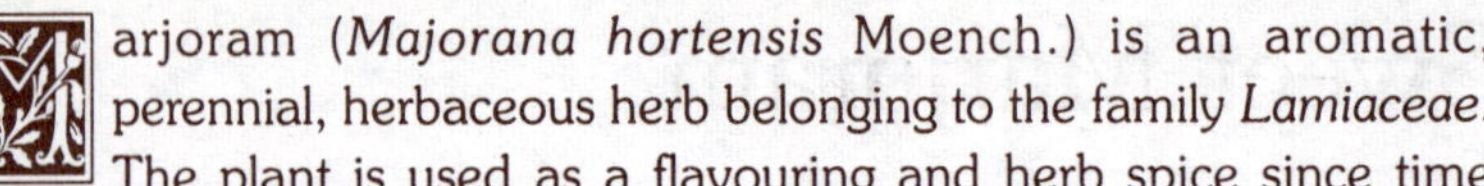

Marjoram (*Majorana hortensis* Moench.) is an aromatic, perennial, herbaceous herb belonging to the family *Lamiaceae*. The plant is used as a flavouring and herb spice since time immemorial. The fresh sprigs of this plant are also used in garlands, bouquets and other kind of wreaths for hair adornment. In the countries traditionally growing the plant the leaves are used fresh or dried and are highly esteemed as a condiment for seasoning food. The fresh leaves are employed as a garnish and incorporated in salads and vinegar. The dried flowering tops are used for sachets and potpourri. The aromatic seeds also find their use in confectionery.

The dry herb contains: water-7%, protein-14.31%, fixed oil-5.60%, volatile oil-1.72%, pentosans-7.68%, fibre-22.06%, ashi-9.69% tannin, an astringent substance, ursolic acid (0.05-0.21%) and minerals like potash (18.3-20.2%), sodium (0.65-0.68%), calcium (17.6–24.8%), phosphorus (8.9-9.1%), iron (6.1-7.3%), silica (19.4-26.5%), magnesium (4.8-6.7%), manganese (trace to 1.05%) and chlorine (1.51-2.05%).

The steam distillation of the leaves and flower-heads yields a volatile oil known in the trade as Oil of Sweet Marjoram. The oil is pale yellow to yellowish-green or colourless, depending upon the material used for distillation. The oil content in the fresh leaves of marjoram ranges from 0.3 to 0.5%. Although the dried plant gives a higher yield of essential oil, up to 1%, it is better to distill the fresh leaves and flowering heads as against dried herbage as there is a possibility of lower yields of oil with a flat note in the case of dried leaves.

The chemical analysis of the essential oil shows that it contains mainly terpene-4-01 (28.973%), linalool + *cis-sabinene* hydrate (27.404%), linalyl acetate (4.972%), alpha terpenol (4.419%), sabinene (2.317%), carvacrol, eugenol, chavicol, methyl chavicol and caryophyllene. The oil is mainly used in flavouring food articles. It is very much in demand in the food processing industries for flavouring canned food. To a small extent, it is also used in high-grade flavour preparations, perfumes and in soaps. Medicinally, it is also useful as an external application for sprains, bruises, stiff and paralytic limbs and tooth-ache, and for hot fomentation in cases of acute diarrhoea. Sweet marjoram is considered to be a carminative, expectorant and tonic. The leaves and seeds are astringent. An infusion of the plant is used as a stimulant, sudorific, emmenagogue and galactagogue. It is reported to be useful in treating asthma, hysteria and paralysis.

The essential oil obtained from the leaves of marjoram has antimicrobial activity against *Bacillus anthracis*, *Proteus vulgaris*, *Salmonella stanley*, *Streptococcus agalactiae*, *S. guneus*, *Salmonella newport and Aspergillus fumigatus*.

Origin and Distribution

The plant is native to Southern Europe, North Africa and Asia Minor. It is an important kitchen herb with a wide range of uses and it is cultivated throughout Europe, in the Mediterranean region as well as North and South America. France and Egypt are the major exporters of marjoram oil in the world. The USA alone consumes about 300-400 t of marjoram oil per annum, but France and Germany have larger markets with over 500 t requirements. In India, this crop is cultivated only on a small scale in several states. It is being grown in home gardens in Karnataka, Andhra Pradesh and Tamil Nadu.

Description of the plant

Sweet marjoram is a perennial plant, which is often grown as an annual, because it does not survive well in wet cold winters. It grows as an upright compact bush, about 20 to 40 cm in height, with a woody main stem and many soft branches. The leaves are round, soft, matte green and have a marvellous sweet smell. In the bud, the flower heads (spicules) are the same grey-green colour as the leaves and resemble little oval knobs. These expand slightly to show the frill of tiny white flowers only just protruding from their overlapping bracts. The flowers are bracteate, sessile, complete and bisexual. Both the calyx and corolla are irregular, zygomorphic and hypogynous. The bract is prominent, spatulate, 4 mm in diameter, hairy along the margin and on the abaxial surface. It encloses the calyx, which has five sepals united to form a single spatulate structure, with a wide and large (5 mm in diameter), bilobed anterior and narrow small posterior parts. It is hairy along the margin and has an abaxial surface. The corolla has five petals. It is gamopetalous, bilobed with 3 petals (posterior) and 2 petals (anterior), with imbricate aestivation. The corolla tube is 4 mm long and 5 mm in diameter. The corolla tube is white in colour, wide (2 mm) above and narrow (1 mm) below, each limb measuring 1.5 to 2.0 mm, the stamens are 4 epipetalous and didynamous, the anterior stamens being shorter (3 mm) than the posterior (4 mm) ones. The staminode is absent. The anthers are two-celled, dehiscing introrsely or longitudinally slit, hypogynous, with fleshy disc. The ovary is superior, bicarpellary, syncarpous, four partite by a secondary division with a single anatropus erect, basal ovule attached to an axile placenta in each locule. The style is gynobasic (5 mm long) and simple with a bifid stigma. The fruit has 1-4 seeded nutlets included within the persistent calyx. The seed is oblong, brown in colour, 3 mm long and 1 mm wide, a their test, a scanty endosperm. The embryo has flat cotyledons parallel to the fruit axis and a short inferior radical.

Anthesis occurs between 8.00 a.m. and 3.00 p.m and the anther dehiscence takes place from 10.00 a.m. to 4.00 p.m. The maximum receptivity of the stigma is found on the day of the anthesis. The pollen grains are dull white in colour, oblong to oval in shape, 43.74 mm in length and 27.70 mm in breadth.

Climate

Marjoram is primarily a warm season plant and, as such, is rather sensitive to cold weather. Yet, it is hardy enough to grow in colder regions, provided that the plants are well mulched with straw or leaves for protection against the winter or still better grown as an annual crop and planted anew each season. It requires a sunny position, although mid-day shade is preferred.

Soil

Majoram prefers well-drained and nutrient-rich soil for its successful growth. Soils which have a tendency to form crusts will retard germination greatly and hence are not suitable for marjoram, unlike most other herbs from the same family. Marjoram has a strong flavour when grown on a rich soil. It is one of the few crops which can be grown even on saline or alkaline soils as well and can withstand high salt content in the soil. The plants are resistant to drought, but prolonged periods of drought are harmful.

Season

The planting time is one of the important factors in obtaining the optimum yields from a unit area. Raising the crop in the proper season is essential because in prevailing optimum conditions, the plants show maximum growth and result in a good yield. Under Bangalore conditions, November planting is found to be superior to its counterparts for all the growth and yield attributes.

Cultivation

Propagation

Marjoram can be propagated by both seeds and cuttings.

i) Nursery raising

The seeds can be sown twice a year, during June-July and September-October. They are sown in well-prepared nursery beds, in rows spaced about 15 cm apart. About 50 g seeds give sufficient seedlings to plant an area of one hectare. As the seeds are tiny, they are mixed with dry sand in the ratio of 1:10 and broadcasted in rows of nicely prepared nursery beds. A thin layer of sand is spread uniformly over the seeds to cover them; the seeds germinate after 8-10 days. Sprinkler irrigation and thorough care of seedlings is taken after germination. The seedlings are ready for transplanting when they are about 2 months old.

As this method is rather cumbersome and the initial growth of the seedlings is slow, it is followed only where growing the crop through seeds is absolutely necessary.

ii) Vegetative propagation

Sweet marjoram is commercially propagated through cuttings. The plants propagated through this method are not only true to type but also grow faster initially. Stem-cuttings obtained from the second ratoon crop are the best for planting. Herbaceous cuttings of 15 cm length are made from healthy plants. Seed pans filled with sand may be used for planting the cuttings. After planting, they should be transferred to the mist propagation chamber or into the shade where they are watered twice a day. The cuttings take about one month to root. Small, individual polythene bags may also be used to raise the cuttings. In this method, the roots do not get damaged during the transplanting. The establishment of rooted cuttings in the main field is about 90-95%.

Planting

Only healthy seedlings or rooted cuttings with a well-developed root system are used for transplanting into the main field at a spacing of 15 cm × 7.5 cm. or 20-25 cm apart in rows, which are spaced 30 cm apart.

Fertiliser application

In an experiment under Bangalore conditions, it was shown that the application of nitrogen at 240 kg/ha, phosphorus at 40 kg/ha and potash at 80 kg/ha in three equal split doses gives the maximum herbage and oil yield in marjoram.

Irrigation and inter-culture

After transplanting, the crop is irrigated by a sprinkler for 5-6 days and then a light surface irrigation is given. Subsequently, the crop may be irrigated once in 4-8 days, depending upon the season and the soil type. The crop is weeded as and when required, followed by hoeing, to provide good aeration in the soil.

Crop rotation/sequencing and intercrops

In many places, sweet marjoram is essentially a short-term winter crop. Of the several possibilities of crop rotation, the raising of coriander, cumin, methi, etc., are promising. These crops can be grown as monsoon crops. After their harvest, the land is rested for three months and then it can be planted with marjoram.

Use of growth regulators

Studies conducted at Bangalore showed that the application of GA_3 at 200 ppm increased the fresh herb recovery, recording the maximum oil yield (0.397% oil content and 55.54 kg/ha of oil yield). Although Cycocel at 500 ppm and Triacontanol at 6 ppm increased the fresh herb recovery and oil yield over the control, they were less successful than GA_3.

Pests and diseases

Marjoram is a comparatively hardy crop and no serious pest and/or disease has been reported on this crop in India. Among the insects, it is found to be sometimes attacked by termites, which may be controlled by the application of Termex 20% EC. With regard to the diseases, sometimes leaf rust disease is found to affect the plants, particularly during the cloudy weather. It can easily be controlled by spraying 0.2% Blitox at weekly intervals.

Harvesting

Under Bangalore conditions, the crop behaves as a perennial. The first harvesting is done after 90 days of planting, when the crop starts flowering. While harvesting, about 10 cm of the shoot above the ground level is left for further growth. The subsequent harvests can be carried out after 45 days of the first harvest. On an average, about 4-5 harvests are taken annually. Also, there are reports that the crop can be retained up to seven cuttings under good management. After every harvest, a top-dressing of nitrogenous fertilisers, preferably urea should be applied to the crop. To obtain oil of export quality, the first harvest should be done only after 90 days of planting. For use as spice, the leaves/tops are dried in the shade and packed in containers and sealed.

Yield

Marjoram may yield about 16-18 t/ha of fresh herbage, which in turn may yield about 35-40 kg marjoram oil by steam distillation. An oil recovery of 0.3 to 0.4% on a fresh weight basis may be considered satisfactory.

Production of essential oil

The fresh herbage of marjoram is steam distilled to obtain the oil. The distillation equipment consists of a boiler, distillation still, condenser and receiver. The distillation still is made of steel. It has a perforated metal sheet at the bottom to support the herbage, which is loaded into the still for distillation.

Loading and unloading can be mechanised with the help of an overhead chain-and-pulley block. The lid of the still can be swung aside during loading and unloading. It is important that the herbage should be evenly packed inside the still, as otherwise steam channels may form during the distillation and result in a poor oil yield. The condenser which cools the hot vapour received from the still, consists of many tubes made of copper or stainless steel, mounted inside a jacket.

The condenser is provided with inlet and outlet for the circulation of cooling water. The hot vapours are cooled in the condenser tubes and the condensate flows out into the receiver. The oil, being lighter than water and insoluble, floats on the top in the receiver and only the water gets drained out. The oil can be drawn-off separately at the end of the distillation. The receiver is fabricated out of two compartments, so that any oil escaping from the first compartment can be retained in the adjoining compartment, although this rarely happens.

Distillation is carried out preferably at atmospheric pressure, as this will prevent any sort of burnt odour in the oil. Usually about four to five hours are required for completing one distillation. The oil should be free from sediment, suspended matter and moisture before storing. It is stored in aluminum containers. The containers should be filled to the brim to exclude all air in order to prevent any deterioration of the oil during storage.

Tamarind

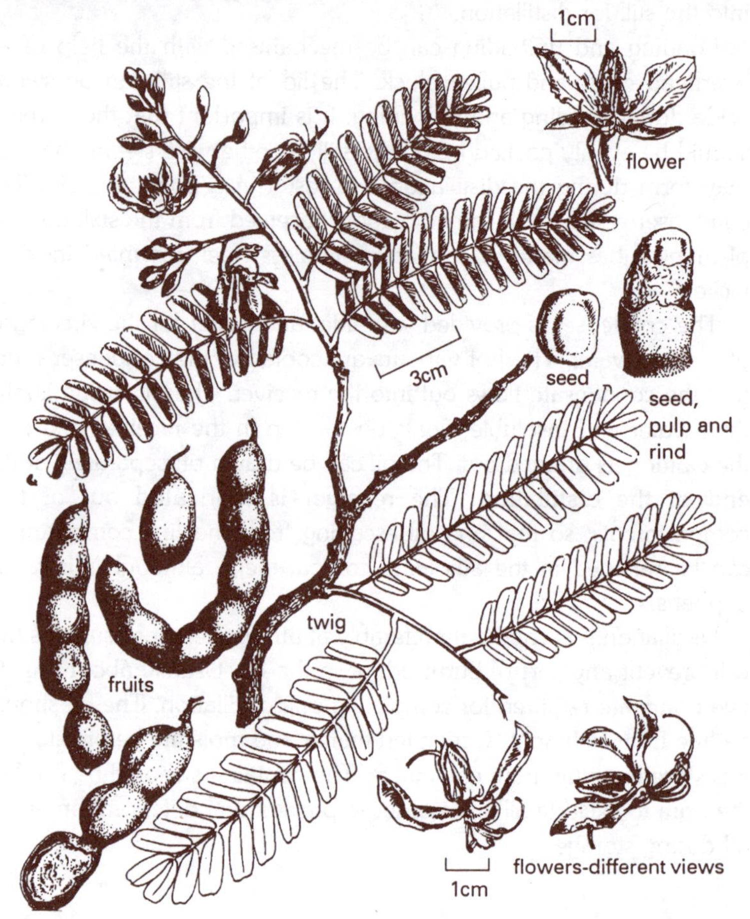

Plant Profile

Family	:	*Caesalpinaceae*
English name	:	Tamarind
Indian name	:	*Imli, Amli* (Hindi), *Hunisehannu, Huli* (Kannada), *Puli* (Malayalam), *Amli, Abdika* (Sanskrit), *Puli* (Tamil), *Chinthappandu, Chinta, Amlika* (Telugu)
Species	:	*Tamarindus indica* Linn.
Distribution	:	Africa, India, Myanmar
Uses	:	Condiment or acidulant

Tamarind (*Tamarindus indica* Linn.), commonly called *imli* in Hindi, is an important tree of semi-arid tropical conditions suited for Dry Land Horticulture. Every part of the plant is used for different purposes. The ripe fruit (pod) on an average comprises about 55% tamarind pulp, 33% seeds and 12% fibre. The tamarind pulp consists of—moisture-18.2%, free acid (tartaric)-9.8 %, combined acids-6.7%, total sugars as invert-38.2%, protein-2.8%, pectin 2.8%, cellulose residue (fibre)-19.4%, mineral matter-2.8%, calcium-0.17%, phosphorus-0.11%, iron-0.011%, calorific value-283 calories per 100 g, vit.A-100 I.U/100 g, niacin-0.2 mg per 100 g. Of the reducing sugars present, 70% is glucose and 30% is fructose (fruit sugar). Only a trace of sucrose or cane sugar is present. The pectin content in the pulp is of good quality having a jelly grade of 180-200. The seed consists of 30% testa and 70% endosperm. The testa contains 40% of water-solubles, 20% of which is tannin and colouring matter. The composition of the seed kernel is—moisture-8.1%, protein-17%, fat-7%, crude fibre-5.6%; non-fibre carbohydrates-65%, other components-5.4% and mineral matter-2.8%.

The fruit pulp is sweetish or acidic in taste and is used as condiment or more precisely as an acidulant like *amchur* (dried unripe mango pieces) or *kokum* (*Garcinia indica*). The pulp is also used for serving curries, *chutneys*, sauces, soups, *sambar, rasam,* etc., which are quite popular in the Southern parts of India.

The pulp is a carminative, laxative, appetiser, warming agent, tonic for the heart, antihelmentic, given as a infusion in bilousness and febrile conditions, cures *vata, pitta* and *kapha,* and wounds and fractures. A gargle of tamarind water is recommended for sore-throat. Because of its antiscorbutic properties, the pulp is used by sailors in place of lime or lemon juice. The tamarind kernel powder (TKP) is used as a sizing material in the textile and leather industries. The polysaccharide, jellose, present in TKP, forms jellies with sugar concentrates and is an excellent substitute for fruit pectins. The seeds are used as a source of carbohydrates for paper and jute products. The seeds also yield a fatty oil which is used in paints and varnishes. Tamarind wood is used for making agricultural implements, tool handles, wheels, mallets and rice pounders. The leaves, flowers, unripe fruits and bark also have many medicinal uses. The plant also acts as a good wind-break.

Original and distribution

Tamarind is probably indigenous to Africa, possibly also to some parts of South India. It is cultivated throughout India and Burma, as far north as the Jhelum. In India, tamarind is grown in Bihar, Orissa, parts of Maharashtra, Madhya Pradesh, Andhra Pradesh, Karnataka and Tamil Nadu. It extends northwards to the sub-Himalayan tract. About 2,50,000 Mt of fruits are marketed within the country and in America. Besides, tamarind is being exported to 46 other countries, of which Pakistan, Saudi Arabia, UAE, Japan and Germany are the major countries importing tamarind from India.

Description of the plant

Tamarindus indica Linn. (2n=24), of the family *Caesalpinaceae,* is a semi-evergreen tree, growing to a height of 24 m with a compact rounded crown with drooping branches which reach to within a few feet of the ground. The trunk is 7 m in girth, stout, and often becomes twisted with a grey, scaly bark. The leaves are alternate, paripinnate, 7-15 cm long, with pulvinous at the base of the petiole. The leaflets are in 10-20 pairs, opposite, entire, almost sessile, oblong, 1-2.5 × 0.5-1 cm, pale green beneath, with an unequal rounded base and a rounded or emarginated apex. The inflorescences are found in small, terminal, drooping racemes and are 5-10 cm long. The flowers are 2-2.5 cm in diameter, zygomorphic; there are two bractioles, reddish, boat-shaped with 3 petals, pale yellow, streaked with red and 1-1.5 cm long. There are three fertile stamens, alternating with staminodes, filaments that are 1 cm long united for about half their length, curving upwards under the petals; the anthers are transverse, reddish-brown and dehiscing longitudinally. The pistil is oblique, curving upwards, green, longer than the stamens, with a small clavate stigma. The pods are usually curved, oblong, 5-10 × 2 cm, constricted, scurfy brown with a brittle shell, 1-10 with seeds and indehiscent. The seeds are obvate, flatt-ended, brown about 1 cm long, joined to each other by tough fibres running through the brown sticky pulp which surrounds the seeds. Tamarind is probably entomophilous.

Climate

The tree is well-adapted to semi-arid tropical regions with alternate wet and dry seasons, but it can also be grown in areas with heavy rainfall if good drainage is provided. It requires a good rainfall, the average being 500-1500 mm during its growth, flowering and fruiting periods, i.e.. from June-October. It needs dry weather from January-March for maturity and ripening of fruits. It is grown in areas where the temperature reaches a maximum of 46°C and a minimum of 0°C. It can withstand drought, but is sensitive to frost. The optimum elevation for tamarind cultivation is 1000 m above MSL. The plants shed their leaves after the fruits are harvested and enter into a short period of dormancy before they starts flowering, in the month of March.

Soil

Tamarind is grown on a variety of soils, ranging from gravelly to deep alluvial soils. However, it thrives best on deep, loamy or alluvial soils. It can tolerate slightly saline and alkaline soils. The plant also adapts to poor soils.

Varieties

The existing cultivated types of tamarind are all of seedling origin. Being highly heterozygous, there are a few recognised varieties based on the pulp colour of the fruit. The local varieties can be classified as 'brown sour types' and 'reddish sweet types'. The popular ones among the local types are Bangalore Tamarind, Tumkur Tamarind, Hosur Tamarind, Krishnagiri Tamarind, Natham Tamarind, Nagarkoil Tamarind, Villupuram Tamarind, Ranchi Tamarind, etc.

A few improved varieties of tamarind are listed below.

PKM-1

A clonal selection from the gene bank, it is an early variety yielding 263 kg pods per tree with a pulp content of 39%. It can give 26 tonnes of pods per hectare if transplanted at a spacing of 10 m × 10 m. The variety is characterised by purple pigmentation in the terminal buds, the stems are dark brown in colour, fruits are borne in clusters of 4-5 and are slightly curved with brown pulp.

Urigam

This is another local type providing very long pods, with a fleshy sweet pulp. It bears 2-3 pods per bunch. The fruit yield is low when compared to PKM-1.

Pratisthan

This is an improved, sweet-pulped, high-yielding variety of tamarind released from the Marathwada Agricultural University, Parbhani.

NTI-19 (DST-1)

This variety is reported to yield well in the Dharwad region of Karnataka. The pods are large and the grafted plants start yielding in 4-5 years of planting. The average yield per plant is about 3 kg the 5th year.

Red tamarind
This variety is recommended for cultivation by the Tamil Nadu Agriculture University. The pulp is red in colour, which has a great demand in Arabian countries.

GKVK-6 and GKVK-33
These two are local types released by the Division of Horticulture, UAS, Gandhi Krishi Vignana Kendra, Bangalore.

Cultivation

Propagation
Tamarind is propagated by seeds, cuttings, gootee, grafts and budding. The seeds are sown in lines 20-25 cm apart and they germinate within a week. The seedlings should be irrigated, if necessary, at regular intervals. Abut 3-6-month-old seedlings are transplanted into the main field. However, 12-15-month-old seedlings of about 75-100 cm in height are preferred for planting.

Seedlings can also be raised in polyethylene bags, but true-to-type plants cannot be produced by seed propagation. The seedling trees are highly erratic in bearing and have an alternate-bearing tendency. They are also low in yield and have a long pre-bearing period. Hence, true-to-type plants are obtained by grafting (approach or soft-wood grafting) and budding (patch budding). Soft-wood grafting is done using 6-12 months old root-stocks.

Land preparation
The land is prepared well by ploughing, levelling and harrowing. The pits (90-100 cm^3) are dug at 10 m × 10 m or 13 m × 13 m for seedlings and at 6.5 m × 6.5 m for buddings and grafts. The pits are filled by mixing the soil with 15-25 kg FYM compost, 100 g of super phosphate and 1 ½ kg neem-cake per pit.

Planting
Planting is done during June-November. While planting, the seedlings/ buddings/ grafts are planted in the centre of the pit. They are provided with suitable support by staking.

Fertiliser application
The fertiliser recommendation for plants of different age groups is as follows:

Age of the plant		Per seedling tree (g)	Per budded/ grafted plant (g)
1st Year	N	73	36
	P	18	09
	K	68	34
2nd Year	N	146	73
	P	38	19
	K	136	68
3rd Year	N	219	109
	P	54	27
	K	204	102
4th Year	N	292	146
	P	72	38
	K	272	136
5th Year	N	365	182
	P	90	45
	K	340	170
6th Year	N	438	219
	P	108	54
	K	408	204
7th Year	N	511	255
	P	126	63
	K	476	238
8th Year	N	584	292
	P	144	72
	K	544	272
9th Year	N	657	328
	P	162	81
	K	612	306
10th Year onwards	N	730	365
	P	180	90
	K	680	340

The recommended dose of fertilisers should be applied to individual plants and forked into the soil in the basin during June-July. In addition to the inorganic fertilisers, it is recommended to apply FYM @ 20 kg, 30 kg and 50 kg per plant/year for plants in the age groups of 1-5 years, 6-8 years and above 9 years, respectively.

After-care

Regular irrigation should be given for the first two years, whenever there is no rainfall, till the plants are established. The field should be kept weed-free during the initial stages. The side shoots are removed from the seedling and the suckers from the budded/grafted plants for the first two years.

The land is ploughed and harrowed between the rows during monsoon during the first 5-8 years in case of seedlings and two years in case of budded plants/grafts. Ragi, jowar, horsegram, field bean and red gram can be raised as inter-crops. Basins should be prepared around each plant. Whenever no cover crops are grown, the inter-spaces are ploughed 2-3 times in a year. This not only helps to check the weeds but also conserves soil moisture. Tamarind is mostly a rain-fed crop, but providing 1-2 irrigations during the flowering and fruit-set periods will give better yields.

Inter-crops

Amla, Pomegranate, *Karonda*, *Anona*, etc., can be grown as inter-crops at a spacing of 5 × 2.5 m up to 10 years. Short-duration vegetables like cucumbers, pumpkin, ashgourd, watermelon and snakegourd could be grown as inter-crops in the basins of tamarind plants up to about 3-4 years.

Pests and diseases

Pests

The fruit borer is the major pest of this crop, which causes heavy crop loss. The damage can be minimised by spraying 40 g Carbaryl or 20 ml Malathion or 20 ml Endosulfan or 13 ml Monocrotophos in 10 litres of water, starting with the first week of August and ending in the third week of August. Tree hopper, mealy bugs and scale insects also attack this crop. They can by controlled by spraying Monocrotophos (0.1%). The other pests include: the beetles and larva of *Lasioderma serricorne, Pachymerus gonagra* and *Calandra linearis* which bore into the seeds; the larvae of *Tribolium castaneum, Argyroploce illepida* and *Virachola isocrates* which attack the fruits and destroy the seeds; and *Aspidiotus* sp, which feed on the sap of fruits. In addition, *Aspidiotus* sp., *and Chionaspis acuminatatricolor* feed on the sap of the young twigs and branches.

Xiphinema citri and *Longidorus elongatus* are the nematodes associated with the roots of the tree.

Diseases

Powdery mildew disease is noticed occasionally in tamarind. The disease can be controlled by spraying Wettable sulphur (0.2%) or Kelthane (0.1%).

The tree is affected by a number of root diseases, such as sap-rot (*Xylaria euglorsa*), brownish sap-rot (*Polyporus calcuttensis*) and white rot (*Trametes floccosa*). Among the other diseases, *Meliola tamarindi* causes sooty mould; *Pholiota gollani* has been recorded on the trunk of living trees; *Polystictus sarbadhikarii* on the dead trunk; and different species of *Hypoxylon* on the decorticated wood, branches and bark. A bacterial leaf-spot has been recorded in trees near Pune.

Harvesting and yield

The seedling plants start yielding in 8-12 years, whereas budded/grafted plants yield 4-5 years after planting. Harvesting is done during January-April, the average yield being 25 tonnes of pods per hectare (180-225 kg/fully grown tree). The budded/grafted plants yield 100 kg/tree /year.

Storage

Tamarind can be stored after removing the seed. Tamarind pulp absorbs moisture from air, thereby reducing the sugars and amino acids present in the pulp which get changed and give a dark colour to the pulp. Therefore, it should be stored in air-tight containers. The pulp and salt can be mixed and packed in palmyrah leaves before sending it to the market. In the West Indies, sugar is mixed with the pulp and it is being exported to Europe.

Value-added products of tamarind

The CFTRI, Mysore in Karnataka has developed a process for the manufacture of tamarind concentrate which is free from fibre, seeds, foreign matter, etc, and is very hygienic. Recently, there has been a growing demand for it from abroad. This concentrate is almost of a jam-like consistency. It is easily dissolved in hot water at the time of its use and is being exported to other countries.

The tamarind seed consists of 30% testa (outer skin) and 70% endosperm. The testa contains 40% of water-solubles, 80% of the seed kernel is—moisture-8.1%, protein-17%, fat-7%, crude fibre-5.6%, non-fibre carbohydrates-65%, other components-5.4% and mineral matter-2.8%. A powder made from the kernel of tamarind seeds is also being exported from India.

The only factors in tamarind pulp and kernel which have food value are the carbohydrates and protein. They also have small amounts of vitamins, carotene, vit. B_1 and nicotinic acid.

An integrated process has also been developed at the CFTRI, Mysore, for the manufacture of tamarind powder, beverages, pectin, tartarates and ethanol (alcohol) from tamarind pulp.

Thyme

Plant Profile

Family	: *Lamiaceae; Labiatae*
English name	: Thyme
Indian name	: *Banajwain* (Hindi), *Thyme* (Kannada)
Species	: *Thymus vulgaris* Linn.; *T. satureioides* Coss; *T. serpyllum* L.; *T. zygus* L.
Distribution	: India, Asia, Australia, Canada, Europe, France, Germany, Greece, Italy, Morocco, Spain, USA
Uses	: Cosmetics, culinary purposes, drugs, insecticides, perfumery

hyme (*Thymus vulgaris* Linn.), commonly called 'Common Thyme' or 'Garden Thyme', belonging to the family *Lamiaceae; Labiatae* is an important herb spice used by man since ancient times. The thyme herb contains—moisture-7.1%, protein-6.8%, fat-4.6%, crude fibre-24.3%, carbohydrates-44.0%, mineral matter-13.2%, calcium-2.1%, phosphorus-0.20%, iron-0.14%, (mg/100 g), sodium-0.08%, potassium-0.9%, vitamin-A-175 IU, vitamin-B_1-51 mg, vitamin B_2-0.4 mg, vitamin-C-12.0 mg, and niacin-4.9 mg, calorific value-340 calories. On distillation, the herb yields the Red Thyme oil of commerce. Apart from oil, the other constituents of thyme are triterpinaoid saponins, flavones, ursolic acid (1.5% in tops), caffeic acid, bitter principle, tannins and resins. In addition, there are several other wild species which yield volatile oil. They are *T. serpyllum* L., *T. zygus* L. (white thyme) and *T. satureioides* Coss and Bal. Thymol, linalool and linalyl acetate are the major compounds present in *T.zygus* and *T. serpyllum. T. zygus* and its var. *gracitis* are also the source for genuine thyme oil. While, thymol, burneol and α-terpineol are the major componds in *T. satureioides*, the oil called the 'oil of thyme' in commerce is a colourless, yellow or red liquid, having a characteristic pleasant odour, and a pungent taste. The composition and also the quality of the oil is reported to vary in different geographical areas. The concentration of the active principle (Thymol) is high (60%) in Spanish oil and is low (20-35%) in French and Moroccon oils, whereas carvacrol is present in minor amounts. The oil also contains amylalcohol, β-y-hexanol, an unidentified terpene with a carrot-like odour, β-pinene, camphene, P-cymene, α-terpinene, linalool, l-borneol, geraniol, caryophyllene, an unidentified sesquiterpene and a sesquiterpinic alcohol, 4-terpineol and *trans*-4-thujanol. The other additional components present in traces are sabinene, verbenene, 1-octen-3-ol, methyl thymol, verbenone, α-muurolen, δ-cadinene and P-cymen-8-ol. The seeds yield 37% of a drying oil which contains mainly linolenic, lindeic and oleic acids. In addition, hydroxy linolenic acid is also present. It is one of the very frequently used spices, in the fresh or dried form, for several European delicacies.

The leaves and flowers find use as food flavourants and in the seasoning of various food items, especially fish and meat preparations and for garnishing. Medicinally, the leaves are said to possess laxative, stomachic and tonic properties, good for the kidney and eye and are blood purifiers. The herb also has an insect-repellent property. It is also used in perfumery and in liquor distillery. The dried leaves and floral tops constituting the thyme of commerce, is known as Thymi Herba in pharmacy, which contains not more than 3% of stems over

1 mm in diameter and 2% of other organic matter and yields 4% acid insoluble ash. The shoot extracts of flowering thyme plants have anti-bacterial action against *Micrococcus pyogenes* var. *aureus* and *Escherichia coli*. The herb, which has a pungent taste, is reported to possess antiseptic, antihelmintic, expectorant, carminative, diuretic, alexiteric, emmenogogue and sedative properties; it is good for liver complaints, pain in the spleen, liver or chest; it is useful in cold, asthma and bronchitis. It thins the phlegm and blood.

Thyme oil has antiseptic, antispasmodic and carminative properties. It is used in mouth-washes and gargles. Formulations containing thyme oil are available for the treatment of whooping cough and bronchitis. The oil is used as a diffusible stimulant in cases of collapse. It is mixed with olive oil and is used as a rubefacient and counter-irritant. It is also used in veterinary medicine. Thyme oil is used in soaps, perfumes and for flavouring food products such as meat, sausages, sauces and canned food. The oil has antifungal and antihelmintic properties and is used as an intestinal antiseptic in treating hookworms. It is also used as a cure for many fungal infections of the skin. The seeds are given as vermifuge. The infusion of the seeds is also used in skin eruptions or diseases. The seed oil has a potential use as a superior drying oil. The plant can be grown for ornamental edging along the paths in gardens, and is also suitable for rockeries.

Origin, distribution and production

Thyme grows wild in almost all the countries bordering the Mediterranean and also over much of Asia and in parts of Central Europe and grows the best in the hills. It is found in the Western temperate Himalayas, from Kashmir to Kumaon, between altitudes of 1,525 m and 4,000 m. Thyme is grown in Europe, Australia and North Asia. It is now cultivated in France, Germany, Spain, Italy, Greece and North Africa, Canada and the USA. In India, thyme may be cultivated in the Himalayas from Kashmir to Kumaon.

Among the countries producing thyme oil, Spain stands first, followed by France, Morocco and the Mediterranean countries. The bulk of the world demand for this oil is met by Spain and Turkey.

The major types of thyme are French thyme, Spanish thyme, American/Californian thyme, etc.

Description of the plant

T. vulgaris is a low, evergreen, perennial under-shrub reaching a height of 20-30 cm whereas, the wild thyme (*T. serpyllum* L) creeps on the soil surface and has broad leaves with a weaker odour. The roots are fairly robust and the stems are branched. The former has oblong-lanceolate, sessile leaves of 10 mm × 3 mm size, with orange-brown, glandular dots, which are coriaceous. The young leaves are slightly woolly. The flowers are small, purplish or bluish to almost white, united in spikes at the tip of the branches and have a bilabiate, tube-like calyx and a bilabiate, tubular corolla with a 3-lobed lower lip. The fruit is a nutlet, brown, 4-sectioned, smooth and is found in the remains of the calyx. The entire plant is aromatic.

Climate

A warm climate is best suited for this crop. It can be grown both in the hills and the plains at an elevation of 1500-4000 m from MSL. Hilly regions are best suited for its cultivation. However, the plants are susceptible to frost.

Soil

The plant prefers a light but fertile and calcareous soil for its good growth and oil content. On heavy, wet soils the aroma of the leaves is less and there are more chances that the plants may dry up soon.

Season

The seeds may be sown in nursery during April. Late summer is the time for transplanting the seedlings or planting rooted cuttings.

Cultivation

Propagation

Thyme can be propagated by seeds and vegetatively by the division of old plants or by cuttings or by layering of the side shoots in March or April.

The seeds are sown directly in rows or are sown in well-prepared nursery beds in good soils. Good soil is preferred for nursery, because the seedlings are very small and remain inconspicuous for several weeks after germination.

Land preparation

The land is prepared well by repeated ploughing or digging and brought to a fine tilth. Then, the land is divided into plots of convenient sizes.

Sowing/planting

When the seeds are sown directly, they can be sown in rows 90 cm apart and later, when the seeds germinate, the seedlings may be thinned out to 30-45 cm within the row. While planting the seedlings or rooted cuttings or layers, etc, they should be planted 30-45 cm apart in rows that are 60 cm apart. A light irrigation is usually provided after planting.

Manures and fertilisers

A light dressing of FYM is given in autumn. Fertilisers are applied @ 100:40:40 kg NPK/ha. A top-dressing of N and FYM in spring is reported to promote the formation of numerous leafy shoots.

Irrigation

No information is available on this aspect. Since the crop is planted late in summer, it requires frequent irrigations during the dry period.

Inter-culture

Inter-culture operations and weeding are done at regular intervals to encourage the good growth of the plants. In the hills, in order to avoid frost injury to plants during winter, mulching is done.

Pests and diseases

The plant is not attacked by any pests of a serious nature, but wilt disease is a major problem in this crop. The disease can be controlled by improving the phytosanitation and by the use of suitable fungicides like Blitox or Dithane M-45, etc, @ 0.3% concentration for soil drenching.

Harvesting and yield

The leaves and flowers, which are used for culinary and medicinal purposes, are harvested five months after sowing/ planting. The leaves and flowers are plucked from the plants or shoots of about 15 cm are cut off from the plants, dried in the shade or in a dryer immediately after harvest and stored in air-tight containers to prevent the loss of flavour. The dried leaves are curled, are brownish-green in colour, and usually not longer than 6-7 mm. The dried shoots may also be powdered and packed. Under favourable conditions, the yield of dry herb is around 1,100-2200 kg/ha. The yield is comparatively low during the first year. The plants become woody, and their replanting becomes necessary after three or four years. For extracting the oil, the fresh herb is collected on dry days. The herb is collected at the stage when it just start flowering. At the time of collection, the lower portions of the stem, along with any yellow or brown leaves, need to be rejected.

Distillation of oil and oil yield

The oil is distilled from the fresh flowering tops by steam distillation. The herb contains about 2% essential oil and the oil recovered by distillation is about 21 kg/ha.

Value-added products

Volatile oil, oleoresin and thymol are the value-added products of thyme.

Turmeric

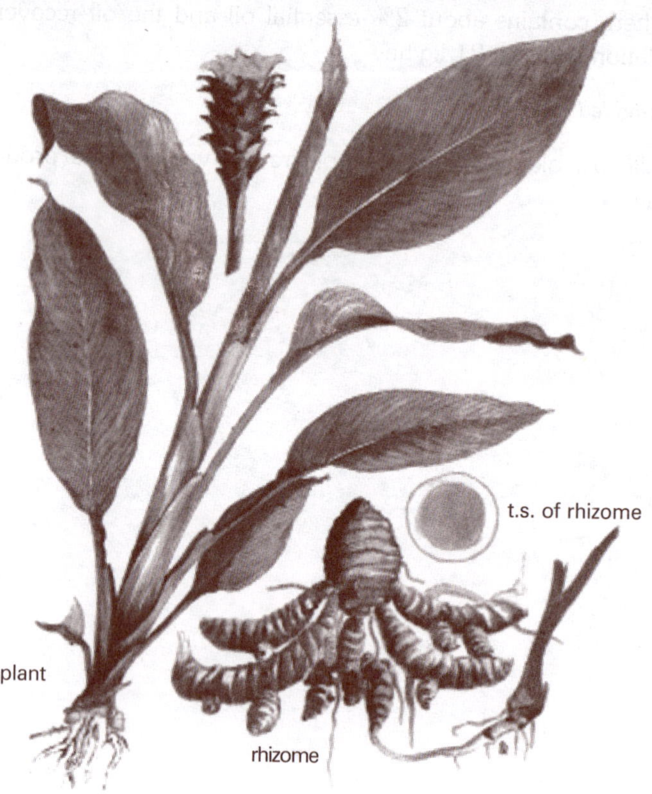

Plant Profile

Family	:	*Zingiberaceae*
English name	:	Turmeric
Indian name	:	*Haldi* (Hindi), *Arishina* (Kannada), *Manjal* (Malayalam), *Haladi, Haridra, Harita* (Sanskrit), *Manjal* (Tamil), *Pasupu* (Telugu).
Species	:	*Curcuma domestica* Val., *C. longa*, C. rotunda, C. amada.
Distribution	:	India, China, Malaysia, Sri Lanka, Pakistan, Indonesia, Haiti, Jamaica, Peru, Bangladesh, El Salvador, Taiwan
Uses	:	Spice, food colourant

urcuma domestica Val., belonging to the family Zingiberaceae, is the plant from which the turmeric of commerce (dried rhizomes) is obtained. Although *C. domestica* is the correct botanical name for turmeric, *C. longa* is sometimes used in literature to describe the finger rhizome and *C. rotunda* for the bulbous central rhizomes. Cured turmeric comprises of moisture-5.8%, protein -8.6%, fat-8.9%, carbohydrates-63.0%, fibre-6.9%, mineral matter-6.8%, calcium-0.2%, phosphorus-0.26%, iron-0.05%, sodium-0.01%, potassium-2.5%, vitamin-A-175 IU/100 g, vitamin-B_1-0.09 mg/100 g, vitamin-B_2-0.19 mg/100g, vitamin-C-49.8 mg/100 g, and niacin-4.8 mg/100 g. It has a calorific value or food energy of 390 calories/100 g. The colouring principle in turmeric is curcumin. Turmeric is an important spice among rice-eating people. Curry powder usually contains about 24% of turmeric powder, with other usual ingredients like coriander, cumin, cardamoms, fenugreek, chillies, ginger, pepper and dill seeds, etc.

In addition to its use as a spice, it has other uses too. Turmeric is a natural source of yellow dye used for dying cotton, silk or wool without a mordant. Turmeric powder and water are used as cosmetics. Turmeric is considered a carminative, tonic, blood-purifier, vermicide and an antiseptic. It is used in folk medicine for intestinal disorders, worms, anaemia, measles, asthma, sore throat, cough and cold, diabetes, sprains, skin disorders, etc., both externally and internally. It is used as a colouring matter in the pharmacy, confectionery and the food industry. Yellow turmeric paper can be used as a test for alkalinity, which turns it brown. It is also used as protection against 'Charam' and considered as a symbol of good fortune. Turmeric oil and oleoresin are also used to impart flavour in food and perfume industries. The essential oil of turmeric is antiseptic. It is used in treating gall stones and gall complaints. The anti-microbial properties of essential oil from turmeric are reported against pathogenic bacteria and fungi.

Origin and distribution

India or China or the Indo-Malayan region is the home of turmeric (a diploid form) from which *C. domestica* is likely to have evolved by a process of triploidy, and subsequent continued selection and cultivation by vegetative propagation. But now, it is not grown in a truly wild state. The crop has now become widely distributed throughout the tropics, but its cultivation as a spice is largely confined to India, Sri Lanka, China, Pakistan, Indonesia, Haiti, Jamaica, Peru, Bangladesh, El Salvador and Taiwan.

Area and production

India is the largest producer and exporter of turmeric in the world having an area of 1,08,000 hectares under turmeric cultivation. Among the turmeric growing states, Andhra Pradesh stands first with 30% of the total annual national production of 2,95,000 tonnes, followed by Maharashtra, Tamil Nadu, Kerala, Karnataka, Orissa, Assam, Bihar and West Bengal. Out of the total production, 12,588 tonnes is being exported to Libya, Moracco, USA, Iran, Japan, UK and Singapore, earning nearly Rs. 171.6 million as foreign exchange.

Description of the plant

Turmeric is an erect perennial herb, but is grown as an annual. The primary tuber at the base of the aerial stem is ellipsoidal, bearing many rhizomes, straight or a little curved, with secondary branches in two rows, which may have tertiary branches, the whole forming a dense clump. The rhizomes have a distinctive taste and smell. They are brownish and scaly outside and bright orange in colour inside. The leafy shoots are erect, less than 1 m in height, bearing 6-10 leaves with the leaf sheaths forming a pseudostem. The thin petiole is rather abruptly broadened from the sheath. The lamina is lanceolate, acuminate and thin, dark green above and pale green beneath, with ellucid dots. The inflorescence is a cylindrical spike, terminal on the leaf shoot, with the scape partly enclosed by the leaf sheaths. The bracts are adnate for less than half their length and are elliptic, lanceolate and acute. The upper sterile bracts are white or white streaked with green, pink-tipped in some cultivars, fading to light-green bracts below. The flowers are borne in cincinni of two in the axils of the bracts, opening one at a time, and are thin-textured and fugacious. The calyx is short, unequally toothed and split nearly half-way down one side. The corolla is tubular at the base, being the upper half-cup-shaped with three unequal lobes inserted on the edge of the cup lip; it is whitish, thin and translucent with a hooded dorsal lobe. There are two lateral staminodes, elliptic-oblong, which are the hood of the dorsal petal. The filament of the stamen is short and broad, united to a versatile anther at about the middle of the parallel pollen sacs. The cylindrical stylodes are about 4 mm long. The ovary is inferior and trilocular with a slender style passing between the anther lobes and held by them. The fruits are few, if ever produced.

Climate

Turmeric prefers a warm and humid climate and can be cultivated in most of the tropics and subtropics, provided that the rainfall is adequate or irrigation facilities are available. An annual rainfall of 100-200 cm is ideal. It can be grown from sea level to 1220 m above MSL at a temperature range of 15-35°C. High temperature and low humidity cause slow emergence of the pseudostem and leaves.

Soil

Turmeric can be grown in various types of soil. However, well-drained loamy or alluvial soils, rich in good organic matter, in a pH range of 5.00 to 7.5 is optimum for the crop. The crop cannot withstand water-logging. Gravelly, stony and heavy clay soils are unsuitable for the crop, because of their interference with the development of rhizomes.

Varieties

Several cultivars are distinguished in our country by the names of localities in which they are grown. The widely grown cultivars in Karnataka are described below.

Kasturi
The core of the rhizome is pale yellow to white in colour. It emits a sweet fragrance after curing.

Mundaga
The rhizomes are large and thick with many fingers.

Balaga
The corns are not so thick, and the number of fingers are also fewer.

Yalachaga
The rhizomes are small and few in number but have a larger number of fingers.

The other popular cultivars grown in different states are Kesari, Kuvvur, Amrutapani, Kothapeta, Duggiala, Tekurpet, Mydukur, Aromoor, Vontimitta, Sugandham, Nandyal, Avanigadda (Andhra Pradesh); Erode, Salem (Tamil Nadu); Aleppey, Mannuthy Local (Kerala) Shillong, Tall Karbi (Assam); Rajpuri, Evaigon (Maharashtra); Duhgi, Jobedi, Katingia (Orissa); and Gorakhpur (Uttar Pradesh).

Some of the improved varieties of turmeric and their important characteristics are given in the table on pg 414.

Variety	Average yield of fresh rhizomes (t/ha)	Duration (days)	Dry recovery (%)	Cur-cumin (%)	Oleore-sin (%)	Recommended for (place)
BSR-1	30.70	285	20.5	4.20	4.0	Tamil Nadu
BSR-2+	37.5	240-245	20.0	3.75	-	Tamil Nadu
Co 1	30.50	285	19.5	3.20	6.7	Tamil Nadu
IISR Prabha	37.47	205	19.5	6.53	15.0	Kerala and Tamil Nadu
IISR Prathiba	39.12	225	18.5	6.21	16.2	Kerala and Tamil Nadu
Kanti	7.90**	240-270	-	7.18	-	-
Krishna	9.20	240	16.4	2.80	3.8	Maharashtra
Megha Turmeric-1	23.0	-	-	-	-	Meghalaya
Rajendra Sonia	4.80	225	18.0	8.40	-	North Bihar
Ranga	29.00	250	24.8	6.30	13.5	Orissa, Tamil Nadu, Andhra Pradesh and Kerala
Rasmi	31.30	240	23.0	6.40	13.4	Orissa, Tamil Nadu, Andhra Pradesh and Kerala
Roma	20.70	250	31.0	9.30*	13.2	Tamil Nadu, Himachal Pradesh, Andhra Pradesh and Kerala
Sobha	5.74**	-	-	7.19	15.9	-
Sudarsana	28.80	190	12.0	5.30*	15.0	Kerala and Andhra Pradesh
Sugandham	15.00	210	23.3	3.10	11.0	Gujarat. Tolerant to pests
Suguna	29.30	190	12.0	7.30*	13.5	Kerala and Andhra Pradesh
Suroma	20.00	253.0	26.0	9.30	13.1	Orissa, Tamil Nadu and Himachal Pradesh
Suvarna	17.40	200	20.0	4.30	13.5	Kerala, Karnataka and Andhra Pradesh

* Provisional, **dry yield   +Resistant to scale insects

The variety RH-10, released as a clonal selection, matures in 288 days and is resistant to leaf-blotch. Pant peetabh is also an improved variety. The variety PCT-13 has been identified as resistant to the rhizome-rot disease. The genetic resources holding of turmeric in India is 1136.

Cultivation

Propagation

Turmeric is generally propagated by using whole or split mother rhizomes. The synseed technology as well as micro-propagation protocols for rapid multiplication have been developed for turmeric.

Land preparation

The land should be ploughed five to seven times thoroughly to a depth of 22-25 cm, followed by two to three plantings, and all the weeds should be cleared away. Well-rotten FYM or compost at the rate of 40 t/ha should be mixed in the soil. After land preparation is completed, the field is laid out either into beds of 1-1.5 m width, 15-25 cm height and of convenient length, or by forming ridges and furrows.

Planting

Healthy and disease-free rhizomes should be stored during the harvest in underground pits. By the beginning of April, when the seed rhizomes start sprouting, they should be taken out from the pits for planting. Care should be taken to protect the sprouting buds from mechanical injury. The planting should coincide with the commencement of the monsoon. Whole or split mother rhizomes as well as finger rhizomes (25-30 g each) are used as planting material (@ 2.5 t/ha). The seed rhizomes are treated with 0.3% Dithane M-45 and 0.1% Quinalphos prior to storage and again at the time of sowing to minimise both rhizome-rot as well as rhizome scales. The recommended spacings are 30 × 25 cm for the beds. In case of ridges and furrows, the ridges should be 45-60 cm apart and the rhizomes are planted at 25 cm along the side of the ridges.

Manures and fertilisers

A basal dose of FYM @ 40 t/ha is applied at the time of land preparation. Under rain-fed conditions, a fertiliser dose of 30:30:60 or 60:30:90 kg of NPK per hectare is recommended, while in irrigated conditions the fertiliser requirement is 60:30:90 kg of NPK. The fertilisers are applied in three split doses. The whole of P_2O_5 and half the dose of K_2O are applied as a basal dose and N is supplied in two split doses after 45 (60% of the recommended N) and 90 (40% of the recommended N) days after planting, along with the remaining half dose of K_2O.

Irrigation

Turmeric can be grown either as a rain-fed crop (Kerala, Orissa and North-Eastern states) or as an irrigated crop (Andhra Pradesh and Tamil Nadu), depending on the location. In the case of the irrigated crop, depending on weather and soil conditions, 15-40 irrigations may be necessary at 7-10 days' intervals.

Mulching

Immediately after planting, the field is mulched with green leaves at 15 t/ha. This is repeated after 50 days of planting when the rhizomes have fully sprouted.

Inter-culture

Weeds should be controlled manually or by the use of herbicides. Usually, weeding is done thrice, at 60, 120 and 150 days after planting, depending upon the weed intensity. Early weeding may be avoided by the use of 2,4-D as a pre-emergent herbicide. Earthing-up is done 60 days after planting.

Inter-cropping and crop rotation

Turmeric also comes up well under sparse shade. It can be grown as an inter-crop in coconut gardens like ginger or as a mixed crop with red gram, chilli, colocasia, vegetables, maize and ragi. In wetlands, it can be rotated with paddy, sugarcane, banana or vegetables. In garden lands, rotation is done with rain-fed paddy or mixed with red gram, maize, groundnut and sunflower.

Pests and diseases

Pests

Shoot borer (*Conogethes punctiferalis*)

As in ginger, the shoot borer is an important pest of turmeric. The larvae bore into the pseudostems and feed on the growing shoots resulting in yellowing and drying of the infested shoots. The presence of bore holes in the pseudostem through which frass is extruded and the withered central shoot (dead heart) are characteristic symptoms of pest infestation. Spraying Monocrotophos (0.075%), or Dimethoate (0.1%) or Quinalphos (0.025%) during July-October at monthly intervals controls the pest. *Hexameris* sp., hymenopteran parasitoids have been identified as bio-control agents for this pest.

Leaf roller (Udaspes folus)
The pest is abundant in the field during August-October. Spraying Carbaryl (0.1%), Dimethoate or Phosphamidon (0.05% each) can control the pest.

Rhizome scale (Aspidiella hartii)
The pest infests the crop both in the field and in storage. In the field, the plants wither and dry during severe infestation. In storage, the shrivelling of the buds and rhizomes is observed due to pest attack. Dipping the seed rhizomes of turmeric in Quinalphos (0.025%) or Fenitrothion (0.05%) for 5 minutes after harvest and before planting/ storage + spraying of Phosphomedon (0.05%) in the field controls the pest.

The larvae of the crane fly (*Libnotes punctipennis*) and rhizome fly (*Calobata albimana*) have been reported to bore into the rhizomes.

Nematodes (Meloidogyne incognita)
The affected plants show stunted growth, yellowing, marginal and tip drying of leaves, reduced tillering and galling, and rotting of the roots. A high population of *M. incognita* causes yellowing and severe stunting and withering in large patches. The plants die prematurely leaving a poor crop stand at harvest. The infested rhizomes lose their bright yellow colour.

Radopholus similes
The infested plants show a tendency to aging and dry faster than healthy plants. The infested rhizomes are a yolk-yellow colour compared with the golden-yellow colour of the healthy rhizomes and have shallow, water-soaked, brownish areas on the surface. The roots show rotting and most of the decayed roots are devoid of the cortex and stylar portions. The application of Thimet @ 10-15 kg/ha or Phorate 10 kg/ha at the time of field prepration or Aldicarb and Carbofuran (1-3 kg a.i/ha) 3 weeks after planting controls the nematode infestation. The varieties Kodour and Chayapasupu are resistant to root-knot nematodes.

The other species of nematodes reported to cause damage to this crop are *Longidorous elongatus, Xiphinema elongatum, Hoplolaimus seinhorstii, Helicotylenchus multicinctus, Tylenchorhychus martini, Pratylenchus delattrei* and *Rotylenchulus reniformis*.

Diseases

Leaf blotch
This disease is caused by *Taphrina maculans*. The disease starts as small, scattered, oily-looking, translucent spots on the lower leaves

when the plants are in the 3-to-4-leaf stage. The leaf-spots later turn dirty yellow and deepen to the colour of gold and sometimes to a hay shade. The adjacent individual leaf-spots of 1-2 mm in diameter coalesce forming reddish-brown blotches and lead to varying degrees of leaf blight. Foliar sprays with Dithane Z-78 (0.2%) or Dithane M-45 (0.25%) at 15-day intervals controls the disease.

Leaf-spot

This is caused by *Collectotrichum capsici*. The infection is usually confined to the leaf-blades and occasionally extends to the leaf sheath. The disease appears as elliptic to oblong spots of variable size on both surfaces, but more on the upper surface. These spots gradually enlarge in size and coalesce to cover the entire leaf area or form large necrotic patches. The spots vary from 4-5 cm in length and 2-3 cm in breadth. A fully-developed spot has a grayish-white centre with a brown margin surrounded by a yellowish colour in the shape of a defined halo. In severe infections, the leaves dry up and defoliate. Spraying of Bordeaux mixture (1%) or Dithane M-45 (0.25%) controls the disease.

Rhizome-rot

This disease is caused by *Pythium graminicolum* and *Pythium aphanidermatum*. This is an important disease, prevalent in all turmeric-growing areas. The disease appears in patches in the field. The leaves of the infected plants become yellow and exhibit gradual drying along the margins. Water-soaked lesions develop on the pseudostem and become soft to the touch. The root system is adversely affected, as the roots rot completely. As the disease progresses, the infection gradually spreads. The infected rhizomes turn brown in colour. In the case of a severe attack, the yield is reduced considerably. Planting healthy rhizomes and seed treatment with Dithane M-45 (0.25%) + Carbendazim (0.1%) for 60 minutes and drying under the shade for 24 hours before sowing/storage are the recommended control measures. The seed rhizomes should be dipped in Blitox 50 (0.3%) before storage or sprayed with Carbendazim or Bordeaux mixture at the first apperance of the disease symptoms to control the disease. *Trichoderma harziarum*, *Trichoderma hamatum* or *Trichoderma viride* can be used as bio-control agents for this disease.

Harvesting and yield

Turmeric comes to maturity in 7-10 months after planting, depending upon the variety. The noticeable maturity indices are the complete yellowing of the leaves and the drying up of plants including the pseudostem. Usually the land is ploughed or dug with a spade and the whole clump is lifted with the plant, including the base of the stem. While doing so, it must be ensured that the rhizomes are not injured. The leafy tops are then cut off, the roots removed, and the rhizomes thoroughly washed with water to remove the adhering earth. The fingers (daughter rhizomes) are separated form the rounds (mother rhizomes) manually.

The yield in turmeric has been observed to vary very greatly. However, a yield of about 20-25 t of raw turmeric per hectare in the case of irrigated and 6-9 t/ha in the case of rain-fed crops is considered satisfactory.

After keeping aside the required quantity of rhizomes for planting, the rest of the bulk is cured and polished.

Processing

The quality of cured turmeric is assessed on the basis of several factors, such as the curcumin content, the organoleptic character, the general appearance, size and physical form of the rhizome.

The method of curing turmeric consists of boiling or steaming the prepared fresh rhizomes in water, drying in the sun and finally peeling or polishing. The following steps are involved in the processing.

Cleaning

The harvested rhizomes are cleaned of other extraneous matter adhering to them and the roots are removed. Only the good fingers separated from the rhizomes are used for curing.

Boiling

The process is aimed at destroying the vitality of the fresh rhizomes and to obviate the raw odour, to reduce the drying time, to gelatinise the starch and give a more uniformly coloured product.

The fingers and mother rhizomes are separated and cured separately, since the mother rhizomes take a little longer time to cook. The boiling is done in galvanized iron vats or pans or other containers of a suitable size. A pan of 1 m length, 0.62 m breadth and 0.48 m depth is found to be suitable. The cleaned fingers (approximately 50 kg) are placed in a trough, which is 0.9 m × 0.55 m × 0.4 m in size and made of GI sheets with an extended parallel handle. This perforated trough in which the fingers are placed is then immersed in the pan.

An alkaline solution prepared by dissolving 100 g sodium bicarbonate ($NaHCO_3$) or sodium carbonate (Na_2CO_3) in 100 ml of water is poured into the trough so as to immerse the turmeric fingers. The whole mass is boiled till the fingers become soft. This can be tested by piercing it with a wooden needle. If the needle passes through the rhizome without much resistance, the softness is confirmed. Usually the boiling is continued for 45 to 60 minutes until frothing occurs and white fumes appear emitting a characteristic turmeric odour.

The cooked fingers are taken out of the pan by lifting the trough and draining the solution into the pan. The alkalinity of the boiling water helps the development of an orange-yellow tinge to the core of the turmeric. The drained solution in the pan can also be used for cooking another lot of turmeric along with the fresh solution prepared for the purpose.

Usually, the cooking of turmeric should be done within 2 or 3 days after harvesting. The colour and aroma of the final product depends greatly on the stage at which the boiling is stopped. Therefore, only skilled persons should be employed for this job. Both delayed cooking and over-cooking will spoil the quality of the final product.

Drying

The cooked fingers are sun-dried by spreading them out on a bamboo mat or on the drying floor for 5-7 days until the moisture reduces to 8-10%, which is fit for storage and sale. The sun-drying of sliced turmeric in particular gives a slight surface bleaching effect on the colour. During the night-hours, the material should be heaped and covered over to protect it from dew. It takes 10-15 days for the rhizomes to become thoroughly dry, when they become quite hard and brittle, and break (with the fingers) with a metallic sound.

Polishing

This operation is done in order to smoothen the rough and hard outer surface of the boiled and dried turmeric. It also improves the colour of the product from a dirty-brown to a bright-yellow. There are two types of polishing; they are Hand polishing, and Machine polishing.

Hand polishing

This method is simple and involves rubbing the turmeric fingers wrapped in several folds of gunny cloth on a hard surface with the hands or by trampling them under the feet or shaking the rhizomes mixed with stones in a long narrow gunny bag or in a bamboo basket. The improved method is through the use of a hand-operated barrel or drum mounted on a central axis, the sides of which are made of expanded metal mesh. When the drum filled with turmeric is rotated, the abrasion of the surface against the mesh as well as by mutual rubbing against each other will polish them effectively.

Machine polishing

The machine used for polishing consists of a hexagonal wooden drum, mounted on a central axis and rotated by power. These power-driven drums are gaining popularity because a higher degree of polishing (smoothness) can be attained by this method. The capacity of these drums is also high compared to the hand-operated ones and, hence, the output is also higher.

Colouring

The colour of turmeric always attracts buyers. Thus, giving the required colour externally to the rhizome is an important step in the processing of turmeric. For this, the boiled, dried and half-polished fingers are placed in a basket, which is shaken continuously in a prepared emulsion. They are later sun-dried. The colour emulsion comprises of alum (0.04 kg), turmeric powder (2 kg), castor seed (0.14 g) or sodium bisulphate (30 g) and concentrated HCl (30 ml).

Grading (Indian standards)

Turmeric is included in the list of spices which must have an 'Agmark' grading before they can be exported from India. Three grades of finger turmeric, two of bulb turmeric and one grade of powdered turmeric are specified, as follows:

a. Finger turmeric, other than the Alleppey variety is sub-graded into 'Special' 'Good' and 'Fair'.
b. Alleppey finger turmeric is sub-graded into 'Good' and 'Fair'.
c. Rajapore finger turmeric is sub-graded into 'Special'. 'Good' and 'Fair'.

The maximum limit for extraneous matter in the prime sub-grades of the above grades of whole turmeric is 1.0%. In the case of Alleppey finger turmeric, which is exported to the United States, the content of extraneous matter according to the specifications of the American Spice Trade Association is usually less then 0.5%. In the case of the turmeric powder, the characteristics are more exacting The maximum percentage limits are specified for moisture, total ash, acid insoluble ash and starch.

The ASTA (1) and FDA (2) chemical and physical specification of turmeric are as follows: (1) Whole dead insects by count 3; mammalian excreta (mg/lb) 5; other excreta (mg/lb) 5; mould (% by weight) 3; insect defiled (% by weight) 2.5 (2) volatile oil/ curcumin (% min) 5; moisture (% max) 10; ash (% max) 8; acid insoluble ash (% max) 1.0, average bulk index (mg/100 g) N/A.

Value-added products

The value-added products of turmeric are curcuminoids, dehydrated turmeric powder, oil and oleoresin.

Estimation of curcumin content

The curcumin content can be estimated adopting the method reported by Manjunath *et al.* (1991). For this, the cured rhizomes should be ground to a fine powder and 0.1 g of the powder should be extracted with 40 ml of distilled alcohol by refluxing over a water-cooled condenser for 2 ½ hours. The extract should then be filtered into a 100 ml volumetric flask and made up to volume with alcohol. An aliquot of 5 ml should be transferred to a 100 ml volumetric flask and again the volume is made up. The absorbance of the solution can be measured at 425 nm with alcohol as the blank. The curcumin content in the sample can be calculated using the absorbance value of a standard solution of curcumin (0.00025 g/100 ml has an absorbance of 0.42), adopting the following formulae:

$$\text{Curcumin content (per cent by weight)} = \frac{0.00025 \times \text{absorbance of sample} \times 100 \times 100}{\text{absorbance of standard} \times \text{weight of sample} \times 5}.$$

Curcumin is highly aromatic with a musky odour and a pungent bitter taste. When it comes into contact with alkali it turns red and is called '*Kum kum*'.

Turmeric oleoresin

This is a highly viscous, orange-brown product containing 30-35% curcumin, 15-20% volatile oil and has a characteristic turmeric aroma. Turmeric oleoresin is prepared by the extraction of a good quality turmeric powder with solvents like acetone, alcohol or ethylene dichloride and subsequent desolventisation to meet the specifications. Since turmeric oleoresin is highly viscous, it is mixed with dilutents like propylene glycol or vegetable oil, to obtain a homogenous pourable product. Though such products have a lower colour value, they are more suitable for specific end uses. New-generation products based on turmeric oleoresin include spice emulsions/aqua resins and encapsulated water dispersible powders with advantages like dispersibility in oil or water, and ease of application. An established curcumin colourant has been prepared by spray drying turmeric oleoresin with an organic acid, a buffer, a dispersant and an encapsulating material. Also, a powdered colouring agent, wherein curcumin is completed with gelatin in acetic acid, has been reported.

Vanilla

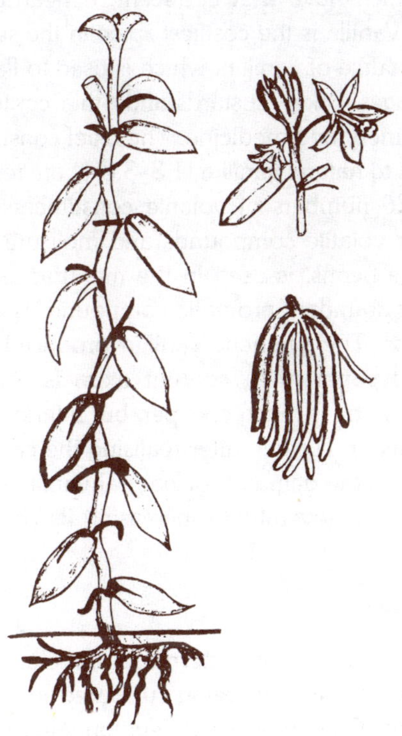

Plant Profile

Family	:	*Orchidaceae*
English name	:	Vanilla
Indian name	:	*Vanilla* (Hindi, Kannada, Malayalam,Tamil).
Species	:	*Vanilla fragrans* (Salisb) syn., *V. planifolia* Andrews *V. pompona* *V. tahitensis*
Distribution	:	Mexico, Guatemala, Central America, Java, Comoro Islands, Reunion, Mauritius, Madagascar, Tahiti, Zanzibar, Uganda, Tonga, Jamaica, West Indies.
Uses	:	Food flavourant

Vanilla fragrans (Salisb) (syn.*V.planifiolia* Andrews) is an orchid belonging to the family Orchidaceae. It is cultivated for its beans, which have a sweet scent, a heartful aroma and a pleasant flavour. Vanilla is the costliest spice in the spice horizon and is the important source of vanillin, which is used to flavour ice-cream, chocolate, beverages, cakes, custards and other confectionery. It also finds use in perfumery and medicine. The chief constituents of vanilla beans, in addition to natural vanillin (1.3–3.8%) are resins, fat, glucose, fructose, about 26 numbers of volatile constituents as well as 144 numbers of other volatile compounds and moisture. 'Vanilla sugar', obtained from the beans, is used in the manufacture of chocolates. Vanillin, the most abundant aromatic compound, is produced during the curing process. The synthetic vanillin produced from the waste sulphite liquor of paper mills, coal-tar extracts and from eugenol obtained from clove oil is much cheaper, but inferior in quality to the natural vanilla flavour. Further, after realising the carcinogenic effects of synthetic vanillin, the demand for natural vanilla in the developed countries has shot up several-fold increasing its prospects for large-scale cultivation.

Origin and distribution

Vanilla is indigenous to South-eastern Mexico, Guatemala and parts of Central America but is now being cultivated in other parts of the tropics such as Java, Comoro Islands, Reunion, Mauritius, Madagascar, Tahiti, Zanzibar, Uganda, Tonga, Jamaica and other islands of West Indies. The history of this orchid spice starts from 1520 AD, however the commercial cultivation started only after 1814. This crop was introduced to India 200 years ago. In India, it is grown in a few estates of Kerala, Tamil Nadu and Karnataka.

Area production and trade

The total area under vanilla in the world during 1999 was 37,525 ha with a production of 4,303 metric tonnes. Presently, Madagascar alone contributes 80% of the world production of vanilla beans. The world market for vanilla beans is highly concentrated in a few developing countries. Three countries, the US, France and Germany, account for 80% of the world imports. The US absorbs 50-60%, France and Germany between 10 and 15% each. These countries are also major exporters of both vanilla beans and processed vanilla products.

The area under vanilla cultivation in India is about 1000 ha and the production was about 6-8 t during 1999-2000. At present, Karnataka has the largest area under vanilla, followed by Kerala. The suitable areas for vanilla cultivation are Kerala, Karnataka, Tamil Nadu, the North-eastern Region, Lakshadweep and the Andaman and Nicobar Islands.

Description of the plant

In addition to *Vanilla fragrans,* there are two other species of the genus *Vanilla*, out of the 50 species, which are occasionally cultivated, though they yield an inferior product. They are listed below.

Vanilla pompona Schiede, known as West Indian vanilla, which occurs wild in Mexico, Central America and Trinidad and is cultivated to a small extent in Guadeloupe. It resembles *V. fragrans,* but the leaves are larger. The flowers are also larger and more fleshy. The cylindrical pods are shorter and thicker.

V. tahitensis (Tahitian vanilla) is indigenous to Tahiti and is cultivated there and also in Hawaii. It is less robust than *V. fragrans* with more slender stems and narrower leaves. The pods taper towards each end.

V. fragrans is a fleshy, herbaceous perennial vine, climbing by means of adventitious roots up onto trees or other supports to a height of 10-15 m, but in cultivation it is trained to a convenient height to facilitate hand-pollination and harvesting. The adventitious roots are long, whitish, aerial, singly opposite to the leaves and adhere firmly appressed to the support upon which the plant climbs. The long, cylindrical, monopodial stem is simple or branched, succulent, flexuose, brittle and has internodes that are 5-15 cm long. They are dark green and bear photosynthethic leaves with stomata. The leaves are large, 8-25 cm long and 2-8 cm broad, flat, fleshy, subsessile and alternate, oblong-elliptic to lanceolate. The tip is acute to acuminate and the base is somewhat rounded. The petiole is short, thick and canalised above.

The protruding inflorescences are axillary, racemose, usually simple, and only rarely branched, borne towards the top of the vine, with up to 20-30 flowers, opening from the base upwards, generally with only 1-3 flowers opening at one time and each lasting one day. The bracts are rigid, concave and persistent.

The flowers are large, waxy, fragrant and a pale-greenish yellow. The pedicil is very short and the ovary is inferior, cylindrical, tricarpellary. There are three sepals, which are obtuse to sub-acute and slightly reflexed at the opening. The two upper petals resemble the sepals in shape, but are smaller. The lower petal is modified as a trumpet-shaped labellum or lip, which is shorter than the other perianth lobes. Darker-coloured papillae form a crest in the median line, with a tuft of hair in the middle of the disc. The column or gynostemium is attached to the labellum. It is hairy on the inner surface, bearing at its tip the single stamen containing the two pollen masses or pollinia covered by a cap, and below is the concave sticky stigma, which is separated from the stamen by the thin flap-like rostellum

The fruit is a capsule, known in the trade as a bean, is pendulous, narrowly cylindrical, obscurely three-angled, 10-20 cm long and 1.5 and 4.5 cm in girth. It is aromatic on drying, containing when ripe, myriads of very minute globose seeds. In commercial production, the capsules are harvested before they are quite ripe.

Varieties

Except in the countries of origin the introduction of material which is propagated vegetatively is likely to be of very limited clonal origin and so little variation can be expected. Thus, there are no named varieties of vanilla as such. However, four major types of vanilla beans are distinguished in the world market, based on the countries of origin. They are: (1) Bourbon vanilla (produced in Madagascar, Comoro and Reunion); (2) Java vanilla (produced in the islands of Java in Indonesia); (3) Bourbon-like vanilla (mainly produced on the islands of Bali in Indonesia) and (4) Mexican vanilla. Among the four types, the first type ranks first in terms of quality.

Climate

Vanilla thrives well in a warm, humid, tropical climate at an elevation of 700 to 1500 m with an annual precipitation of 150-300 cm and a mean day temperature range of 25° to 32°C. The rainfall should be well distributed for a period of nine months and the rest of the three months should be a dry period. Although it thrives best under moist conditions, excessive humidity and moisture lead to diseases, while a very dry climate and direct sunlight may cause yellowing by sun scorching. Vanilla can be grown successfully as an inter-crop in coconut and arecanut gardens.

Soil

It grows well in loamy soils as well as in laterite soils with plenty of organic manure. Any soil which has poor drainage should be avoided as the plant does not withstand water-logging. It prefers land with a gentle slope.

Cultivation

Propagation

Commercially, vanilla is always propagated by planting *in situ* stem cuttings. Cuttings of 60-120 cm length and 4 to 5 cm diameter, with 8-10 internodes should be taken from healthy vigorous plants and may be cut from any part of the vine. The plants raised from 2-3.5 m-long cuttings will come to the bearing stage within 1 to 2 years, while those from shorter cuttings take 3 to 4 years. However, cuttings with less than 5-6 internodes and 60 cm length should not be used directly for planting. Vanilla can also be propagated by single-node cuttings using growth regulators. IBA at 10,000 ppm followed by NAA at 2000 ppm concentrations produced the maximum rooting. For raising rooted cuttings, the month of August is the most suitable giving the best results even without any hormone treatment. The vines are coiled around and buried inside the soil. Tissue culture technology from the shoot meristem has been standardised and the tissue-cultured plant can also be used for planting. The protocols for seed and embryo culture have also been standardised.

Planting

The ideal time for planting vanilla is when the weather is neither too rainy nor too dry; planting in August or September can also be done under South Indian conditions. At the time of planting, the three or four basal leaves are removed from the cuttings and dipped in 1% Bordeaux mixture and kept in the shade to dry for about a week. They are planted at the rate of one per support at a spacing of 1.5 × 3.0 m. About 1,600 to 2,000 plants are accommodated in one hectare area as a pure crop.

After planting the cuttings, frequent irrigations are given and the basins are mulched with leaves and trash. While planting, about 3-5 cm of the cutting is projected above the ground and stakes are provided for the cuttings to climb and grow. Vanilla flourishes well in partial shade of about 50% sunlight. Low-branching trees with rough barks and small leaves are grown for support. Under Indian conditions, Glyricidia has been found to be a good standard. The growth of the live standard should be adjusted so as to make them branch at a height of 120-250 cm. The standards are planted at a spacing of 2.5-3 × 2 m (1600-2000 trees/ha). The best time for planting supports is with the onset of rains after the summer break, and it should be atleast six months before the vanilla planting. In one or two weeks, extension growth appears along with roots in the axils, which cling on to the support for climbing. Frequent inspections are necessary for turning of the shoots so as to enable proper climbing. The vines may be trained on trellises, or on other live standards like Bauhinia and silver oak. In the early stages, lateral shade may be provided by bananas and windbreaks should be planted if necessary. On account of the very superficial roots, no cultivation around the vanilla vines is recommended after planting.

Manuring

Vanilla loves a lot of organic matter and decomposed mulch, as they are the main sources of nutrients. A thick layer of organic debris also helps to retain enough moisture and provides a loose soil structure for the roots to spread. Hence, it is advisable that early decomposable organic material is applied around the plant base atleast 3-4 times in a year. In addition, the application of 40-60 g N as Ammonium sulphate or Calcium ammonium nitrate, 30 g P_2O_5 and 60-100 g of potash in the form of muriate of potash per vine in two or three split doses promotes vegetative growth, leading to higher pod yields. The organic manures can also be applied in the form of leaf compost, starting from June-July. The application of fresh cattle manure has an adverse effect on the growth of the vines. The manures should be applied 20-30 cm away from the plant. Vanilla responds well to the foliar application of nutrients and therefore it is recommended that part of the fertiliser dose be given as a foliar spray before flowering. Spraying 1% 17:17:17 NPK complex on the foliage and stems has been found to be beneficial in enhancing the growth of the vines.

Irrigation

Timely irrigation should be provided in the initial two to three years of growth. During this period, a weekly irrigation of 2-3 litres per plant is sufficient.

Weeding

The weeds growing at the base of the support tree around the plant, within a radius of 50 cm, should be pulled out and left *in situ* to decompose.

Pruning

The height of the vines should be restricted to facilitate hand-pollination and harvesting. Care should be taken not to tear or bruise the leaves, branches or roots of the plant. The top 7.5–10 cm of the vine is pinched off 6-8 months before the flowering season to encourage the production of inflorescence in the vines.

Flowering, pollination and fertilisation

The vanilla flower has a peculiar morphology. The pollen sacs of the flower are covered by a hood-like structure called 'anther cap', and the stigma is actually prevented from coming into contact with stamens due to the presence of a lip-like projection called the labellum or rostellum. Because of this separation, self-pollination of the individual flower is impossible. In Mexico, natural pollination is effected by bees of the genus *Melapona* and hummingbirds, but in the absence of those pollinators in our local conditions, hand-pollination is the only alternative. While doing hand-pollination, a contact between the stigma and stamens is brought about by lifting the rostellum and tilting the anther cap by means of a small bamboo splinter, the size of a tooth-pick. A trained person can complete pollination of about 1200-2000 flowers per day.

V. fragrans usually flowers only once a year over a period of about two months, starting from the third year after planting. The flowers open from the base of the raceme upwards and usually one, or more rarely two or three flowers, are open on the inflorescence at the same time. The flowers open early in the morning and remain receptive for eight hours and wither away within 2-3 days if pollination and fertilisation do not take place successfully. The fruit set is highest (97.5-100%) when pollination is done early on a bright morning (6 a.m.–1 p.m.), following a shower of rain. When a fertile cross-pollinated seed is required, it is necessary to obtain pollen from another plant.

Usually on vigorous plants, 8-10 flowers in 10-12 inflorescences are pollinated, of which 4-8 capsules are allowed to grow to maturity on each raceme. When the desired number of fruits have set, the remaining buds are removed by clipping off the tip of the inflorescence. All the damaged and malformed capsules are removed during growth. The final number of beans per vine varies greatly and is usually between 30 and 150.

It has been found that growth regulators like 2,4-D, IAA and IBA will induce the development of parthenocarpic fruits and give a higher percentage of fruit set.

Pests and diseases

Pests

Among the insect pests, a few small longicorn beetles (*Hoplia retusa* and *Saula ferruginea*) and an ash-weevil (*Cratopus retuse*) bite holes in the flowers and others destroy the column. In addition to these, there are a few caterpillars and certain earwigs, snails and slugs which live on the tender parts of the plant. Grasshoppers and crabs are also found cutting the growing tips of plants during the establishment stage of the plantation. Poultry birds sometimes cause considerable damage to the roots while scratching the mulch placed around the plants. Regular surveillance and the removal of pests reduces their damage to a great extent.

An aphid, *Cerataphis variabilis,* has been observed occasionally infesting the leaves which can be controlled by spraying 0.02% Dimethoate @ 250 l/ha. The insecticide may be sprayed as soon as the pest appears. If necessary, the spray can be repeated at fortnightly intervals or Demecron @ 200 ml/ha can be sprayed at monthly intervals.

During the processing and storage, the pods are attacked by *Tyrophagus* sp. of mites.

Diseases

Diseases are more serious than pests, as they sometimes cause heavy losses in crop production. The important diseases of this crop are described below.

Root-rot (Fusarium batatis* var. *vanillae)

This is the most serious problem, severely limiting vanilla production. The pathogen affects the roots first causing a browning and later the death of the roots as well as the underground part of the stem. The plant ceases to grow, the stems and leaves become flaccid and the stem begins to shrivel, leading to the ultimate death of the planet. Drenching with 0.1% Bavistin or Methyl Benzymidazole carbomate as a soil drench and foliar spray helps in controlling the disease, besides following suitable phytosanitary measures.

Stem-rot or wilt (Fusarium oxysporum* var. *vanillae)

This disease can be controlled by removal of the diseased plants along with the surrounding soil. The control strategy for this disease is as follows:

- Removal of weeds around the plants
- Mulching the base of the stem with dry leaves before and after the monsoon.
- Avoiding injury to roots
- Drenching the base of the stem with 0.5% Bordeaux mixture or 0.2% Copper oxychloride.
- Pre-treatment with non-pathogenic *Fusarium* is reported to suppress the disease incidence. The extracts of clove leaves have been found to be antifungal *in vitro* and may be useful as a soil amendment to control the disease in the field.

Anthraconse (Calospora vanillae)

This is a very serious disease attacking the stem apex. The wilting of the leaves and roots, dropping off of the beans after turning black are the symptoms. The disease is aggravated by excess moisture, a prolonged rainy season, excess shade and over-crowding. Phytosanitation, removal of the excess shade, avoiding over-crowding coupled with spraying 0.1% Bavistin help to check the disease.

Fruit-rot and flower-rot are the other diseases that have been observed on this crop. During processing and storage, the pods are susceptible to moulds (mainly *Penicillium* and *Aspargillus* sp.).

Harvesting and yield

Almost 9 -11 months are required for the fruit to mature after successful pollination and fertilisation, depending on the temperature and elevation. The pods or sticks are harvested rotationally when they become fully yellow. The immature pods produce an inferior product and, if it is picked too late, it will split during curing.

The yield of the crop is variable. A good vanillery may yield about 300 kg/ha of cured beans per year. About 6 kg of green pods produce 1 kg of cured beans. The curing should be done within a week of harvesting the beans.

After fruiting, the old and weak branches are pruned off. The tree supports or shade should be pruned to provide 30-50% of full sunlight and to induce branches at the correct height for training the vines. The yield may decline after the 12th or 13th year of planting, and so replanting should be done at this stage.

Curing

There are different methods of curing the beans, viz., Mexican process, Bourbon process, Peruvian process and Guyana process. Under Indian conditions, either the Mexican or Bourbon process can be followed. Every procedure of curing of vanilla beans is characterised by these four phases

Killing or wilting

This process stops further vegetative development in the fresh beans and initiates the enzymatic (β-glycosidase) reactions responsible for producing the aroma and flavour. Generally for killing, the harvested mature beans are sorted according to their size viz., those above 15 cm and those below 15 cm are separately taken in an open basket, and are dipped in hot water with a temperature of 65°C for 3 minutes. Later, the basket is taken out and the hot water is drained off. Killing is indicated by the development of a brown colouration in the beans.

Sweating or fermentation

The killed beans are then subjected to sweating. This results in a rise in the temperature of the killed beans, which promotes the desired enzymatic reactions and hastens the drying period to prevent harmful fermentations. The sweating operation is carried out for two days, depending on the weather conditions. In warm weather, the pods are spread over blankets and exposed to the sun. During mid-day the blanket is folded over and bundled and left in the open for the rest of the day. They are wrapped in blankets in the night to maintain continuous fermentation and sweating. The pods should be wrapped in blankets when they are hot to touch. This process is repeated for 7-12 days till they become dark brown in colour, soft and flexible.

Slow drying

This is done at an ambient temperature in the shade, till the beans have reached about one-third of their original weight. The beans are dried for 17-22 days inside a room and will have 25-30% moisture in them.

Conditioning or aging

Here, the beans are stored in closed boxes in bundles of 25-30 tied together from tip to top with cotton threads and these bundles according to their size are wrapped in tissue paper. These wrapped bundles are again wrapped in woollen blankets and kept for a period of three months or longer in closed wooden boxes to permit the full development of the desired aroma and flavour.

Grading

In India, the beans are graded into four groups. The first (best) grade beans consist of those above 16 cm in length, soft to touch without any blemishes and containing 2.5% vanillin. The second-grade beans consist of those of 12-15 cm length, without blemishes and with more than 1.8% vanillin. The third-grade beans are shorter, with a length of 8-12 cm and with some blemishes. The rest of the beans are taken as fourth (last) grade and are of poor quality.

The grades or types of vanilla (beans) entering the trade are:

i. Mexican vanilla
ii. Bourbon vanilla (from Madagascar)
iii. Indonesian vanilla
iv. Tahiti vanilla (from Tahiti)
v. Vanillons (from West Indies)

Extraction of vanillin

For the first time in 1858, Gobley isolated the crystalline substance vanillin ($C_8H_8O_3$) from the vanilla beans. The vanillin extract is taken from the cured beans by hydro-alcoholic entraction. The finely-chopped beans are put in large stainless steel baskets and are immersed in a warm alcohol-water solvent in glass-lined or stainless steel percolators. The solvent, which has been piped in from the top of the vat, percolates down through the beans and on reaching the bottom is recirculated to the top to continue the percolation, until the maximum vanilla flavour is extracted. The freshly-percolated extract is drawn off, filtered, aged for about three weeks and bottled.

Vanilism (Vanilla poisoning)

The vanilla plant produces a fair quantity of oxalic acids in the leaves, stems and pods, on account of which the plant has to be handled with care. Too much of contact with vanilla creepers is reported to bring about an ailment called vanillism, which is characterised by heaviness of the head, skin irritation, fever and intestinal disorders.

Vanilla poisoning may occur while planting and harvesting the green beans and also later during curing, especially during the conditioning operation. Therefore, it is better to wear rubber gloves while handling the vanilla beans.

Products of vanilla

The various types of vanilla products employed in flavouring are vanilla extract, vanilla flavouring, vanilla tincture (for pharmaceutical use), concentrated vanilla extract and concentrated vanilla flavouring, vanilla oleoresin, vanilla–vanillin extract and flavouring, vanilla powder, vanilla-vanillin powder, perfumery vanilla, vanilla absolute, etc.

Bibliography

Abdus-Samad, Luchon-Saikia, A. Samad, and L. Saikia (1997) "Performance of black cumin (*Nigella sativa* L.) in Assam", *Indian Cocoa, Arecanut and Spices J.*, **21** (4): 108–109.

Abraham, C.C. and B.P. Skaria (1995) "New record of the red borer *Zeuzera coffeae* Nietner as a pest of allspice (*Pimenta diocia* Linn.) Merril", *Insect Environment*, **1** (1): 6.

Abraham, V.A. (1991) "Curry leaf cultivation in Andhra Pradesh", *Spice India*, **4** (11): 13–14.

Agarwal, N.L. (1995) "Production of coriander in Rajasthan", *Spice India*, **8** (6): 17.

Ahmed, A., A.A. Farooqi and K.M. Bojappa (1988) "Effects of nutrients and spacing on growth, yield and essential oil content in Fennel (*Foeniculum vulgare* Mill.)", *Indian Perfum.*, **32** (4): 301–305.

Akbari, L.F. and I.U. Bhruj (1995) "Chemical control of *Alternaria* blight of cumin", *J. Spices and Aromatic Crops*, **4** (1): 82–83.

Akhtar Hussain, O.P., Virmani, Ashok Sharma, L.N. Anupkumar, Mishra (1988) *Major Essential Oil Bearing Plants of India*, Central Institute of Medicinal and Aromatic Plants, Lucknow.

Amit Krishna D. (2000) "The wonders of Ajowan 2000", *Spice India*, **10** (1): 9–10.

Anandaraj M. and K.V. Peter (1996) *Biological control in Spices*, IISR, Calicut, 52 pp.

Anandaraj M. and V.R. Sarma (1999) "How to manage *Phytophthora* in black pepper", *Planter's Chronicle*, **95** (6): 277–280.

Anandaraj M. and Y.R. Sarma (1995) "Diseases of Black pepper (*Piper longum* L.) and their management", *J. of Spices and Aromatic Crops*, **4** (1): 17–23.

Anonymous (1956) *The Wealth of India–Raw Materials*, Council of Scientific and Industrial Research, New Delhi.

Anonymous (1974) "A new variety of fennel", *Arecanut and Spices Bull.*, **6** (1): 10–11.

Anonymous (1975) "Report of the VI workshop on All India Crop Improvement Project on Medicinal and Aromatic Plants"

Anonymous (1983) *Chemistry of Natural Products*, Kalyani Publishers, New Delhi.

Anonymous (1985) Ginger: package of practices for ginger (Pamphlet 9E 94th Revised), CPCRI, Kasargod, Kerala, pp. 1–8.

Anonymous (1985) "Report on Herbal Industry", Industrial and Technical Consultancy Organization of Tamil Nadu Ltd., Greames Road, Chennai.

Anonymous (1986) *Export of spices and spice products from India during 1985–86 and 1984–85. Ibid*, **20** (8): 20.

Anonymous (1990) *Magic and Medicine of Plants*, The Reader's Digest Association, Inc, New York.

Anonymous (1991) "Processing of chillies–a scientific method", *Spice India*, pp:7.

Anonymous (1994) "*Kokke kandu*–virus disease of cardamom and its management", *Planters' Chronicle*: 31–32.

Anonymous (1999) *Annual Report*, Central Institute of Medicinal and Aromatic Plants, Lucknow.

Anonymous (2000) *Annual Report*, Central Institute of Medicinal and Aromatic Plants, Lucknow.

Anonymous (2001) *AICRPS–a profile, Indian Institute of Spices Research*, Marikannu P.O., Calicut, Kerala.

Anonymous (2002) "Dominating global markets", *The Hindu Survey of Indian Agriculture*, pp: 87–95.

Anu, A. and K.V. Peter (2000) "The chemistry of paprika", *Indian Spices*, **37** (2): 15–18.

Atal, C.K. and B.M. Kapoor (1982) *Cultivation and Utilization of Aromatic Plants*, Regional Research Laboratory, Jammu.

Aykroyd, W.R., V.N. Wardhan, and S. Ranganathan (1991) "The nutritive value of Indian foods and the planning of satisfactory diets", Manager of Publications, Govt. of India, New Delhi.

Azgar Sheriff, P. (1992–1996) Spices bibliography from 1990 to 1995, *J. Spices and Aromatic Crops*, I to V.

Babu, K.N., D. Minoo, S.P. Geetjha, K. Samsudeen, J. Rema, P.N. Ravindran, and K.V. Peter (1998) "Plant biotechnology–its role in improvement of spices", *Indian J. of Agric.Sci.*, **68** (8): 533–547.

Badiyala, D. and K. Saroch (1997) "Effect of seed corm size and planting geometry on saffron under dry temperate conditions of Himachal Pradesh" *Indian Perfum*, **41** (4): 167–69.

Bajaj, K.L. and Kaur (1979) "Colorimetric determination of capsaicin in capsicum fruits with the Folin ciocalten", *Mikrochimica Acta*, **1**: 81–86.

Bakhru, R.K. (1992) "Sweet flag–calamus", *Spice India*, **5** (1): 9–10.

Bathakur, M.P. and D.N. Bordoloi (1992) "Micro-propagation of *Curcuma amada* (Roxb.)", *J. of Spices and Aromatic Crops*, **1** (2): 154–159.

Bhandoria, B.B.S. and D.V.S. Chauhan (1995) "Efficacy of herbicides in the control of weeds infesting Indian Mustard", *Indian J. Agron.*, **40** (2): 327–29.

Bhat, V.R., G.S. Sulikeri, and N.C. Hulamani (1992) "Coriander–Effect of NPK on herbage yield of coriander", *Spice India*, **5** (10): 11–13.

Bhati, D.S. (1993) "Effect of irrigation and phosphorus on seed yield of fenugreek and its attributes", *Indian J. Agron.*, **38** (3): 449–453.

Blank, I., A. Sen, and W. Ghosh (1992) "Sensory study on the character impact on flavour compounds of dill herbs", *Fd. Chem.*, **43** (5): 337–43.

Boatto, G., G. Pintore, M. Palomba, F. De Simone, E. Ramundo, and C. Iodice (1994) "Composition and antimicrobial activity of *Rosmarinus officinalis* essential oils", *Fitoterapia*, **65** (3): 279–80.

Bose, T.K., S.K. Mitra, A.A. Farooqi and M.K. Sadhu (Eds) (2000) *Tropical Horticulture*–Vol.1. Eds., Pub. Nayaprakosh, Calcutta.

Chadha, K.L., (1998) "Improvement in tree fruit and plantation crops", *Indian J. of Hort.*, **55** (4): 265–296.

Chadha, K.L. and P. Rethinam (Eds.) (1994) *Advances in Horticulture*, Vol. 9–Plantation and Spice Crops Part 1.

Chadha, K.L. (Ed.) (2001) *Hand Book of Horticulture*, Pub. Indian Council of Agricultural Research, New Delhi.

Charalambous, G. (1994) *Spices, herbs and edible fungi*. In: Development in Food Science, pp. 34.

Chauhan, H.S. and Kamala Singh (1998) "Response of mint species to varying levels of irrigation schedules in Uttar Pradesh, foothills of the Himalayas", *Indian Perfum*, **42** (1): 27–32.

Chezhiyan, N., M. Ananthan, and G.B.P. Vedaonuthu (1996) "Standardization of seed propagation techniques in nutmeg and clove", *Spice India*, **9** (16): 21–22.

Chopra, R.N., I.C. Chopra,and B.S. Verma (1969) *Supplement to Glossary of Indian Medicinal Plants*, Council of Scientific and Industrial Research, New Delhi.

Curzio, O.A. and A.M. Urioste (1994) "Sensory quality of irradiated onion and garlic bulbs", *Food Processing and Preservation*, **18** (2): 149–159.

Dake, G.N. (1995) "Diseases of ginger (*Zingiber officinale* Rosc.) and their management", *J. of Spices and Aromatic Crops.* **4** (1): 40–48.

Dange, S.R.S. (1995) "Diseases of cumin and their management", *J. Spcies and Aromatic Crops*, **4** (1): 57–60.

Deepu Mathew, K.C. Shivakumar,and C. Kathiresan (2001) "Biological control of Important Sucking Pests of Chilli", *Spice India,* **38** (6): 5.

Devasahayam, S. (1992) "Insect pests of black pepper and their control" *Planter's Chronicle*, pp. 153–155.

Eapen, S.J. (ed) and K.V., Ramana (2000) *Annual Report*, Indian Institute of Spices Research, Calicut.

Edison, S. (1987) "Fungal, bacterial and viral diseases of spices and condiments–their control measures, constraints and future R.D. needs, In: *Souv. Nat. Symp" On Spice Industries* (Ed. J.S. Pruthi), AFST(I) Delhi, pp. 17–19.

Edison, S. and J.A. Kallupurackal (1990) "New varieties to improve productivity of seed spices", *Indian Cocoa, Arecanut and Spices J.*, **13** (4): 121–123.

Edison, S., A.A. Johny, K. Nirmal Babu and A. Ramadasan (1991) *Spices Varieties*, National Research Centre for Spices, Calicut, Kerala.

Elizabeth Peplow, (1984) *The Herb Book*, W.H. Allen and Co., PLC, London.

Elizabeth, K.G. and K.G. Thomas (1989) "New concepts in pepper cultivation", *Indian Cocoa, Arecanut and Spices J.*, **13** (1): 7–11.

Farooqi, A.A. and B.S. Sreeramu (2001) *Cultivation of Medicinal and Aromatic Crops, Universities Press*, (India) Ltd., Hyderbad.

Farooqi, A.A., K.A. Devaiah, M. Shridharayya and M. Vasundhara (1994) "Effect of nitrogen and phosphorus on growth yield and essential oil content of Marjoram (*Marjoram hortensis* Moench.)", *Indian Perfum.*, **38** (1): 9–14.

Farooqi, A.A., M.M. Khan and M. Vasundhara (2001) *Production Technology of Medicinal and Aromatic Crops*, Natural Remedies Pvt., Ltd., Bangalore.

Fateh Singh, Vinay Singh and R.S. Malik (1995) "Effect of different doses and sources of sulphur on yield and uptake of sulphur by garlic", *J. Indian Soc. Soil Science*, **43** (1): 130–131.

Frederic, Jr. Rosengarten (1969) *The Book of Spices*, Livingston Publishing Co., Wynnewood, Pennsylvania, USA.

Geetha, V. (1995) "Urigam tamarind—A unique variety", *Spice India*, **8** (8): 8–9.

George, C.K. (1993) "Vanilla cultivation in India and its economics", *Vatika*, No.**2**: 43–45.

Gill, H.S. (1989) "Improved technologies for chilli production", *Indian Cocoa, Arecanut and Spices J.*, **12** (4): 118–119.

Gopinath, G. and N.M. Poonacha (1992) Bishop's weed, *Spice India*, **5** (9): 9–14.

Guenther, E. (1950) *Essential Oils*, Van Hostrand Co., Inc., New York.

Guenther, E. (1980) *Essential Oils*, Van Hostrand Co., Inc., New York.

Gupta, S.C. (1996) "Phenological observations on basil", *Indian Perfum*, **40** (1): 17–22.

Haldankar, P.M., D.D. Nagwekar, A.G. Desai, J.L. Patil, and R.T. Gunzate (1996) "Airlayering in allspice", *Indian Cocoa, Arecanut and Spices J.*, **19** (2): 56–57.

Handa, S.S. and M.K. Kaul (1997) *Supplement to cultivation and utilization of Aromatic plants*, Regional Research Laboratory, Council of Scientific and Industrial Research, Jammu.

Hegde, S.G. and S.H. Anahosur (1998) "Integrated management of foot-rot of black pepper", *Karnataka J. of Agric Sci.*, **11** (1): 78–82.

Hore, A. (1979) "Improvement of minor umbelliferious speices in India", *Econ. Bot.*, **33** (3): 290–297.

Hosamani, R.M. (2000) "Paprika in North Karnataka–Status, Research and Development", *Indian J. of Arecanut, Spices and Medicinal Plants*, **2** (2): 50–57.

Jackson, S.A.L. and R.K.M. Hay (1994) "Characteristics of varieties of thyme for use in the UK: oil content, composition and related characters", *J. Hort. Sci.*, **69** (2): 275–83.

Jacob, S. (1995) "Pests in stored tamarind", *Spice India*, **8** (7): 9–10.

James Verghese (2000) *Piprum nigrum* L. Essential Oil Specifications, *Indian Spice*, **37** (2): 8–9.

Jarnail Singh, Singh, J., Chadha, K.L. and G. Kalloo (Ed) (1993) "Improvement of chillies" *Advances in Horticulture*; vegetable crops–volume 5: 69–86.

Jasvir Singh, M.R. Sudarshan, D. Kumerasan,and J. Singh (1994) "Cardamom foliar pest, lace–wing bug (*Stephanitis typicus*) (Distant) (Heteroptera: Tingidae)", *Annals of Entomology*, **12** (2): 81–84.

Johny Manithottam (1994) "Vegetative propagation in nutmeg", *Spice India*, **7** (2): 5–7.

Joshi, M.S. and S.P. Raut (1995) "Occurrence of foliar diseases in the clove nursery", *Indian Cocoa, Arecanut and Spices J.*, **19** (1): 19–20.

Joshi, B.S., S. Ramanujam and M.B.L. Saxena (1963) "Improvement of some essential oil bearing spice plants", *Information Bull.* Regional Research Laboratory, Jammu, **18**: 94–100.

Kalpana, M., H.L. Prabhakara and R.V. Hegde (2001) *In vitro* propagation of curry leaf, *Indian Spices*, **38** (1): 5–7.

Kanwar, J.S. and M.S. Saimbhi (1993) "Response of seed crop of Kasuri fenugreek to nitrogen, potassium and leaf cutting", *Indian J. Agric. Sci.*, **63** (5): 299–301.

Kaul, V.K. and S.S. Nigam (1997) "Antibacterial and antifungal studies on some essential oils", *J. Res. Indian Med. Moga and Pomeo*, **12** (3): 132–135.

Kempe Gowda, K. and P. Mudappa Gowda (1988) "Effects of dates of planting on yield of green and red chillies", *Spices Newsletter*, **22** (8): 10, 23.

Kennedy, R.R., V. Swaminathan, and K. Hariharan (1995) "Problems and prospects in tamarind cultivation", *Spice India*, **8** (5): 14, 19.

Kirtikar, K.R. and B.D. Basu (1963) *Indian Medicinal Plants*, **2**: 1345, L.M.Basu, Allahabad.

Korikanthimath, V.S., D. Prasanth and Ashok Kumar Gaddi (2001) "Profitable cultivation of vanilla", In: *Proc. National Seminar on Transfer of Technology of Medicinal and Aromatic Crops*, 20–22, February, Bangalore; pp: 200–225.

Korikanthimath, V.S., G.M. Hiremath, M.S. Venugopal, Rajendra Hegde and R. Hegde (1999) "Feasibilty of vanilla cultivation with coconut", *J. of Medicinal and Aromatic Plants Sci.*, **21** (4): 1033–1039.

Krishnamoorthy, B. and J. Rema (1991) "Cassia and cinnamon", *Spice India*, **4** (3): 10–12.

Krishnamoorthy, B. and J. Rema (1994a) "Airlayering in cassia", *J. Spices and Aromatic Crops*, **3** (1): 48–49.

Krishnamoorthy, B. and J. Rema (1994b) "Tree spices–crop improvement" *Spice India*, **9** (11): 24–30.

Krishnamoorthy, N. and S.R. Sampathu (1987) "Application of garcinol as a colourant for butter and ghee and method of its estimation", *J. Fd. Sci. Technol.*, **24** (6): 312–13.

Krishnamoorthy, N. and S.R. Sampathu (1988) "Anti–oxidant properties of Kokum rind", *J. Fd. Sci. Technol.*, **25** (1): 44–45.

Krishnamoorthy, V., M.B. Madalageri and C. Kannan (2000) "Ajowan (Bihshop weed)–A crop for dry land", *Agro India* (8): 7.

Krishnamurthy, B. and K. Sivaraman (1999) "Quality planting materials of tree species–an overview", *Indian J. of Arecanut, Spices and Medicinal Plants*, **1** (3): 98–99.

Krishnarajan, J. (1991) "Curry leaf cultivation", *Spice India*, **4** (3): 5–6.

Kshirasagar, P.J., P.M. Haldankar, B.P. Patil and H.R. Nadakarni (2001) "Kokum–Present status and future prospects", In: *Proc. National Seminar on Transfer of Technology of Medicinal and Aromatic Crops*, 20–22, February, Bangalore: pp.49 55.

Kuttappa, K.M. and S. Eddy (1988) "Cardamom whitefly and its predator", *Spices Newsletter*, Spices Board, Cochin, **22** (8): 16.

Lawrence, B.M. (1989) Parsley leaf and seed oils progress in essential oils–A review, *Perfumer and Flavorist*, **14** (5): 54–55.

Lawrence, B.M. (1993) "Essential oils, 1988–91 (including ajowan oil)" Allured Pub. Corpn., Carol Stream, Illinois 60188, USA, Vol. 4, p. 226.

Lewis, Y.S. and S. Neelakantan (1964) "The Chemistry, Biochemistry and Technology of Tamarind", *J.Sci. and Industr. Res.*, **23** (5): 204–206.

Madan, M.S. and M. Tamil Selvan (2001) "Globalisation and spice economy of India", *Indian J. of Arecanut, Spices and Medicinal Plants*, **3** (4): 262–269.

Madan, C.L., B.M. Kapur,and U.S. Gupta (1966) Saffron, *Econ.Bot.*, **20** (4): 377–385.

Madhusudana Rao, J. (1988) "Analysis of essential oils by nuclear magnetic resonance (NMR) spectroscopy", *Indian Perfum*, **32** (1): 109–12.

Mangalakumari, C.K. and A.G. Mathew (1986) "Localisation of significant constitutents of turmeric", *J. Fd. Sci. Technol.*, **23** (2): 93–96.

Manjunath, M.N., V.V. Sattigeri, and K.V. Nagaraj (1991) Curcumin in Turmeric, *Spice India,* **14** (3): 7–9.

Mathew, A.G. (2000) "Spices oils and other essential oils from Kerala", *Indian J. Arecanut, Spices and Medicinal Plants,* **2** (4): 150–152.

Mathew, N.M.(ed)., C . Kuruvilla Jacob, (ed), T. Joseph, (ed), J.R. Meenattoor, (ed) and K.K. Thomos (1998) "Developments in plantation crops research", *Proc. of the 12th Symposium on Plantation crops*, PLACROSYM XII, 27–29 November, 1996. India, 343 pp.

Mc Gimpsey, J.A., M.H. Douglas, and A.R. Wallace (1997) "Evaluation of Saffron (*Crocus sativus* L.) Production in New Zealand" *New Zealand J. of Crop and Hort.Sci.,* **25** (2): 159–168.

Mehta, K.G., P.G. Patel, B.T. Kachhadia, and S. Edison (1995) "Varietal response to leaf cutting management on seed and green leaf yields of fenugreek", *Indian Cocoa, Arecanut and Spices J.,* **19** (3): 91–93.

Menon, P.S. (2000) "IISR Scientists develop new clove varieties", *J. of Herbs, Spices and Medicinal Plants*, **7** (1): 103–105.

Mohd. Abdulkhader (1987) "Improved varieties of annual spice crops (turmeric, coriander, fennel and fenugreek) released by the Tamil Nadu Agricultural University. Abstract No. 1.20" In: poster session abstracts, (Ed. J.S. pruthi), *Nat. Symp. On Spice Industries, AFST (India),* New Delhi, pp. 16–17.

Morris, G.E.L. and G.W. Tucker (1985) "Study of components of growth of the spring sown onions and their relationships and meteorological data", *J. Hort. Sci.,* **60** (3): 341–51.

Munshi, A.M. (1992) "Genetic variability for important traits in Saffron (*Crocus sativus* L.)", *Crop Res.,* **5** (2): 326–332.

Muthulaxmi, P. and T. Sarah, George (1999) "Kodampuli a multipurpose fruit tree", *Indian Horticulture.* **3**: 7–8.

Narasimha Rao, B. and K. Rama Subbaiah (1987) "Dissipation of monocrotophos and carbaryl in green and ripe fruits of chilli and removal of the insecticides by different processing procedure", *Indian Cocoa, Arecanut and Spices J.,* **11** (2): 53–57.

Nawale, R.N. and M.S. Karmarkar (1997) "Soft wood grafting technique for rapid multiplication of high yielding clones in Kokum (*Garcinia indica* Choisy)", *Indian Cocoa, Arecanut and Spices J.,* **21** (3): 76–77.

Nawale, R.N., Y.R. Parulekar, and M.B. Magdum (1997) "Kokum (*Garcinia indica* Choisy) cultivation in *Konkan* region of Maharashtra", *Indian Cocoa, Arecanut and Spices J.,*. **21** (2): 42–43.

Nybe, E.V. and V.S. Sujatha (2001) "A wonder Drug", *Indian J. Arecanut, Spices and Medicinal Plants,* **3** (4): 270–272.

Nybe, E.V. (Ed.) (2001) *Three Decades of spices Research at KAU,* Kerala Agricultural University, Vellinikkara–680 656, Thrissur, Kerala.

Pandey, A.K. (1999) "Response of mango ginger to nitrogen and plant spacing under terrace system of cultivation in mid-altitude areas of Mizoram", *J. of Hill Res.*, **12** (1): 77–78.

Parameswaran, M. (1997) "Curry leaf: a plant with much potential", *Food Australia.* **49** (12): 590–591.

Parkash (1998) "Paprika", *Agro India.* **6** (3): 31.

Parry, J.W. (1969) *Handbook of Spices*, Chem. Pub. Co., Inc, New York, N.Y., USA.

Parry, J.W. (1969) Spices II. *Their Morphology, History and Chemistry*, Chem.Pub.Co.Inc., New York, N.Y., USA.

Patil, J.D., G.Y. Desale, and P.N. Kale (1986) "Genetic variability studies in onion (*Allium cepa* L.)", *J. Maharashtra Agri. Univ.*, **12**: 114–15.

Perez, G.F. (1995) "Effect of light, temperature seed priming on germination of celery seeds", *Seed Sci. and Technol.*, **23** (2): 377–85.

Peter, K.V. (1998) "Spices research in India", *Indian J. of Agric. Sci.*, **68** (8): 527–532.

Peter, K.V. (1999) "Bioinformatics on Tree Spices", *Indian Spices*, **36** (1): 16–18.

Peter, K.V. (2000) "Informatics on Turmeric and Ginger", *Indian Spice*, **38** (2 & 3): 12–14.

Poornima, K., Sivagami–Vadivelu and S. Vadivelu (1999) "Occurrence of seasonal population behaviour of phytonematodes in turmeric (*Curcuma longa* L.)", *Pest Management in Hort. Ecosystems.*, **5** (1): 42–45.

Potty, S.N. and V.Krishnakumar, (2000) "Indian Spice production problems and prospects", *Plant Hort. Tech.*, **1** (4): 50–55.

Prakash, V. (1990) *Leafy Species*, pp. 1–114. ICRC Press, Roca Raton, Ann Arbor, Boston, USA.

Prasanna Kumar, G.K., M.M. Khan, A.A. Farooqi and S.S. Joshi (1994) "Studies on floral biology of Marjoram (*Marjorana hortensis* Moench.)", *Indian Perfum.*, **38** (1): 5–8.

Pruthi, J.S., (Ed.) (1987) "Bibliography on 57 spices and condiments", 1972–86, AFST(I), Delhi chapter, pp. 168 (1.617 references including some on aniseed).

Pruthi, J.S. (1985) "Recent advances in the packaging of spices and spice products", *Souv. Semin. Innovations in packaging of processed foods*, Small Industries Service Institute, 17–18 May, New Delhi, pp. 1–7.

Pruthi, J.S. (1988) "Recent developments in processing technology of Pepper", presented at the meeting of International Pepper Community, Goa, November, 1998.

Pruthi, J.S. (1989a.) "Need for modernization and partial mechanization of fruit and vegetable murabba industry with special reference to ginger preserve and candy" Paper presented at Nat. Semin. On

Fruit and Veg. Murabba Industry in India, 16th April, 1989, IIC, New Delhi, pp. 1–8., *Indian Fd Packer,* **44** (6): 51–62 (1990).

Pruthi, J.S. (1989b) "Indian fruit and vegetable preserve industry–A retrospect and prospect", *Souv. Nat. Semin. On Fruit and Murrabba Industry,* 16th April, 1989, IIC, New Delhi, pp. 1–10.

Pruthi, J.S. (1989b) "Processing technology of spices–present world scenario", Invited paper, *Indian Cocoa, Arecanut and Spices J.,* November, XII (4).

Pruthi, J.S. (1998) Aniseed, In: *Spices and condiments,* National Book Trust, India, New Delhi, 5th Edn., pp. 19–22.

Pruthi, J.S. (1998) Celery, In: *Spices and Condiments,* 5th Edn., National Book Trust, India, New Delhi, pp. 80–83.

Pruthi, J.S. (1998) Galangal, In: *Spices and Condiments,* 5th Edn. National Book Trust (India), New Delhi 1, pp. 1–325.

Pruthi, J.S. (1998) Long pepper, In: *Spices and condiments,* 5th Edn., pp. 191–93, National Book Trust, India, New Delhi.

Pruthi, J.S. (1998) Parsley, In: *Spices and Condiments,* 3rd Edn. National Book trust (India), New Delhi, pp. 177–80.

Pruthi, J.S. (1998) Rosemary, In: *Spices and Condiments,* 5th Edn. National Book Trust India, New Delhi.

Pruthi, J.S. (1998) Saffron, In: *Spices and condiments,* 5th Edn., National Book Trust, India, New Delhi, pp. 203–6.

Pruthi, J.S. (1998) "Sage–description, distribution, composition and end uses", In: *Spices and Condiments,* 5th Edn., pp. 205–8.

Pruthi, J.S. (1998) *Spice and Condiments* (chapter on Thyme) 5th Edn., pp. 1–325. National Book Trust India, New Delhi.

Pruthi, J.S. (1998) "Sweet flag or calamus", In: *Spices and condiments,* 5th Edn., National Book Trust, India, New Delhi, pp. 217–19.

Pruthi, J.S. (1998) Tamarind, In: *Spices and Condiments,* Edn. 5th, pp. 217–19, National Book Trust, India, New Delhi–16.

Pruthi, J.S. (1998) Vanilla, In: *Spices and condiments,* 5th Edn., National Book Trust, India, New Delhi, pp. 230–34.

Pruthi, J.S. (1980c) "Spices as antioxidants in foods", *J.Bev. Food World,* **7** (3): 13–16, 19.

Pruthi, J.S. (1970) "Packaging of Spices and Condiments", *J. packaging, India,* **3** (1): 11.

Pruthi, J.S. (1976) *Spices and Condiments,* National Book Trust, India, New Delhi.

Pruthi, J.S. (1980) "Spices–Chemistry, Microbiology and Technology", (Ist ed), *Advances in Food Research,* **22**: 1–450, Academic Press, New York, USA.

Pruthi, J.S. (1980b) "Spices as flavourants", *J. Bev. and Food World,* **7** (3): 7–11.

Pruthi, J.S. (1980d) "Spices as food colourants", *J. Bev. and Food World.* **7** (4).

Pruthi, J.S. (1981) "Manufacture and use of Spices, Essential Oils in the perfumery and cosmetics", *Pafai J.,* **2** (2): 85–94.

Pruthi, J.S., (1986) "Spices and condiments–crop management and post–harvest technology", *Tech. Monograph*, Indian Council of Agricultural Research, New Delhi.

Pruthi, J.S. (1998) *Major Spices of India*, Indian Council of Agricultural Research, New Delhi.

Pruthi, J.S. (2000) *Minor Spices and Condiments*, Indian Council of Agricultural Research, New Delhi.

Purseglove, J.W. (1984) *Tropical Crops–Dicotyledons*, Longmans, London.

Pushpangadan and B.L. Bradu (1995) Basil, In: *Advances in Hort.*, K.L.Chadha and Rajendra Gupta (Eds.), **11**: 623–658.

Rajagopal, V. (1997) *A Handbook on Drought Tolerant Crop Species of India*, Central Plantation Crops Research Institute, Kasargod: 228 pp.

Rajagopalan, A., T.V. Vishwanathan, and P.K. Gopalakrishnan (1989) "Phytochemical analysis and nutrient uptake studies on *Kaempferia galanga* L", *South Indian Hort.,* **37** (1): 34–38.

Rajendra Gupta (1972) "Cultivation and distillation of Japanese mint in India", *Indian Fmg.,* **22** (3): 18–23.

Rajendra Gupta (1997) "Studies on cultivation and improvement of dill (*Anethum graveolens*) in India (Part I)–Floral biology and biochemical studies *CIMAP Symp*", Jammu, March 1976, In: *Cultivation and Utilization of Medicinal and Aromatic plants* (Eds. Atal & Kapur), pp. 337–43.

Rajinder Singh, B.S. Gill and J.N. Handal (2001) "Studies on Extraction of Chilli Oleoresin", *Spice India,* **38** (5): 13–15.

Ramana, K.V. and S.J. Eapen (1995) "Parasitic nematodes and their management in major Spices", *J. of Spices and Aromatic Crops,* **4** (1): 1–16.

Ramanujam, S. and V.P. Tewari (1986) "Production of chillies in various states of India and their commercial classification", *Spices Bull. Chillies Seminar*, Spices Export Promotion Council, Cochin, **6** (3): 13–21.

Ranadive, A.S. (1994) "Vanilla–cultivation, Curing, Chemistry, technology and commercial products", In: *Spices, Herbs and Edible Fungi* (pp. 517–79) Charalanibus, G. (Ed.).

Ranaware, V.S., R.N. Nawale, R.G. Khandekar, and M.B. Magdum (1995) "Effect of season on air-layering of cinnamon", *Indian Cocoa, Arecanut and Spcies J.,* **19** (3): 81–84.

Randhawa, G.S. and A. Singh (1991) "Effect of sowing time and harvesting stage on oil content, herbage and oil yield of dill (*A.graveolens*)", *Indian Perfum.* **35** (4): 204–18.

Randhawa, G.S. and R.K. Mahey (1988) "Advances in the agronomy and production of turmeric in India", In: *Herbs, Spices and Medicinal Plants,* (USA), V.3P–71–101.

Rao, Y.S., Mary, K. Mathew, and K.J. Madhusoodanan (2001) "Small cardamom research", *Indian J. of Arecanut, Spices and Medicinal Plants,* **3** (4): 225–245.

Rao, T.G.N. (1995) "Diseases of turmeric (*Curcuma longa* L.) and their management", *J. of Spices and Aromatic Crops,* **4** (1): 49–56.

Rao,Y.S., K.M. Mathew and S.N. Potty (1999) "Tamarind (*Tamarindus indica* L.) research–a review", *Indian J. of Arecanut, Spices and Medicinal Plants,* **1** (4): 127–145.

Ravindran, T.C. and K. Manojkumar (2001) "Production, export and marketing of spices in India", *Indian J. of Arecanut, Spices and Medicinal Plants,* **3** (4): 246–257.

Raychaudhuri, S.P. (1991) "Recent advances in medicinal, aromatic and spice crops (Vol.1)", *Proc. Int. Conference,* held on 28–31, January, 1989, New Delhi.

Regeena, S. and S. Ravi (1995) "Curry leaf cultivation in the homesteads of South Kerala", *Spice India,* **8** (9): 11–13.

Rekhi Singh, Kamlesh Kumar, R.C. Sagwal, R. Singh, and K. Kumar (1997) "Research gaps in scientific cultivation of Saffron (*Crocus sativus* L.) in Kashmir", *Agric. Sci., Digest*–Karnal, **17** (3): 161–164.

Rema, J. and B. Krishnamoorthy (1990) "Vegetative propagation of tree spices", *Spice India,* **2** (5): 10–13.

Rema, J. and B. Krishnamoorthy (1994a) "Approach grafting of clove", *Indian Hort.*, **39** (3): 53–55.

Rema, J. and B. Krishnamoorthy (1994b) "Vegetative propagation of clove" *Eugenia caryophyllus* (Sprengel), B & H" *Tropical Agri.*, **71** (2): 144–46.

Rema, J. (1995) "Production technology of tree spices and propagation technique", In: *Conf. on Spices,* 11th January, 1995, CPCRI, Kasargod.

Rethinam, P., S. Edison, A.K. Sadanandan, and A.K. Johny (1994) "Major achievements of AICRP on spices", *Indian Cocoa, Arecanut and Spices, J.* **18** (1): 16–21.

Rosebrook, D.D., C.C. Prolze and J.E. Barney (1968) "Improved method for determination of extractable colour in *Capsicum* Species", *J. Assoc. Anal. Chem.* **51** (6): 37–43.

Rosemary Hemphill (1966) *The Penguin Book of Herbs and Spices*, Penguin Books Ltd., Harmonds Worth, Middlesex, England.

Rostogi, J.M.S. (1997) "Response of clary sage (*Salvia sclarea* L.) to nitrogen levels and row spacing" *Indian Perfum*. **41** (3): 113–16.

Sanjeev Agarwal (1995) "Volatile oil constituents in different entries of fruits of fennel", *Indian Perfum.*, **39** (2): 70–71.

Sanjeev Agarwal, R.K., M.P. Sharma, Jain and S. Edison (2001) "Cumin wilt and its management", *Indian J. of Arecanut, Spices and Medicinal Plants*. **3** (2): 56–59.

Sankaracharya, N.B. and C.P. Natarajan (1971) "Cardamom chemistry, technology and uses", *Indian Food Packer*, **XXV** (5): 28–36.

Sarma,Y.R. and M. Anandraj (2000) "Diseases of spice crops and their management", *Indian J. of Arecanut, Spices and Medicinal Plants*. **2** (1): 6–20.

Sarma, Y.R., M. Anandaraj, K.V. Ramana, M.N. Venugopal and S.M.P. Khurana (1998) "Major diseases of black pepper and cardamom and their management", *Pathological problems of economic crop plants and their management:* 281–293.

Sasikumar, B., P.N. Ravindran and K.J. George (1994) "Breeding ginger and turmeric", *Indian Cocoa, Arecanut and Spices J.*, **18** (1): 10–12.

Sasikumar, B., J. Rema, and P.N. Ravindran (1992) Vanilla, *Indian Cocoa, Arecanut and Spices J.*, **16** (1): 6–10.

Sathessan, K.V. and A. Ramadasan (1987) "Curcumin and essential oil contents of three turmeric (*Curcuma domestica* Val.) cultivars grown in monoculture and as an intercrop in coconut garden", *J. Plant Crops*, **15**(1).

Shadakshari, Y.G., H.M. Chandrappa, and K. Umesha (1996) "Flowering and fruit set in vanilla (*V.planifolia*)", *J. Spices and Aromatic crops*, **5** (2): 148.

Shankaracharya, N.B. and C.P. Natarajan (1971) "Chemical composition and uses of cumin", *Indian Food Packer*. **XXV** (6): 22–28.

Shankaracharya, N.B. and C.P. Natarajan (1971) "Coriander–Chemistry, Technology and Uses", *Indian Spices,* **8** (8): 4.

Shankaracharya, N.B. and C.P. Natarajan (1973) "Vanilla–Chemistry, Technology and Uses", *Indian Food Packer*, **XXVI** (3): 20–36.

Sharma, S.K. (1996) "Garlic–A gift of nature: Medicinal values", *Spice India*, **9** (4): 9–10.

Shashikumar, B., J. Rema, and P.N. Ravindran (1993) "Vanilla, The much fancied export crop", *Indian Perfum.*, **37** (4): 330.

Shivaraman, K.M., Tamil Selvan and K.G. Elizabeth (2001) "Value addition in seed spices", *Indian J. of Arecanut, Spices and Medicinal Plants*, **3** (3): 187–188.

Simon, J.E. and J. Quinn (1988) "Characterization of essential oil of parsley", *J. Agric. Fd. Chem.*, **36**: 467–72.

Singh, A.K. and S. Sharma (1994) "Quick estimation of menthol concentration in the essential oil of *Mentha arvensis*: A ready reckoner", *Indian Perfum.*, **38** (3): 93–97.

Singh, B., Y.S. Babel, B.D. Yadav, and D.R. Sharma (1995) "Effect of size of onion sets and planting distance on the yield and economics of *kharif* onion", *Indian J. Hort.*, **52** (3): 215–217.

Singh, B.P. and Ravindranath (1971) "Celery–A profitable crop enterprise for Punjab", *Indian Fmg.*, **21** (4): 26–27, 41.

Sir George Watt (1966) *The Commercial Products of India*, Today and Tomorrows Printers and Publishers, Delhi.

Sivarajan and Indira Balachandran (1994) *Ayurvedic Drugs and their Plant Sources*, Oxford and IBH publishing Company, New Delhi.

Sivaraman Nair, P.C. (1970) "Tree Spice Crops–Clove and Cinnamon" *Arecanut and Spices Bull.*, **1** (3): 1–4.

Small, E. (1997) *Culinary Herbs: Agriculture and Agri–Food*, Canada, 710 pp.

Souret, F.F. and P.J. Weathers (1999) "Cultivation *in vitro* culture, secondary metabolites production, and phytopharmacognosy of Saffron (*Crocus sativus* L.)", *J. of Herbs, Spices and Medicinal Plants*, **6** (4): 99–116.

Spices Board, Govt. of India (1994) *Spices statistics*, 2nd Edn., (Update), Kochi.

Srivastava, A.K. (1978) *French Basil and its Cultivation in India*, Central Institute of Medicinal and Aromatic Plants, Lucknow.

Steffen Arctander (1960) *Perfume and Flavour Materials of Natural Origin*. Elizabeth, N.J., U.S.A.

Subpdjkpsjo, P.C. Thakur and T.S. Verma (1999) "Kt–pl 19: A paprika line for commercial cultivation", *Indian Horticulture*, **4**: 53–54.

Subramanian, S. and Md. Abdulkhader (1987) "Quality parameters of turmeric rhizomes harvested at different stages of growth", In: *Proc. Nat. Symp. On spice industries* (Ed. J.S. Pruthi), AFST, India, 9–10, April 1987, Abstract No. 2.07, pp. 39–40.

Sudheendra, S., A.A. Farooqi, Thilak Subbaiah and B.V. Bhat (1993) "Influence of nitrogen, phosphorus and potassium on growth, seed yield and essential oil content in Celery (*Apium graveolens* L.)", *Indian Perfum.*, **37** (4): 315–317.

Sungha, S.K. (1995) "Management of purple blotch (*Alternaria porri*) of garlic with fungicides", *Indian J. Agric. Sci.*, **65** (6): 455–59.

Tamil Selvan, M. (2001) "Eco–friendly insect pest management using bio–control agents", *Indian J. Arecanut, Spices and Medicinal Plants*, **3** (4): 258–261.

Tanaka (1976) *Encyclopedia of Edible Plants of the World*, Keigaku Publishing Co., Tokyo, Japan.

Tankamani,C.K., K. Sivaraman, K. Kandiannan, and K.V. Peter (1994) "Agronomy of tree species (clove, nutmeg, cinnamon and allspice)–A review", *J. of Spices and Aromatic Crops*, **3** (2): 105–123.

Tetenyi,P. (1997) Opium poppy (*Papaver somniferum*): Botany and horticulture, *Hort. Reviews*, **19**: 373–408.

Thamburaj and Narendra Singh (2001) *Textbook of Vegetables*, Tuber Crops and Spices.

Thekkayam, S.B., (1988) "Garcinia–The neglected acidic fruit of Kerala", *Indian Cocoa, Arecanut and Spices*, **11** (3): 100.

Thomas, J. and R.S. Bhai (1995) "Fungal and bacterial diseases of cardamom (*Elittaria cardamomum* Maton) and their management", *J. of Spices and Aromatic Crops*. **4** (1): 24–31.

Thomas, J., P.P. Joy, Samuel Mathew, Baby P. Skaria, P.P. Duethi and T.S. Joseph (2000) *Agronomic Practices for Aromatic and Medicinal Plants*, Directorate of Arecanut and Spices Development, Calicut.

Umrao Singh, A.M., Wadhwani and B.M. Johri (1983) *Dictionary of Economic Plants of India*, Indian Council of Agricultural Research, New Delhi.

Upadhaya, S.P., M. Kirihata, and I. Ichimoto (1994) "Cinnamon leaf oil from *Cinnamomum tamala* grown in Nepal", *J. Japanese Society of Fd. Sci. and Tech.*, **41** (7): 512–15.

Varier, P.S., (1988) "Ginger (ayurvedic uses)", *Spices Newsletter*, March 1988, p. 13.

Vasanthakumar, G.K. (2000) Garcinia–"The wonder plant for obesity reduction", In: *Proc. National Seminar on Transfer of Technology of Medicinal and Aromatic Crops*, 20–22, February, Bangalore, pp: 56–61.

Vasundhara, M., A.A. Farooqi, K.A. Devaiah and M. Sridharayya (1992) "Influence of some growth regulators on the growth, herbage and oil yield in Majoram (*Majorana hortensis* Moench.)", *Indian Perfum.*, **36** (3): 171–174.

Venugopal, M.N. (1995) "Viral diseases of cardamom (*Elitteria cardamomum* Maton) and their management", *J. of Spices and Aromatic Crops*, **4** (1): 32–29.

Verma, T.S. and S. Joshi (2000) "Status of paprika development in India", *Indian J. of Arecanut, Spices and Medicinal Plants*, **2** (2): 39–44.

Virmani, O.P. and S.C. Datta (1968) "Oil of spearmint", *Perfum. and Essen. Oil, Rec.*, London, pp: 1–11.

Vishwanathan, T.V. (1995) "Long pepper", In: *Advances in Horticulture*, (Eds): K.L.Chadha and Rajendra Gupta, **11**: 373–383.

Woodbury, J.E. (1994) "Spices and other condiments", *J. AOAC International*, **77** (1): 184.

Yadav, N.D. (1995) "Influence of irrigation methods and time of nitrogen application on growth, seed yield and water use of Indian Mustard", *Indian J. Agron.*, **40** (1): 120–22.

Yadav, T.D. (1987) "Turmeric–great potential source of *Lasioderma serricorne* (Fabr.) infestation", In: *Proc. Nat. Symp. on Spice Industries* (Ed. J.S. Pruthi), AFST (I) New Delhi, Abstract No. 1.22, pp. 17–18.

Annexure I: Spice Research Stations in India

Table 1 *AICRPS centres and mandate crops*

Sl. No.	Name of centre and location	Crops handled	State	Year of establishment
1.	Chinthapalle (ANGRAU)	Black pepper & Ginger ***	Andhra Pradesh	1981 VI plan
2.	Coimbatore (TNAU)	Coriander, Fenugreek* & Turmeric ***	Tamil Nadu	1975 V plan
3.	Dapoli (KKV)	Black pepper, Nutmeg, Clove & Cinnamon	Maharashtra	1995 VII plan
4.	Dholi (RAJU)	Turmeric, Ginger, Coriander & Fenugreek	Bihar	–
5.	Gangtok (ICAR Research Complex for NEH Region)	Large Cardamom	Sikkim	1986 VII plan
6.	Guntur (ANGRAU)	Coriander & Fenugreek	Andhra Pradesh	1975 V plan
7.	Hissar (CCSHAU)	Coriander, Fennel & Fenugreek	Haryana	1993 VII plan
8.	Jagital (ANGRAU)	Coriander & Fenugreek	Andhra Pradesh	1986 VII plan
9.	Jagudan (GAU)	Cumin, Coriander, Fennel & Fenugreek	Gujarat	1975 V Plan
10.	Jobner (RAU)	Cumin, Coriander, Fennel and Fenugreek	Rajasthan	1975 V Plan
11.	Kumarganj (NDUAT)	Turmeric, Ginger, Fennel, Coriander & Fenugreek	Uttar Pradesh	1995 VII plan

(contd...)

(...contd.)

Table 1

Sl. No.	Name of centre and location	Crops handled	State	Year of start
12.	Mudigere (UAS-B)	Cardamom & Black pepper ***	Karnataka	1971 IV plan
13.	Pampadumpara (KAU)	Cardamom & Black pepper **	Kerala	1971 IV plan
14.	Panniyur (KAU)	Black pepper	Kerala	1971 IV plan
15.	Pottangi (OUAT)	Turmeric & Ginger	Orissa	1975 V plan
16.	Pundibari (UBKVV)	Black pepper, Turmeric & Ginger	West Bengal	1996 VIII plan
17.	Raigarh (IGKV)	Coriander, Turmeric & Ginger	Chattisgarh	1996 VIII plan
18.	Sirsi (UAS-D)	Black pepper	Karnataka	1981 VI plan
19.	Solan (YSPHUF)	Ginger & Turmeric *	Himachal Pradesh	1971 IV plan
20.	Yercaud (TNAU)	Clove **, Nutmeg **, Cinnamon **, Cardamom *, & Black pepper	Tamil Nadu	1981 VI pla

Vittal centre for pepper and Gonicopal for cardamom—established in 1971, discontinued in 1980. Vellanikkara (KAU) established in 1981 (for ginger and turmeric)—discontinued with effect from 1. 4. 1990.

ICAR R. C. NEHR, Gangtok started in 1986 to work on large cardamom, discontinued in 1988.

* Crop discontinued, ** Work initiated in 1993 – 1994 *** Work initiated in 1995 – 1996.

(**Source:** Spice Board, Cochin)

Table 2 *Voluntary / Participating AICRPS Centres*

Sl. No.	Name of centre and location	Crops handled	State	Year of start
1.	Ambalavayal (KAU)	Black pepper, Clove & Cinnamon	Kerala	1992
2.	Bhavanisagar (TNAU)	Turmeric	Tamil Nadu	1993
3.	Myladumpara (ICRI)	Cardamom	Kerala	1993
4.	Pechiparai (TNAU)	Clove, Nutmeg & Cinnamon	Tamil Nadu	1993
5.	Sakleshpur (ICRI-RS)	Cardamom	Karnataka	1993
6.	Thadiyankudisai (TNAU)	Clove, Nutmeg, & Cinnamon	Tamil Nadu	1993
7.	Thadiyankudisai (ICRI-RS)	Cardamom	Tamil Nadu	1993
8.	Gangtok (ICRI, RS)	Large cardamom	Sikkim	1999

(**Source:** Spice Board, Cochin)

Mandates

- Evolving high-yielding varieties, resistant/tolerant to pests and diseases and their multilocation testing.
- Developing agrotechnologies through integrated nutrient management for increasing sustainable productivity under various agro-ecological situations.
- Evolving suitable pests and disease management technologies through integrated approaches.
- Collection, conservation, study and sustainable utilisation of the biodiversity of spices in the country.
- To act as an interface between ICAR/SAU's/IISR, DASD systems.

Table 3 *Mandatory crops under AICRPS (State-wise)*

Sl. No.	State	Crops
1.	Andhra Pradesh	Turmeric, Coriander, Fenugreek, Black pepper, Ginger
2.	Bihar	Turmeric, Ginger, Coriander, Fenugreek
3.	Chattisgarh	Ginger, Turmeric, Coriander
4.	Gujarat	Coriander, Cumin, Fennel, Fenugreek
5.	Haryana	Coriander, Fennel, Fenugreek
6.	Himachal Pradesh	Ginger
7.	Karnataka	Black pepper, Cardamom
8.	Kerala	Black pepper, Cardamom
9.	Maharashtra	Black pepper, Cinnamon, Nutmeg, Clove
10.	Orissa	Ginger, Turmeric
11.	Rajasthan	Coriander, Cumin, Fennel, Fenugreek
12.	Sikkim	Coriander, Cumin, Fennel, Fenugreek
13.	Tamil Nadu	Black pepper, Turmeric, Coriander, Clove, Nutmeg, Cinnamon
14.	Uttar Pradesh	Turmeric, Ginger, Coriander, Fennel, Fenugreek
15.	West Bengal	Black pepper, Ginger, Turmeric

(**Source:** Spice Board, Cochin)

Spectrum of Activities

1. Germplasm collection, conservation, evaluation
2. Evolving high-yielding spices varieties
3. Developing production technology in spices
4. Evolving control measures against pests and diseases
5. Developing a package of practices for different spices
6. Planting material production
7. Post-harvest technology studies
8. Extension and training programmes
9. Linkages
10. Publications
11. Monitoring cess fund/*ad hoc* schemes
12. Management and allotment of funds of AICRPS units

Table 4 *Coordinating Centres (Crop-wise) of AICRPS*

Crops	No. of centres	Name of centres
Black pepper	7	Panniyur, Sirsi, Chinthapalle, Pampadumpara, Yercaud, Pundibari, Dapoli
Cardamom	2	Mudigere, Pampadumpara
Ginger	6	Solan, Pottangi, Raigarh, Kumarganj, Pundibari, Dholi
Turmeric	6	Pottangi, Jagtial, Dholi, Raigarh, Kumarganj, Pundibari
Clove	5	Ambalayal, Pechiparai, Thadiyankudisi, Yercaud, Dapoli
Nutmeg	2	Yercaud, Dapoli
Cinnamon	5	Ambalayal, Pechiparai, Thadiyankudisi, Yercaud, Dapoli
Coriander	8	Coimbatore, Jobner, Jagudan, Guntur, Hisar, Dholi, Raigarh, Kumarganj
Cumin	2	Jobner, Jagudan
Fennel	3	Jobner, Jagudan, Hisar
Fenugreek	5	Guntur, Jobner, Jagudan, Hisar, Dholi
Cardamom (Large)	1	ICRI, Gangtok (Voluntary Centre)

Genetic Resources of AICRPS

The AICRPS centres have a good assemblage of valuable germplasm collection of cultivated types, endangered and wild species. About 5000 accessions in 12 spices have been conserved, evaluated and characterised.

Table 5 *Genetic Resources Holding in Various Spices*

Crop	Holding	Locations
Black pepper	367	Panniyur, Sirsi, Chinthapalle, Yercaud, Pundibari, Dapoli
Cardamom	336	Mudigere, Pampadumpara
Ginger	406	Solan, Pottangi, Kumarganj, Pandibari, Raigarh, Dholi
Turmeric	1136	Jagtial, Dholi, Pottangi, Raigarh, Pundibari, Kumarganj
Clove	42	Yercaud, Dapoli, Pechiparai
Cinnamon	41	Yercaud, Dapoli, Pechiparai
Nutmeg	42	Yecaud, Dapoli, Pechiparai
Coriander	1467	Coimbatore, Jobner, Jagudan, Guntur, Hisar, Dholi, Raigarh, Kumarganj
Cumin	495	Jobner, Jagudan
Fennel	420	Jobner, Jagudan, Hisar
Fenugreek	944	Guntur, Jobner, Jagudan, Hisar, Dholi

(**Source:** Spice Board, Cochin)

Annexure II

Table 1 *Average Composition of Manures and fertilisers*

Manures and Fertilizers	Percentage of Nutrients		
	N	P_2O_5	K_2O
A) Fertilizers			
Ammonium sulphate	20.5	–	–
Ammonium sulphate nitrate	26.0	–	–
Ammonium nitrate	33.5	–	–
Ammonium phosphate	16.0	20.0	–
Calcium ammonium nitrate	20.5	–	–
Nitrate of soda	16.5	–	–
Urea	46.0	–	–
Superphosphate-single	–	18.0	–
Superphosphate-double	–	35.0	–
Superphosphate-triple	–	45.0	–
Ultraphosphate	–	28.3	–
Mussoorie-Phosphate	–	20.24	–
Rock phosphate	–	28.4	–
Bone-meal	3.5	21.0	–
Muriate of potash	–	–	50.0 or 60
B) Organic Manures			
Poultry manure	1.2 – 1.5	–	–
Sheep manure	0.8 – 1.6	–	–
Farm Yard Manure (FYM)	0.4	0.3	0.2
Compost	0.5	0.25	0.5

Annexure III

Table 1 *Conversion of pure nutrients to various N, P and K Fertilizers: Quantities of fertilisers required for the supply of specified doses of nutrients*

Rate of Application (kg / ha)	Ammonium Sulphate (20% N)	Urea (46 % N)	Superphosphate (18 % P_2O_5)	Muriate of potash	
				(50% K_2O)	(60% K_2O)
10	50	22	56	20	17
20	100	44	112	40	34
30	150	66	168	60	51
40	200	88	224	80	68
50	250	110	280	100	85
60	300	132	336	120	102
70	350	154	392	140	119
80	400	176	448	160	136
90	450	198	504	180	153
100	500	200	560	200	153
110	550	242	616	220	187
120	600	264	672	240	204
140	700	308	784	280	238
150	750	330	840	300	255

Annexure IV
Preparation of Bordeaux Mixture (%)

Dissolve 1 kg of powdered Copper sulphate crystals in 50 litres of water. In another lot of 50 litres of water, prepare milk of lime with 1 kg of quicklime. Pour the copper sulphate solution into the milk of lime, slowly stirring the mixture all the while. Test the mixture before use for the presence of free copper, which is harmful to the plants, by dipping a polished knife in it. If the blade turns a reddish colour, add more lime till the blade is not stained on dipping.

Important Precautions

Always use wooden, earthen or copper vessels for the preparation of Bordeaux mixture that is mentioned above.

Universities Press
BIOTECHNOLOGY
Agriculture and Intellectual Property Rights
Economic, Institutional and Implementation Issues in Biotechnology
Edited by
V Santaniello • R E Evenson
D Zilberman • G A Carlson

Universities Press
Joint Forest Management in India
Spread, Performance and Impact
N H Ravindranath
P Sudha

Universities Press
Nursery Manual for Forest Tree Species
D M Bhat • Vidya S Swamy • N H Ravindranath

Universities Press
Universities Press Dictionary of
BIOLOGY
Robert Hine
Third Edition

Universities Press
BIOTECHNOLOGY
Travels in the
Genetically Modified Zone
Mark L Winston

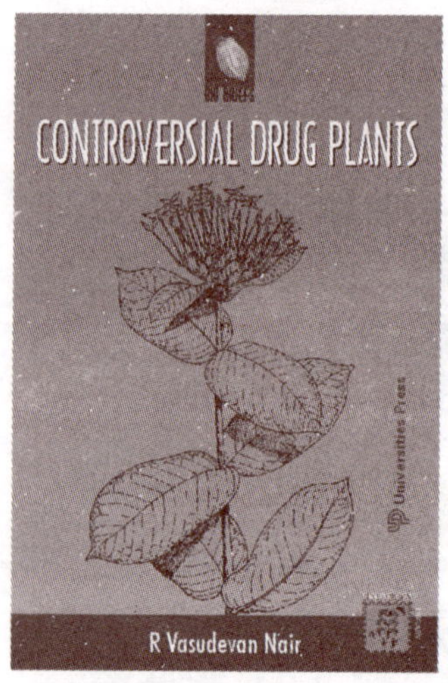
CONTROVERSIAL DRUG PLANTS
Universities Press
R Vasudevan Nair

Universities Press
BIOTECHNOLOGY
Plant
Tissue Culture
Basic and
Applied
Timir Baran Jha
Biswajit Ghosh